LINEAR SYSTEMS

Non-Fragile Control and Filtering

LINEAR SYSTEMS

Non-Fragile Control and Filtering

Guang-Hong Yang • Xiang-Gui Guo
Wei-Wei Che • Wei Guan

CRC Press
Taylor & Francis Group
Boca Raton   London   New York

CRC Press is an imprint of the
Taylor & Francis Group, an **informa** business

CRC Press
Taylor & Francis Group
6000 Broken Sound Parkway NW, Suite 300
Boca Raton, FL 33487-2742

First issued in paperback 2017

© 2013 by Taylor & Francis Group, LLC
CRC Press is an imprint of Taylor & Francis Group, an Informa business

No claim to original U.S. Government works

Version Date: 20130129

ISBN 13: 978-1-138-07206-0 (pbk)
ISBN 13: 978-1-4665-8035-0 (hbk)

Visit the Taylor & Francis Web site at
http://www.taylorandfrancis.com

and the CRC Press Web site at
http://www.crcpress.com

Contents

Preface

Digital control systems design has become an important field in electrical engineering and in systems and control theory. One of the important and fundamental issues in digital control systems design is the filter or controller coefficient sensitivity because even vanishingly small perturbations in controller or filter coefficients may destabilize the resulting systems. In the actual engineering systems, the controllers or filters realized by microprocessors/microcontrollers do have some uncertainties due to limitations in available microprocessor/microcontroller memory, effects of finite word length (FWL) of digital processors, quantization of the A/D and D/A converters, and so on. Therefore, non-fragile (insensitive) control is becoming popular in many fields of engineering and science, and there is a vast amount of literature on design and analysis of non-fragile control problems using rigorous methods based on different performance criteria.

In order to obtain non-fragile (insensitive) controllers, numerous works in the filtering and control theory are devoted to solving such problems. The previous results were mainly developed in the framework of robust control theory, that is, non-fragile controller/filter design methods have been proposed to obtain the non-fragile controllers/filters which can be insensitive or non-fragile with respect to controller/filter gain uncertainties by considering controller/filter gain uncertainties directly. There are two main types of gain uncertainties considered in the design methods. One is known as norm-bounded gain uncertainty, the other is known as interval-bounded coefficient variations. It is worth mentioning that the type of norm-bounded uncertainty cannot exactly reflect the uncertain information due to the FWL effects, while the type of interval-bounded coefficient variations may result in numerical problems because the number of linear matrix inequalities (LMIs) involved in the design conditions grows *exponentially* with the number of uncertain parameters. On the other hand, sensitivity analysis techniques in performance assessments are important in operations research as well as in the practical design of control systems because sensitivity analysis provides valuable insights into the influence of parameter variations on the dynamic behavior of systems. However, they mainly consider the optimal realization of a controller or filter via minimizing the coefficient sensitivity.

In this book, the aim is to present our recent research results in designing non-fragile controllers/filters for linear systems. The main feature of this text is that the algebraic Riccati equation technique is successfully introduced to solve the type of additive/multiplicative norm-bounded controller/filter gain uncertainty, while a structured vertex separator is proposed to approach the numerical problem by considering interval-bounded coefficient variations. Moreover, sensitivity theory is always used to characterize the phenomenon of trivial deviations, which motivates us to design insensitive controllers/filters in the framework of coefficient sensitivity theory because the controller/filter coefficient variations resulting from limitations of the available computer memory are of trivial deviations. This book provides a coherent approach and contains valuable reference materials for researchers wishing to explore the area of non-fragile control/filtering. Its contents are also suitable for a one-semester graduate course.

The text focuses exclusively on the issues of non-fragile control/filtering in the framework of algebraic Riccati equations, LMI techniques, structured vertex separator methods, and coefficient sensitivity methods. The book begins with the development and main research methods in non-fragile control, while offering a systematic presentation of the newly proposed methods for non-fragile control/filtering of linear systems with respect to additive/ multiplicative controller/filter gain uncertainties. The tools for design and analysis presented in the book will be valuable in understanding and analyzing parameter uncertainties.

This work was partially supported by the Funds of the National 973 Program of China (Grant No. 2009CB320604), the Funds of the National Science of China (Grant Nos. 60974043, 61273148, 61104106, 61104029, 61203087), the Funds of Doctoral Program of Ministry of Education, China (20100042110027), the Foundation for the Author of National Excellent Doctoral Dissertation of PR China (No. 201157), the Natural Science Foundation of Liaoning Province (Grant No. 201202156), and by the Program for Liaoning Excellent Talents in University (LNET)(LJQ2012100).

Guang-Hong Yang
Northeastern University, China
State Key Laboratory of Synthetical Automation for Process Industries
(Northeastern University), China

Xiang-Gui Guo
Tianjin Key Laboratory for Control Theory Application
in Complicated Systems
School of Engineering
Tianjin University of Technology, China

Wei-Wei Che
Shenyang University, China

Wei Guan
Shenyang Aerospace University, China

MATLAB^{TM} is registered trademark of The MathWorks, Inc. For product information, please contact:

The MathWorks, Inc.
3 Apple Hill Drive
Natick, MA 01760-2098 USA
Tel: 508-647-7000
Fax: 508-647-7000
E-mail: info@mathworks.com
Web: www.mathworks.com

Symbol Description

$\in$	belongs to	$\|e\|_2$	L_2-norm of signal e
R	field of real numbers	$L_2[0, \infty)$	space of square integrable functions on $[0, \infty)$
R^n	n-dimensional real Euclidean space	$l_2[0, \infty)$	space of square summable infinite vector sequences over $[0, \infty)$
$R^{n \times m}$	set of $n \times m$ real matrices		
I	identity matrix	$*$	symmetric terms in a symmetric matrix
I_n	$n \times n$ identity matrix		
X^T	transpose of matrix X	$He\{M\}$	$He\{M\} := M + M^T$
$P \geq 0$	symmetric positive semi-definite matrix $P \in R^{n \times n}$	$\otimes$	Kronecker product
		diag$\{...\}$	block diagonal matrix
$P > 0$	symmetric positive definite matrix $P \in R^{n \times n}$	$\bigoplus\limits_{i=1}^{n} A_i$	block diagonal matrix with blocks $A_1, A_2, \cdots, A_n$, i.e., diag$\{A_1, A_2, \cdots, A_n\}$
$P \leq 0$	symmetric negative semi-definite matrix $P \in R^{n \times n}$		
$P < 0$	symmetric negative definite matrix $P \in R^{n \times n}$	CCLM	cone complementarity linearization method
P^{-1}	the inverse of matrix P	SLPMM	sequential linear programming matrix method
rank$(\cdot)$	rank of a matrix		
trace$(\cdot)$	trace of a matrix	LMI	linear matrix inequality
$\| \cdot \|$	Euclidean matrix norm	BLMI	bilinear matrix inequality

1

Introduction

With the rapid development of computer and automation technologies, more and more attention is paid to the *digital control system* which has been considered as one of the most important and active fields in the research. A typical configuration of the digital control system is shown in Figure 1.1, in which limitation in available microprocessor memory, effects of *finite word length* (FWL) of the digital processor, errors for truncation and quantization of the A/D and D/A converters, and so on, always cause the controller parameters *trivial deviations* from the original design values [47]. Keel and Bhattacharyya [77], by means of numerical examples, have shown that the controllers designed by using weighted H_∞, μ, and L_1 synthesis techniques may be very sensitive, or fragile with respect to relatively small perturbations in controller parameters. Therefore, a significant issue is how to design a filter or controller for a given plant such that the filter or controller is insensitive to some errors with respect to its coefficients, that is, the designed filter or controller is *insensitive* or *non-fragile*.

The configuration shown in Figure 1.2 simply describes the *robust control*. Figure 1.3 shows the non-fragile control, while the robust *non-fragile control* is shown in Figure 1.4, where P denotes the plant, ΔP stands for the uncertainties of the plant P, K denotes the controller, and ΔK denotes the inaccuracies or uncertainties in the implementation of a designed controller. There are two main types of coefficient uncertainties considered in the designed methods. One is of a *norm-bounded* type, the other is of an *interval-bounded* type. Furthermore, the above two types also can be divided into *additive* case and *multiplicative* case. Then, the models of the uncertainty ΔK are given as follows:

- Norm-Bounded Uncertainty:

$$\begin{cases} \Delta K = H_a \Delta_a E_a, \Delta_a^T \Delta_a \leq I & \text{Additive Case} \\ \Delta K = H_m \Delta_m E_m K, \Delta_m^T \Delta_m \leq I & \text{Multiplicative Case} \end{cases} \quad (1.1)$$

where H_a, E_a, H_m, and E_m are known constant matrices of appropriate dimensions, and Δ_a and Δ_m are the uncertain parameter matrices.

- Interval-Bounded Uncertainty:

$$\begin{cases} \Delta K = [\theta_{ij}] & \text{Additive Case} \\ \Delta K = [\theta_{ij} k_{ij}] & \text{Multiplicative Case} \end{cases} \quad (1.2)$$

where k_{ij} denotes the (i,j)th element of the matrix K, and θ_{ij} ($|\theta_{ij}| \leq \theta$) is

1

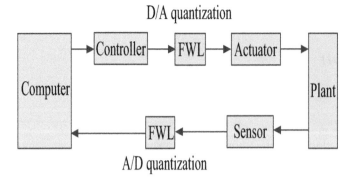

FIGURE 1.1
A typical digital control system configuration.

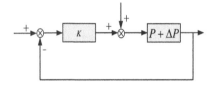

FIGURE 1.2
A robust control configuration.

used to describe the magnitude of the deviation of the matrix coefficient k_{ij}, where θ denotes the maximum possible deviation.

In recent years, the type of norm-bounded uncertainty (1.1), which is investigated in Chapters 3–5, has received wide attention, however, it cannot exactly reflect the uncertain information due to implementation imprecision. Therefore, the type of interval-bounded uncertainty (1.2) is introduced in Chapters 6 and 7. Yet, it has a numerical problem because the number of the linear matrix inequalities (LMIs) involved in the design conditions grows *exponentially* with the number of uncertain parameters, which make it difficult to apply the results to systems with high orders. Although the *structured vertex separator method* is proposed to deal with the numerical problem, the

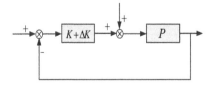

FIGURE 1.3
A non-fragile control configuration.

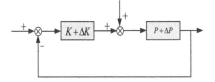

FIGURE 1.4
A robust non-fragile control configuration.

number of LMI constraints involved in the design conditions is still large. Furthermore, another important problem is that the computational efficiency is critical in real-time applications, so it is highly desirable for a controller to have a *sparse structure*, namely containing many trivial parameters (trivial parameters mean that they are 0 and ±1, which can be digitally implemented exactly and cause no rounding errors). Other parameters are, therefore, referred to as nontrivial parameters [92]. The problem of finding sparse controller realizations has been considered by several researchers [3, 57, 58, 92]. These results consider the sparse structure problem from the point of view the controller realization. How to design a controller with sparse structure is a valuable problem.

On the other hand, *sensitivity analysis* allows us to assess the effects of changes in the parameter values [12, 13, 23, 80]. Hence, it is very useful to understand how changes in the parameter values influence the design [12, 13, 23, 50, 80]. After the hard work of many researchers in more than one decade, fundamental results have been obtained for the study of sensitivity analysis and performance limitations in automatic control systems ([see, for example, 21, 59, 60, 129, 140, 141], and the references therein), and many different definitions of sensitivity have been used for sensitivity analysis [61–64, 87, 91, 100, 125, 131, 132, 136, 152]. One of the effective synthesis methods is the *coefficient sensitivity method*, which describes the variations in performance due to variations in the parameters that affect the system dynamics [see 90, 92, 99, 121]. It is well known that very small perturbations in the coefficient of the designed controller or filter may result in the serious deterioration of the system performance, including instability. Therefore, the controller or filter should be designed to be insensitive to some amount of error with respect to its coefficients. Sensitivity theory is always used to characterize the phenomenon of *trivial deviations*, which motivates us to design insensitive controllers and filters in the framework of coefficient sensitivity theory in Chapters 8–11 because the coefficient variations resulted from the limitation of the available computer memory are of trivial deviations.

The main contribution of this book is that the algebraic Riccati technique, the *linear matrix inequality* technique, and the sensitivity analysis method have been successfully combined to establish a set of new non-fragile (insensitive) control methods [19, 47–49, 142–144, 147–150]. The proposed method can optimize the closed-loop system performances and simultaneously make the designed controllers or filters tolerant of coefficient variations in controller or filter gain matrices. Parts of the developed theories are applied to the simulation studies of the F-404 engine model and the F-18 aircraft model, which show intuitively the feasibility and superiority of the newly proposed methods.

A summary of the rest of the chapters of this book is given below.

Chapter 2 presents some preliminaries about the considered problem. Some lemmas to be used to derive the main results of this book are also given.

Chapter 3 investigates the problem of *guaranteed cost control* of discrete-time linear systems subject to *additive/multiplicative* controller gain uncertainties, respectively. First, an optimal guaranteed cost control design method is presented by using the algebraic Riccati equation technique. It is worth mentioning that the standard optimal control design for the same system can be obtained by modifying the cost function. Under a bound condition for the gain uncertainties, an optimal guaranteed cost control design method is also given for the case of the multiplicative gain uncertainties. The numerical example has shown the effectiveness of the proposed design procedures.

Based on the results in Chapter 3, Chapters 4 and 5 deal with the corresponding non-fragile controller and filter design problems. The procedures of designing non-fragile dynamic output-feedback controllers that can tolerate some *additive/multiplicative* controller gain uncertainties are presented in Chapter 4 in terms of symmetric positive-definite solutions of algebraic Riccati inequalities. Chapter 5 presents a robust *non-fragile Kalman filter* design method corresponding to the filter gain uncertainties in terms of solutions to *algebraic Riccati* equations, which depend on two design parameters, one from the system uncertainty and another from the state estimator gain uncertainty. When the controller/filter gain uncertainties are not considered, the results are reduced to those for the standard control. Finally, the effectiveness of the proposed methods is validated by numerical examples.

Chapter 6 studies the full parameterized and sparse structured non-fragile H_∞ controller design problems. The type of the additive interval-bounded coefficient variations, which less conservative than the type of norm-bounded controller gain uncertainties, is considered. First, a *two-step procedure* is adopted to solve the full parameterized controller design problem for the discrete-time and continuous-time systems, respectively. In addition, a structured vertex separator is proposed to approach the numerical computational problem resulting from the interval type of coefficient variations, and exploited to develop sufficient conditions for the non-fragile H_∞ controller design in terms of solutions to a set of LMIs. Second, for the sparse structured controller design problem, a class of sparse structures is specified. Then, a three-step procedure for non-fragile H_∞ controller design under the restriction of the sparse structure is provided. The contribution of this method is that it not only reduces the number of nontrivial parameters but also designs the *sparse structured controllers* with non-fragility. The resulting designs of the two cases guarantee that the closed-loop system is *asymptotically stable* and the H_∞ performance from the disturbance to the regulated output is less than a prescribed level. Finally, the effectiveness of the proposed design methods is illustrated by numerical examples.

Based on the results of Chapter 6, Chapter 7 deals with the problem of non-fragile H_∞ filter design subject to the additive interval-bounded filter co-efficient variations. The full parameterized and sparse structured filter design problems are investigated simultaneously. For the full parameter filter design, the structured vertex separator proposed in the previous chapter is exploited to solve the numerical computational problem and to further develop sufficient conditions for the non-fragile H_∞ filter design in terms of solutions to a set of LMIs. For the sparse structured filter design, first, a class of *sparse structures* is specified. Then, an LMI-based procedure for non-fragile H_∞ filters design under the restriction of the sparse structure is provided. The effectiveness of the proposed methods are illustrated via some numerical examples and their simulations.

Chapter 8 investigates the problem of designing multi-objective coefficient insensitive H_∞ filters for linear continuous-time systems. Parameter sensitivity functions of transfer functions with respect to filter *additive/multiplicative* parameter variations are defined first, and the H_∞ norms of the sensitivity functions are used to measure the sensitivity of the transfer functions with respect to filter parameters. In addition, in order to deal with the filter design problem for the multiplicative filter coefficient variation case, new measures based on the average of the sensitivity functions are also defined. Based on the above two types of sensitivity measures, novel methods for designing insensitive H_∞ filters subjected to *additive/multiplicative* filter coefficient variations, respectively, are given in terms of LMI techniques. Furthermore, an indirect method for solving the multiplicative variations is also proposed. In comparison with the existing method, the new proposed method has less computational burden. In addition, it is difficult to use the techniques developed in Chapter 7 to obtain convex conditions for the filter design problem with respect to the interval multiplicative parameter variation case, while this problem can be resolved well by using the new proposed method. The simulation examples have also shown the effectiveness of the proposed method.

Based on the results in Chapter 8, Chapters 9 and 10 focus on the problems of designing multi-objective coefficient insensitive H_∞ filters and an *output tracking controller* for delta operator discrete-time systems, respectively. The designed filters/controllers are insensitive to the filter/controller parameter variations. Being different from using a common Lyapunov matrix of Chapter 8, the design conservatism is reduced by introducing slack variables in these two chapters. It is worth mentioning that the delta operator approach offers better parameter sensitivity than the traditional shift operator approach at a high sampling rate. Finally, some numerical examples including a linearized model of an F-404 *engine* and an F-18 *aircraft* are given to show the effectiveness and superiority of the proposed approaches in the above two chapters.

Chapter 11 studies the problem of designing multi-objective coefficient insensitive H_∞ *dynamic output feedback* controllers for linear discrete-time systems. Two different design methods with different degrees of conservativeness and computational complexity are proposed for this problem. The designed controllers are insensitive to the controller parameter variations. The first method presents a necessary and sufficient condition for the existence of the insensitive controller. The problem of designing multi-objective dynamic output feedback controllers is a *non-convex problem* itself, an LMI-based procedure which is a *sequential linear programming matrix method* (SLPMM) is proposed to solve this non-convex problem. However, the search for satisfactory solutions may be difficult when the SLPMM algorithm acts on a module of very high dimension. To overcome the above difficulty, the non-fragile controller design method is adopted to obtain an initial solution for the SLPMM algorithm for the first time. In the second method, a sufficient condition is provided for the multiplicative parameter variation case based on a new type of sensitivity measures. Finally, the effectiveness of the proposed method is validated by numerical examples.

2

Preliminaries

In this chapter, *non-fragile control* and *filtering* problems for *linear systems* are investigated under both H_∞ and *guaranteed cost performance index*, using the *linear matrix inequality (LMI)* technique and the *coefficient sensitivity method*. For the convenience of discussion in the rest of this chapter, some preliminaries, including a few definitions, notions, and lemmas, are presented in this chapter.

2.1 Delta Operator Definition

A definition of *delta operator* or Euler operator is introduced as follows:

Definition 2.1 *[44,45] For a continuous-time signal $x(t)$, the discrete-time sequence by sampling the continuous-time signal is $x(nh)$ where h is the sampling period and $n = 0, 1, 2, \cdots$. We assume that $h = 0$ signifies $x(nh) = x(t)$. For $h \neq 0$ we denote $x(nh) = x_q(k), x((n+1)h) = x_q(k+1), k = 0, 1, 2, \cdots$. Then, the definition of an incremental difference operator (or delta operator for short) is given out as follows:*

$$\delta x(k) \triangleq \begin{cases} \frac{d}{dt}x(t) & h = 0 \\ (x(k+1) - x(k))/h & h \neq 0 \end{cases}$$

where the delta representation converges to the continuous-time representation as $h = 0$, and it converges to the discrete-time representation as $h \neq 0$. Obviously, the delta operator provides a theoretically unified formulation of continuous-time and discrete-time systems.

In addition, from the above definition, the delta operator and the traditional forward *shift operator* (q operator) are related as

$$\delta x(k) = \delta[x(nh)] = \frac{x(nh+h) - x(nh)}{h} = \frac{q[x(nh)] - x(nh)}{h}, \text{ for } h \neq 0 \quad (2.1)$$

where q is a forward shift operator ($qx_q(k) = q[x(nh)] = x_q(k+1)$) with $x_q(k)$ being sampled by using the forward shift operator approach. The above equation can be rewritten as

$$\delta = \frac{q-1}{h}.$$

In view of this, δ as a dynamic operator provides the same flexibility and implementability as a shift operator [44, 45]. However, it is well known that the usual shift operator approach suffers from numerical ill-conditioning at sufficiently small sampling periods. Therefore, in order to solve this problem, the delta operator instead of the traditional *shift operator* was constructed to study sampling continuous-time systems by Goodwin et al. [44, 45]. Two major advantages are known for the use of delta operator parameterization: a theoretically unified formulation of continuous-time and discrete-time systems, and better numerical properties in FWL implementations when compared with traditional z-transform at high sampling periods [90, 92]. Therefore, the delta operator is widely applied in many fields such as high-speed digital signal processing [36], system modeling [35,81], robust control/filtering [135], reliable control [116], and non-fragile control/filtering [47, 96].

2.2 H_∞ Performance Index

A popular performance measure of a stable linear time-invariant system is the H_∞ norm of its *transfer function*. It is defined as follows.

Definition 2.2 *[154] Consider a linear time-invariant continuous-time system*

$$\dot{x}(t) = Ax(t) + B_1\omega(t)$$
$$z(t) = Cx(t) + D_1\omega(t) \tag{2.2}$$

where $x(t) \in R^n$ is the state, $\omega(t) \in R^s$ is an exogenous disturbance in $L_2[0,\infty]$, that is,

$$\|\omega(t)\|_2^2 = \int_0^\infty \omega^T(t)\omega(t)dt < \infty$$

and $z(t) \in R^r$ is the regulated output, respectively. A, B_1, C, D_1 are known constant matrices of appropriate dimensions.

   Let $\gamma > 0$ be a given constant, then the system (2.2) is said to be with an H_∞ performance index no larger than γ, if the following conditions hold:
(1) Systems (2.2) are asymptotically stable
(2) Subject to initial conditions $x(0) = 0$, the transfer function matrix $T_{\omega z}(s)$ satisfies

$$\|T_{\omega z}(s)\|_\infty := \sup_{\|\omega\|_2 \leq 1} \frac{\|z\|_2}{\|\omega\|_2} \leq \gamma \tag{2.3}$$

Equation (2.3) is equivalent to

$$\int_0^\infty z^T(t)z(t)dt \leq \gamma^2 \int_0^\infty \omega^T(t)\omega(t)dt, \quad \forall \omega(t) \in L_2[0,\infty) \tag{2.4}$$

It is easy to see that the inequality (2.4) describes the restraint disturbance ability. Moreover, the smaller the value of γ is, the better the system performance is.

In addition, the definition of the H_∞ performance index for the z-domain or δ-domain is similar to Definition 2.2, therefore it is omitted here.

2.3 Operations on Systems

In this section, some facts about system interconnection are introduced, which will be used to obtain the sensitivity functions.

For brevity, the state-space models in the s-, z-, and δ-domains are unified as

$$
\begin{aligned}
\rho x(t) &= A_\rho x(t) + B_\rho u(t) \\
y(t) &= C_\rho x(t) + D_\rho u(t)
\end{aligned}
\tag{2.5}
$$

where

$$
\begin{cases}
\rho x(t) &= \dot{x}(t) & s\text{-domain} \\
\rho x(t) &= x(t+1) & z\text{-domain} \\
\rho x(t) &= \delta x(t) & \delta\text{-domain}
\end{cases}
$$

The state-space of the transfer function is described by

$$
T(\rho) = \left[\begin{array}{c|c} A_\rho & B_\rho \\ \hline C_\rho & D_\rho \end{array} \right] = C_\rho(\rho I - A_\rho)^{-1} B_\rho + D_\rho
$$

Then, the transpose of the transfer matrix $T(\rho)$ (or the dual system) is defined as

$$
T^T(\rho) = \left[\begin{array}{c|c} A_\rho^T & C_\rho^T \\ \hline B_\rho^T & D_\rho^T \end{array} \right] = B_\rho^T(\rho I - A_\rho^T)^{-1} C_\rho^T + D_\rho^T
$$

or equivalently

$$
\left[\begin{array}{c|c} A_\rho & B_\rho \\ \hline C_\rho & D_\rho \end{array} \right] \longmapsto \left[\begin{array}{c|c} A_\rho^T & C_\rho^T \\ \hline B_\rho^T & D_\rho^T \end{array} \right]
$$

Further, suppose that $T_1(\rho)$ and $T_2(\rho)$ are two subsystems with state-space representations:

$$
\begin{cases}
\rho x_1(t) &= A_\rho^1 x_1(t) + B_\rho^1 u_1(t) \\
y_1(t) &= C_\rho^1 x_1(t) + D_\rho^1 u_1(t)
\end{cases}
$$

$$
\begin{cases}
\rho x_2(t) &= A_\rho^2 x_2(t) + B_\rho^2 u_2(t) \\
y_2(t) &= C_\rho^2 x_2(t) + D_\rho^2 u_2(t)
\end{cases}
$$

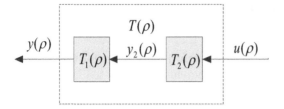

$$u(\rho) = u_2(\rho), u_1(\rho) = y_2(\rho), y(\rho) = y_1(\rho)$$

FIGURE 2.1

Two subsystems in a series.

The state-space of their transfer functions can be described by

$$T_1(\rho) = \left[\begin{array}{c|c} A_\rho^1 & B_\rho^1 \\ \hline C_\rho^1 & D_\rho^1 \end{array}\right] = C_\rho^1(\rho I - A_\rho^1)^{-1}B_\rho^1 + D_\rho^1$$

$$T_2(\rho) = \left[\begin{array}{c|c} A_\rho^2 & B_\rho^2 \\ \hline C_\rho^2 & D_\rho^2 \end{array}\right] = C_\rho^2(\rho I - A_\rho^2)^{-1}B_\rho^2 + D_\rho^2$$

Then the series or cascade connection of these two subsystems is a system with the output of the second subsystem as the input of the first subsystem as shown in the following.

$$u(\rho) = u_2(\rho), u_1(\rho) = y_2(\rho), y(\rho) = y_1(\rho)$$

The diagram is shown in Figure 2.1.

This operation in terms of the transfer matrices of the two subsystems is essentially the product of two transfer matrices. Hence, the representation for the series system can be obtained as

$$
\begin{aligned}
T(\rho) &= T_1(\rho)T_2(\rho) \\
&= \left[\begin{array}{c|c} A_\rho^1 & B_\rho^1 \\ \hline C_\rho^1 & D_\rho^1 \end{array}\right]\left[\begin{array}{c|c} A_\rho^2 & B_\rho^2 \\ \hline C_\rho^2 & D_\rho^2 \end{array}\right] \\
&= \left[\begin{array}{cc|c} A_\rho^1 & B_\rho^1 C_\rho^2 & B_\rho^1 D_\rho^2 \\ 0 & A_\rho^2 & B_\rho^2 \\ \hline C_\rho^1 & D_\rho^1 C_\rho^2 & D_\rho^1 D_\rho^2 \end{array}\right] \\
&= \left[\begin{array}{cc|c} A_\rho^2 & 0 & B_\rho^2 \\ B_\rho^1 C_2 & A_\rho^1 & B_\rho^1 D_\rho^2 \\ \hline D_\rho^1 C_\rho^2 & C_\rho^1 & D_\rho^1 D_\rho^2 \end{array}\right]
\end{aligned}
$$

Similarly, the parallel connection or the addition of $T_1(\rho)$ and $T_2(\rho)$ can

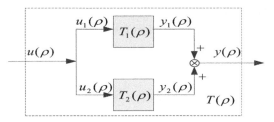

$$u(\rho) = u_1(\rho) = u_2(\rho), y(\rho) = y_1(\rho) + y_2(\rho)$$

FIGURE 2.2
Two subsystems in parallel.

be obtained as

$$
\begin{aligned}
T(\rho) &= T_1(\rho) + T_2(\rho) \\
&= \left[\begin{array}{c|c} A_\rho^1 & B_\rho^1 \\ \hline C_\rho^1 & D_\rho^1 \end{array} \right] + \left[\begin{array}{c|c} A_\rho^2 & B_\rho^2 \\ \hline C_\rho^2 & D_\rho^2 \end{array} \right] \\
&= \left[\begin{array}{cc|c} A_\rho^1 & 0 & B_\rho^1 \\ 0 & A_\rho^2 & B_\rho^2 \\ \hline C_\rho^1 & C_\rho^2 & D_\rho^1 + D_\rho^2 \end{array} \right]
\end{aligned}
$$

The diagram is shown in Figure 2.2.
More system operations can be found in Zhou, Doyle, and Glover [154].

2.4 Some Other Definitions and Lemmas

Some other definitions and lemmas that will be used in this chapter are presented as follows.

Definition 2.3 *[152] For a matrix $M \in R^{n \times m}$, m_{ij} denotes the (i,j)th element of the matrix M. Then, $\frac{\partial M^{-1}}{\partial m_{ij}}$ can be evaluated by*

$$\frac{\partial M^{-1}}{\partial m_{ij}} = -M^{-1} \frac{\partial M}{\partial m_{ij}} M^{-1}$$

Definition 2.4 *[43] Let m_{ij} denote the (i,j)th element of the matrix M with M being an $m \times n$ real matrix and let $f(M)$ be a matrix function of M. Then, the coefficient sensitivity function of f with respect to the (i,j)th element of M is given by*

$$S_{ij} = \frac{\partial f}{\partial m_{ij}}$$

Definition 2.5 *[111, 151] Let $V(x_\delta(k))$ be a Lyapunov functional in the delta-domain. A delta operator system is asymptotically stable, if the following conditions hold:*

(i) $V(x_\delta(k)) \geq 0$, with equality if and only if $x_\delta(k) = 0$;

(ii) $\delta V(x_\delta(k)) = [V(x_\delta(k+1)) - V(x_\delta(k))]/h < 0$.

Remark 2.1 *For Lyapunov functional $V(\bullet)$ both in the z-domain and the s-domain, the condition (i) in Definition 2.5 can always be given. In condition (ii), when $h = 1$, there exists*

$$\delta V(x_\delta(k))|_{h=1} = \frac{V(x(nh+h)) - V(x(nh))}{h}|_{h=1} = \Delta V(x_q(k)) < 0.$$

On the other hand, when $h \to 0$, referring to Equation (2.1) there exists

$$\lim_{h \to 0} \delta V(x_\delta(k)) = \lim_{h \to 0} \frac{V(x(nh+h)) - V(x(nh))}{h} = \frac{dV(x(t))}{dt} < 0.$$

The above results imply that the Lyapunov functional in the δ-domain can be reduced to the traditional Lyapunov functional in the z-domain and s-domain when the sampling period is 1 or tends to be 0.

Now, some important lemmas are introduced, which will be useful in this chapter.

Lemma 2.1 *[11] (Schur Complement Lemma) For any given symmetric matrix $S = \begin{bmatrix} S_{11} & S_{12} \\ S_{12}^T & S_{22} \end{bmatrix}$, where $S_{11} \in R^{r \times r}$. Then the following three conditions are equivalent:*
(i) $S < 0$
(ii) $S_{11} < 0, \quad S_{22} - S_{12}^T S_{11}^{-1} S_{12} < 0$
(ii) $S_{22} < 0, \quad S_{11} - S_{12} S_{22}^{-1} S_{12}^T < 0$

Lemma 2.2 *[113] Let matrices $Q = Q^T$, G, and a compact subset of real matrices $\mathbf{H}$ be given. Then the following statements are equivalent:*

(i) for each $H \in \mathbf{H}$

$$\xi^T Q \xi < 0 \text{ for all } \xi \neq 0 \text{ such that } HG\xi = 0;$$

(ii) there exists $\Theta = \Theta^T$ such that

$$Q + G^T \Theta G < 0, \mathbf{N}_H^T \Theta \mathbf{N}_H \geq 0 \text{ for all } H \in \mathbf{H}.$$

Lemma 2.3 *[154] Let $T_{azw} = C_a(sI - A_a)^{-1}B_a$, then A_a is Hurwitz and $\|T_{azw}\| < \gamma$ for some constant $\gamma > 0$ if and only if there exists a symmetric matrix $X > 0$ such that*

$$A_a^T X + X A_a + \frac{1}{\gamma^2} X B_a B_a^T X + C_a^T C_a < 0.$$

Lemma 2.4 *[46] Let $G_{azw}(z) = C_a(zI - A_a)^{-1}B_a$, then A_a is Shur stable and $\|G_{azw}(z)\| < \gamma$ for some constant $\gamma > 0$ if and only if there exists a symmetric matrix $X > 0$, such that*

$$\begin{bmatrix} -X & 0 & XA_a & XB_a \\ * & -I & C_a & 0 \\ * & * & -X & 0 \\ * & * & * & -\gamma^2 I \end{bmatrix} < 0 \qquad (2.6)$$

Denote

$$G_{0zw}(z) = C_{e0}(zI - A_{e0})^{-1}B_{e0}, \qquad (2.7)$$

where

$$A_{e0} = \begin{bmatrix} A & B_2 C_k \\ B_k C_2 & A_k \end{bmatrix}, \quad B_{e0} = \begin{bmatrix} B_1 \\ B_k D_{21} \end{bmatrix}, C_{e0} = \begin{bmatrix} C_1 & D_{12}C_k \end{bmatrix}, \qquad (2.8)$$

with $A_k \in R^{n \times n}$.

Then, we have the following lemma.

Lemma 2.5 *Let $\gamma > 0$ be a given constant. Then the following statements are equivalent:*

(i) A_{e0} is Shur stable, and $\|G_{0zw}(z)\| < \gamma$;

(ii) there exists a symmetric positive matrix $X > 0$ such that

$$\begin{bmatrix} -X & 0 & XA_{e0} & XB_{e0} \\ * & -I & C_{e0} & 0 \\ * & * & -X & 0 \\ * & * & * & -\gamma^2 I \end{bmatrix} < 0 \qquad (2.9)$$

(iii) there exists a symmetric positive matrix $X > 0$ and a matrix G such that

$$\begin{bmatrix} X - G - G^T & 0 & G^T A_{e0} & G^T B_{e0} \\ * & -I & C_{e0} & 0 \\ * & * & -X & 0 \\ * & * & * & -\gamma^2 I \end{bmatrix} < 0 \qquad (2.10)$$

(iv) there exists a nonsingular matrix T and a symmetric matrix $P > 0$ with

$$P = \begin{bmatrix} Y & N \\ N & -N \end{bmatrix}, \qquad (2.11)$$

such that

$$\begin{bmatrix} -P & 0 & PA_{ea} & PB_{ea} \\ * & -I & C_{ea} & 0 \\ * & * & -P & 0 \\ * & * & * & -\gamma^2 I \end{bmatrix} < 0 \qquad (2.12)$$

where

$$A_{ea} = \begin{bmatrix} A & B_2 C_{ka} \\ B_{ka} C_2 & A_{ka} \end{bmatrix}, \quad B_{ea} = \begin{bmatrix} B_1 \\ B_{ka} D_{21} \end{bmatrix}, \quad (2.13)$$
$$C_{ea} = \begin{bmatrix} C_1 & D_{12} C_{ka} \end{bmatrix}$$

and

$$A_{ka} = T^{-1} A_k T, \quad B_{ka} = T^{-1} B_k, \quad C_{ka} = C_k T.$$

(v) there exist a symmetric matrix $X > 0$ and a matrix G with structure

$$G = \begin{bmatrix} Y & N \\ N & -N \end{bmatrix}, \quad (2.14)$$

such that

$$\begin{bmatrix} X - G - G^T & 0 & G^T A_{ea} & G^T B_{ea} \\ * & -I & C_{ea} & 0 \\ * & * & -X & 0 \\ * & * & * & -\gamma^2 I \end{bmatrix} < 0 \quad (2.15)$$

holds, where A_{ea}, B_{ea}, and C_{ea} are defined by (2.13).

Proof 2.1 *(i) $\Longleftrightarrow$ (ii). From Lemma 2.3, the equivalence of (i) and (ii) is immediate.*
(ii) $\Longleftrightarrow$ (iii). On the one hand, let $G = G^T = X$, then (iii) holds if (ii) holds. On the other hand, if (iii) holds, we have $X - G - G^T < 0$; obviously, G is invertible. It is known to all that

$$G^T X^{-1} G > G^T + G - X. \quad (2.16)$$

In fact, $(G^T - X) X^{-1} (G - X) > 0$. Now, according to (2.16), if (2.10) holds, then the following inequality holds:

$$\begin{bmatrix} -G^T X^{-1} G & 0 & G^T A_{e0} & G^T B_{e0} \\ * & -I & C_{e0} & 0 \\ * & * & -X & 0 \\ * & * & * & -\gamma^2 I \end{bmatrix} < 0. \quad (2.17)$$

Let $T = diag\{G^{-1} X, I, I, I\}$ perform a transformation with T on (2.17), resulting in (2.9), which establishes that (iii) implies (ii).
(ii) $\Longleftrightarrow$ (iv). Notice the fact that, for any square matrix E and scalar $\eta > 0$, there exists an $\epsilon > 0$ with $\epsilon < \eta$ such that $E + \epsilon I$ is nonsingular, which implies statement (ii) if and only if there exists a symmetric matrix $X = \begin{bmatrix} X_{11} & X_{12} \\ X_{12}^T & X_{22} \end{bmatrix} > 0$ with X_{12} nonsingular such that (2.9) holds. Let $A_{ka} = (X_{12}^{-1})^T X_{22} A_k X_{22}^{-1} X_{12}^T, B_{ka} = -(X_{12}^{-1})^T X_{22} B_k, C_{ka} = -C_k X_{22}^{-1} X_{12}^T, Y = X_{11}$, and $N = -X_{12} X_{22}^{-1} X_{12}^T.$

Denote $\bar{\Gamma} = diag\{\Gamma, I, \Gamma, I\}$, *where* $\Gamma = \begin{bmatrix} I & 0 \\ 0 & -X_{12}X_{22}^{-1} \end{bmatrix}$. *Then*

$$P = \Gamma X \Gamma^T = \begin{bmatrix} Y & N \\ N & -N \end{bmatrix}$$

and

$$
\begin{bmatrix}
-P & 0 & PA_{ea} & PB_{ea} \\
* & -I & C_{ea} & 0 \\
* & * & -P & 0 \\
* & * & * & -\gamma^2 I
\end{bmatrix}
$$

$$
= \bar{\Gamma}
\begin{bmatrix}
-X & 0 & XA_{e0} & XB_{e0} \\
* & -I & C_{e0} & 0 \\
* & * & -X & 0 \\
* & * & * & -\gamma^2 I
\end{bmatrix}
\bar{\Gamma}^T < 0,
$$

so inequalities $X > 0$ and (2.9) are equivalent to $P > 0$ and (2.12), respectively.
(iv) $\Longleftrightarrow$ (v). On the one hand, let $G = G^T = X = P > 0$ with the structure (2.11), then (2.15) holds if (2.12) holds.
On the other hand, let $X = P$ with the structure (2.11), according to (2.16), then we have

$$
\begin{bmatrix}
-G^T P^{-1} G & 0 & G^T A_{ea} & G^T B_{ea} \\
* & -I & C_{ea} & 0 \\
* & * & -P & 0 \\
* & * & * & -\gamma^2 I
\end{bmatrix} < 0,
\qquad (2.18)
$$

which holds if (2.15) holds. Let $\Upsilon = diag\{G^{-1}P, I, I, I\}$ perform a transformation with Υ on (2.18), resulting in (2.12), which establishes that (v) implies (iv).
Thus, the proof is complete.

Denote

$$T_{0zw} = C_{e0}(sI - A_{e0})^{-1}B_{e0}.$$

Let controller gain matrices A_k, B_k, and C_k be given, and such that

$$\|T_{0zw}\| = \|C_{e0}(sI - A_{e0})^{-1}B_{e0}\| < \gamma. \qquad (2.19)$$

Lemma 2.6 *Let $T_{azw} = C_a(sI - A_a)^{-1}B_a$, then A_a is stable and $\|T_{azw}\| < \gamma$ for some constant $\gamma > 0$ if and only if there exists a symmetric matrix $X > 0$, such that*

$$A_a^T X + X A_a + \frac{1}{\gamma^2} X B_a B_a^T X + C_a^T C_a < 0.$$

Lemma 2.7 *Let $\gamma > 0$ be a given constant. Then the following statements are equivalent:*

(i) A_{e0} *is Hurwitz, and* $\|T_{0zw}\| < \gamma$.

(ii) There exists a symmetric matrix $X > 0$ *such that*

$$A_{e0}^T X + X A_{e0} + \frac{1}{\gamma^2} X B_{e0} B_{e0}^T X + C_{e0}^T C_{e0} < 0. \qquad (2.20)$$

(iii) There exist a nonsingular matrix T *and a symmetric matrix* $P > 0$ *with structure (2.14) such that*

$$A_{ea}^T P + P A_{ea} + \frac{1}{\gamma^2} P B_{ea} B_{ea}^T P + C_{ea}^T C_{ea} < 0 \qquad (2.21)$$

where $A_{ea}, B_{ea},$ *and* C_{ea} *are defined by (2.13).*

Lemma 2.8 *Let matrices* Q, F_1, *and* F_2 *be constant matrices with appropriate dimensions. Then the following statements are equivalent:*
(i)

$$Q + F_1 \Delta F_2 + (F_1 \Delta F_2)^T < 0, \ \text{for} \ |\theta_i| \le \theta_a, \ i = 1, \cdots, s,$$

where $\Delta = diag[\theta_1, \ \cdots, \theta_s].$
(ii)

$$Q + F_1 \Delta F_2 + (F_1 \Delta F_2)^T < 0, \ \text{for} \ \Delta \in \Delta_v,$$

where $\Delta_v = \{\Delta : \ \theta_i \in \{-\theta_a, \ \theta_a\}, i = 1, \cdots, s\}.$
(iii) There exists a symmetric matrix $\Theta \in R^{2s \times 2s}$ *such that*

$$\begin{bmatrix} Q & F_1 \\ F_1^T & 0 \end{bmatrix} + \begin{bmatrix} F_2 & 0 \\ 0 & I \end{bmatrix}^T \Theta \begin{bmatrix} F_2 & 0 \\ 0 & I \end{bmatrix} < 0, \qquad (2.22)$$

$$\begin{bmatrix} I \\ \Delta \end{bmatrix}^T \Theta \begin{bmatrix} I \\ \Delta \end{bmatrix} \ge 0, \ \text{for all} \ \Delta \in \Delta_v. \qquad (2.23)$$

Proof 2.2 *(i)* $\Longleftrightarrow$ *(ii). It is immediate.*
(i) $\Longleftrightarrow$ *(iii). Let* $f(\Delta) = \bar{\xi}^T[Q + F_1 \Delta F_2 + (F_1 \Delta F_2)^T]\bar{\xi}$. *Then (i) is equivalent to* $f(\Delta) < 0$ *for all* $\bar{\xi} \ne 0$, *which further is equivalent to*

$$\bar{\xi}^T Q \bar{\xi} + 2\bar{\xi}^T F_1 \bar{y} < 0 \qquad (2.24)$$

which holds for all $\begin{bmatrix} \bar{\xi} \\ \bar{y} \end{bmatrix} \ne 0$ *such that*

$$HF \begin{bmatrix} \bar{\xi} \\ \bar{y} \end{bmatrix} = 0, \qquad (2.25)$$

where $H = \begin{bmatrix} \Delta & -I \end{bmatrix}$ *and* $F = \begin{bmatrix} F_2 & 0 \\ 0 & I \end{bmatrix}$. *It is easy to see that* $\begin{bmatrix} I \\ \Delta \end{bmatrix}$ *forms* *a basis of N_H. By Lemma 2.2, and Equations (2.24) and (2.25), it follows that $f(\Delta) < 0$ for all $\bar{\xi} \neq 0$ is equivalent to the symmetric matrix $\Theta = \begin{bmatrix} \Theta_{11} & \Theta_{12} \\ \Theta_{12}^T & \Theta_{22} \end{bmatrix} \in R^{2s \times 2s}$ with $\Theta_{22} \in R^{s \times s}$ such that (2.22) holds and*

$$\begin{bmatrix} I \\ \Delta \end{bmatrix}^T \Theta \begin{bmatrix} I \\ \Delta \end{bmatrix} \geq 0, \text{ for all } |\theta_i| \leq \theta_a. \tag{2.26}$$

From (2.22), it follows that $\Theta_{22} < 0$, which implies that $\begin{bmatrix} I \\ \Delta \end{bmatrix}^T \Theta \begin{bmatrix} I \\ \Delta \end{bmatrix}$ is convex for each θ_i. Hence condition (2.26) is equivalent to condition (2.23). Thus, the proof is complete.

Lemma 2.9 *(Finsler's lemma) [25] Let $X \in R^n$, $P = P^T \in R^{n \times n}$, and $M \in R^{m \times n}$ such that $\text{rank}(M) = r < n$. The following two statements are equivalent:*
  (1) $X^T P X < 0, \forall M X = 0, X \neq 0$;
  (2) $\exists N \in R^{n \times m} : P + He\{NM\} < 0$.

Lemma 2.10 *[71] There exists a positive-definite matrix $P > 0$ that satisfies*

$$A^T P + PA < 0$$

if and only if there exists a matrix $P > 0$ and a sufficiently small positive scalar $\varepsilon > 0$ such that

$$\begin{bmatrix} -\varepsilon P^{-1} & I + \varepsilon A \\ \varepsilon A^T & -\varepsilon^{-1} P \end{bmatrix} < 0.$$

Lemma 2.11 *[71] Let $\lambda(P)$ be a matrix expression that may relate to P, or may have nothing to do with P. Then, there exists a positive-definite matrix $P > 0$ such that*

$$\begin{bmatrix} -\lambda(P) & A^T P \\ PA & -P \end{bmatrix} < 0$$

if and only if there exists a positive matrix $P > 0$ and a matrix Z such that

$$\begin{bmatrix} -\lambda(P) & A^T Z \\ Z^T A & P - Z - Z^T \end{bmatrix} < 0.$$

Lemma 2.12 *[19, 107, 109] Given matrices Y, M, and N with the appropriate dimensions, then the following statements are equivalent:*
  (i)
$$Y + M\Delta N + N^T \Delta^T M^T < 0$$

  holds for all Δ satisfying $\Delta^T \Delta \leq \theta I$.
  (ii)
$$Y + M\Delta N + N^T \Delta^T M^T + \theta I < 0$$

holds for all Δ satisfying $\Delta^T \Delta \le \theta I$ and some $\theta > 0$.

(iii)   There exists a constant $\epsilon > 0$ such that

$$Y + \epsilon M M^T + \frac{\theta}{\epsilon} N^T N < 0$$

Lemma 2.13 *For a given constant $\gamma > 0$, the linear system with the transfer function matrix $T(\rho) = C_{cl}(\rho I - A_{cl})^{-1} B_{cl} + D_{cl}$ is asymptotically stable and satisfies $\|T(\rho)\|_\infty \le \gamma$ if and only if there exists a positive symmetric matrix P such that*

1. *s-domain [11]*

$$\begin{bmatrix} He\{PA_{cl}\} & PB_{cl} & C_{cl}^T \\ * & -\gamma^2 I & 0 \\ * & * & -I \end{bmatrix} < 0 \qquad (2.27)$$

2. *z-domain [11]*

$$\begin{bmatrix} -P & * & * & * \\ 0 & -\gamma^2 I & * & * \\ PA_{cl} & PB_{cl} & -P & * \\ C_{cl} & D_{cl} & 0 & -I \end{bmatrix} < 0 \qquad (2.28)$$

3. *δ-domain [86]*

$$\begin{bmatrix} PA_{cl} + A_{cl}^T P & PB_{cl} & C_\rho^T & \sqrt{h}A_{cl}^T P \\ * & -\gamma^2 I & D_{cl}^T & \sqrt{h}B_{cl}^T P \\ * & * & -I & 0 \\ * & * & * & -P \end{bmatrix} < 0 \qquad (2.29)$$

where h denotes the sampling period.

3

Non-Fragile State Feedback Control with Norm-Bounded Gain Uncertainty

3.1 Introduction

This chapter is devoted to the study of the problem of *non-fragile* guaranteed cost control for discrete-time *linear systems* under state feedback controller gain uncertainties. It is well known that even small perturbations in controller parameters could destabilize the *closed-loop system* [129]. Keel and Bhattacharyya [77] have also shown by a number of examples that the controllers designed by using weighted H_∞, μ and l_1 synthesis techniques may be very sensitive, or fragile, with respect to errors in the controller coefficients, although they are robust with respect to plant uncertainty. Therefore, it is important to design a controller for a given plant with uncertainty such that the controller is insensitive to some amount of error with respect to its gains, that is, the controller is non-fragile. Recently, researchers have increasingly paid attention to the non-fragile controller design problem [8, 27, 34, 54, 68, 74, 78, and the references therein]. In this chapter, two classes of uncertainties are considered, namely, *additive* and *multiplicative*. The state feedback control designs for optimal guaranteed cost control under the two classes of gain uncertainties are given in terms of solutions to algebraic Riccati equations. The designs are such that the cost of the closed-loop system is guaranteed to be within a certain bound for all admissible uncertainties. The proposed approach in this chapter also provides a basis for solving other related problems that are to be studied in the rest of the book.

3.2 Problem Statement

Consider a discrete-time linear system described by the equation

$$x_{k+1} = Ax_k + Bu_k \tag{3.1}$$

where $x_k \in R^n$ is the state and $u_k \in R^m$ is the control input, and A and B are known constant matrices. The cost function associated with this system is

$$J = \sum_{k=0}^{\infty} (x_k^T Q x_k + u_k^T R u_k) \tag{3.2}$$

where $Q > 0$ and $R > 0$ are given weighting matrices. For a given controller $u_k = K x_k$, the actual controller implemented is assumed to be

$$u_k = (K + \Delta K) x_k \tag{3.3}$$

where K is the nominal controller gain and ΔK represents the gain uncertainties. In this chapter, the following two classes of uncertainties are considered.
(a) ΔK is of the additive form

$$\Delta K = H_1 F_1 E_1, \quad F_1^T F_1 \leq \rho I, \quad \rho \geq 0 \tag{3.4}$$

with H_1 and E_1 being known constant matrices, and F_1 the uncertain parameter matrix.
(b) ΔK is of the *multiplicative* form

$$\Delta K = H_2 F_2 E_2 K, \quad F_2^T F_2 \leq \rho I, \quad \rho \geq 0 \tag{3.5}$$

with H_2 and E_2 being known constant matrices, and F_2 the uncertain parameter matrix.

Remark 3.1 *The controller gain uncertainties can result from the actuator degradations, as well as from the requirement for readjustment of controller gains during the controller implementation stage [see 1, 27, 77, 126]. These uncertainties in the controller gains are modeled here as uncertain gains that are dependent on uncertain parameters. The models of additive uncertainties (3.4) and multiplicative uncertainties (3.5) are used to describe the controller gain variations in Famularo et al. [34] and Haddad and Corrad [53], respectively. The multiplicative model can also be used to describe degradations of actuator effectiveness [1].*

For the actual controller implemented, we introduce the following definition, which is similar to that in Xie [137].

Definition 3.1 *Consider the system (3.1) with the cost function (3.2). The control law (3.3) with controller gain uncertainties (3.4) or (3.5) is said to be a guaranteed cost control with matrix $P > 0$ if*

$$[A + B(K + \Delta K)]^T P [A + B(K + \Delta K)] - P + (K + \Delta K)^T R(K + \Delta K) + Q < 0$$

for all uncertainties ΔK satisfying (3.4) or (3.5).

Definition 3.2 *[37] The closed-loop uncertain system*

$$x_{k+1} = [A + B(K + \Delta K)] x_k \tag{3.6}$$

is said to be quadratically stable if there exists a matrix $P > 0$ such that

$$[A + B(K + \Delta K)]^T P[A + B(K + \Delta K)] - P < 0$$

for all uncertainties ΔK satisfying (3.4) or (3.5).

The following result shows that a guaranteed cost control for the system (3.1) will guarantee the quadratic stability of the closed-loop system (3.6) and defines an upper bound on the cost function (3.2).

Lemma 3.1 *Consider the system (3.1) with the cost function (3.2). Suppose that the control law (3.3) with controller uncertainties (3.4) or (3.5) is a guaranteed cost control with matrix $P > 0$. Then the closed-loop uncertain system (3.6) is quadratically stable and*

$$J = \sum_{k=0}^{\infty} x_k^T [Q + (K + \Delta K)^T R(K + \Delta K)] x_k \leq x_0^T P x_0 \qquad (3.7)$$

for all uncertainties ΔK satisfying (3.4) or (3.5).

Proof 3.1 *The quadratic stability of system (3.6) is immediate from Definition 3.1 and Definition 3.2. Let $V(x_k) = x_k^T P x_k$. Then, along the state trajectory of (3.6), we have*

$$V(x_{k+1}) - V(x_k) = x_k^T ([A + B(K + \Delta K)]^T P[A + B(K + \Delta K)] - P)x_k$$
$$\leq -(u_k^T R u_k + x_k^T Q x_k).$$

It follows that

$$J = \lim_{N \to \infty} \sum_{k=0}^{N-1} (u_k^T R u_k + x_k^T Q x_k) \leq \lim_{N \to \infty} [V(x_0) - V(x_N)] = V(x_0).$$

Thus, the proof is complete.

In this chapter, the problem under consideration is to design a state feedback gain K such that the control law (3.3) with (3.4) or (3.5) is a guaranteed cost control associated with a cost matrix P. In particular, the optimal guaranteed cost control will be pursued.

It should be noted that the results in Kaminer, Khargonekar, and Rotea [75] can provide sufficient conditions for the guaranteed cost control problem under controller gain uncertainties. In fact, if the closed-loop cost function J in (3.7) is bounded by $\bar{J} = \sum_{k=0}^{\infty} x_k^T (Q + K^T R_0 K) x_k$, where $R_0 \geq (I + H_2 F_2 E_2)^T R(I + H_2 F_2 E_2)$ for any F_2 satisfying (3.5), then for the multiplicative case, applying the results in Kaminer, Khargonekar, and Rotea [75] to the system $x_{k+1} = A x_k + B(I + H_2 F_2 E_2)u$ and the cost function $\bar{J}$, a sufficient condition for the guaranteed cost control problem under the multiplicative gain uncertainties is obtained. The case for the additive gain uncertainties is similar. In this chapter, we will provide necessary and sufficient conditions for the guaranteed cost control problem under controller gain uncertainties.

Definition 3.3 *[83] A symmetric matrix P is said to be a stabilizing solution to the Riccati equation*

$$A^T P A - P - A^T P B (B^T P B + R)^{-1} B^T P A + N = 0$$

if it satisfies the Riccati equation and the matrix $A - B(B^T P B + R)^{-1} B^T P A$ is stable.

3.3 Non-Fragile Guaranteed Cost Controller Design

In this section, the problem of designing non-fragile guaranteed cost controllers is considered. The designed controller is insensitive to additive controller gain uncertainties and *multiplicative* controller gain uncertainties, respectively.

3.3.1 Additive Controller Gain Uncertainty Case

In this section, we consider the guaranteed cost control problem under additive gain uncertainties of the form (3.4). We first give the following theorem.

Theorem 3.1 *Consider the system (3.1) with the cost function (3.2). There exists a state feedback gain K such that the control law (3.3) with additive uncertainty (3.4) is a quadratic guaranteed cost control with a cost matrix P if and only if there exists a constant $\epsilon > 0$ such that*

$$R_2 = R_2(P, \epsilon) \overset{\triangle}{=} I - \epsilon H_1^T (B^T P B + R) H_1 > 0 \qquad (3.8)$$

and

$$S_a(P, \epsilon) \overset{\triangle}{=} A^T P A - P + \frac{\rho}{\epsilon} E_1^T E_1 + Q - A^T P B (B^T P B + R)^{-1} B^T P A < 0. \quad (3.9)$$

Furthermore, if (3.8) and (3.9) are satisfied, then a guaranteed cost control law is given by (3.3) with

$$K = - \left(B^T P B + R \right)^{-1} B^T P A. \qquad (3.10)$$

Proof 3.2 *Let the control law (3.3) with controller gain uncertainty (3.4) be a quadratic guaranteed cost control with a cost matrix P. Then from Definition 3.1, it follows that*

$$[A + B(K + \Delta K)]^T P [A + B(K + \Delta K)] - P + (K + \Delta K)^T R(K + \Delta K) + Q < 0$$

for all uncertainties ΔK of the form (3.4). By Lemma 2.1 and (3.4), this *inequality is equivalent to the following inequality,*

$$
\begin{bmatrix}
-P^{-1} & 0 & A+B(K+\Delta K) \\
* & -R^{-1} & K+\Delta K \\
* & * & Q-P
\end{bmatrix}
$$

$$
= \Theta + He\left\{ \begin{bmatrix} BH_1 \\ H_1 \\ 0 \end{bmatrix} F_1 \begin{bmatrix} 0 & 0 & E_1 \end{bmatrix} \right\} < 0
$$

where $\Theta = \begin{bmatrix} -P^{-1} & 0 & A+BK \\ * & -R^{-1} & K \\ * & * & Q-P \end{bmatrix}$.

By Lemma 2.12, the above inequality holds if and only if there exists a constant $\epsilon > 0$ such that

$$
\Theta + \epsilon \begin{bmatrix} BH_1 \\ H_1 \\ 0 \end{bmatrix} \begin{bmatrix} BH_1 \\ H_1 \\ 0 \end{bmatrix}^T + \frac{\rho}{\epsilon} \begin{bmatrix} 0 \\ 0 \\ E_1^T \end{bmatrix} \begin{bmatrix} 0 & 0 & E_1 \end{bmatrix}
$$

$$
= \begin{bmatrix}
-P^{-1}+\epsilon BH_1H_1^TB^T & \epsilon BH_1H_1^T & A+BK \\
* & -R^{-1}+\epsilon H_1H_1^T & K \\
* & * & Q-P+\frac{\rho}{\epsilon}E_1^TE_1
\end{bmatrix} < 0.
$$

By the Schur complement and completing the square, it follows that the above inequality is equivalent to

$$
M = \begin{bmatrix} M_{11} & M_{12} \\ M_{12}^T & M_{22} \end{bmatrix} \triangleq \left(\begin{bmatrix} P^{-1} & 0 \\ 0 & R^{-1} \end{bmatrix} - \epsilon \begin{bmatrix} BH_1 \\ H_1 \end{bmatrix} \begin{bmatrix} BH_1 \\ H_1 \end{bmatrix}^T \right)^{-1} > 0 \tag{3.11}
$$

and

$$
\begin{aligned}
\Delta_1 &= \begin{bmatrix} (A+BK)^T & K^T \end{bmatrix} M \begin{bmatrix} A+BK \\ K \end{bmatrix} - P + \frac{\rho}{\epsilon}E_1^TE_1 + Q \\
&= (A+BK)^T M_{11}(A+BK) + K^T M_{12}^T(A+BK) + (A+BK)^T M_{12}K \\
&\quad + K^T M_{22}K - P + \frac{\rho}{\epsilon}E_1^TE_1 + Q \\
&= A^T M_{11}A - P + \frac{\rho}{\epsilon}E_1^TE_1 + Q - A^T \mathcal{M}R_1^{-1}\mathcal{M}^T A \\
&\quad + \begin{bmatrix} K^T + A^T\mathcal{M}R_1^{-1} \end{bmatrix} R_1 \begin{bmatrix} K^T + A^T\mathcal{M}R_1^{-1} \end{bmatrix}^T < 0, \tag{3.12}
\end{aligned}
$$

where

$$
\begin{aligned}
\mathcal{M} &= M_{11}B + M_{12}, \\
R_1 &= B^T M_{11}B + M_{22} + M_{12}^TB + B^T M_{12}. \tag{3.13}
\end{aligned}
$$

It is easy to show that $M > 0$ holds if and only if the inequality (3.8) holds.

By computing directly, we have

$$
M = \begin{bmatrix} P & 0 \\ * & R \end{bmatrix} + \epsilon \begin{bmatrix} P & 0 \\ * & R \end{bmatrix} \begin{bmatrix} BH_1 \\ H_1 \end{bmatrix} R_2^{-1} \begin{bmatrix} BH_1 \\ H_1 \end{bmatrix}^T \begin{bmatrix} P & 0 \\ * & R \end{bmatrix}
$$

$$
= \begin{bmatrix} P + \epsilon PBH_1 R_2^{-1} H_1^T B^T P & \epsilon PBH_1 R_2^{-1} H_1^T R \\ * & R + \epsilon RH_1 R_2^{-1} H_1^T R \end{bmatrix}. \tag{3.14}
$$

Thus, from (3.13) and (3.14), it follows that

$$
\begin{aligned}
R_1 &= B^T PB + \epsilon B^T PBH_1 R_2^{-1} H_1^T B^T PB + R + \epsilon RH_1 R_2^{-1} H_1^T R \\
&\quad + \epsilon B^T PBH_1 R_2^{-1} H_1^T R + \epsilon RH_1 R_2^{-1} H_1^T B^T PB \\
&= X + \epsilon X H_1 R_2^{-1} H_1^T X, \tag{3.15}
\end{aligned}
$$

$$
\begin{aligned}
M_{11}B + M_{12} &= PB + \epsilon PBH_1 R_2^{-1} H_1^T B^T PB + \epsilon PBH_1 R_2^{-1} H_1^T R \\
&= PB(I + \epsilon H_1 R_2^{-1} H_1^T X), \tag{3.16}
\end{aligned}
$$

$$
\begin{aligned}
R_1^{-1}(M_{11}B + M_{12})^T &= (X + \epsilon X H_1 R_2^{-1} H_1^T X)^{-1}(I + \epsilon H_1 R_2^{-1} H_1^T X) B^T P \\
&= X^{-1} B^T P, \tag{3.17}
\end{aligned}
$$

where

$$
X = B^T PB + R. \tag{3.18}
$$

By (3.11), (3.12), and (3.14)-(3.18), it follows that

$$
\begin{aligned}
\Delta_1 &= A^T PA - P + \frac{\rho}{\epsilon} E_1^T E_1 + Q - A^T PB \left[(I + \epsilon H_1 R_2^{-1} H_1^T X) X^{-1} \right. \\
&\quad \left. - \epsilon H_1 R_2^{-1} H_1^T \right] B^T PA + \left[K^T + A^T PBX^{-1} \right] R_1 \left[K^T + A^T PBX^{-1} \right]^T \\
&= S_a(P, \epsilon) + \left[K^T + A^T PBX^{-1} \right] R_1 [K^T + A^T PBX^{-1}]^T. \tag{3.19}
\end{aligned}
$$

From (3.11) and (3.19), the necessity is obvious. For the sufficiency, the proof is completed by substituting K in (3.10) into Equation (3.19).

Theorem 3.1 provides a necessary and sufficient condition for the solution to the quadratic guaranteed cost control problem. But it remains unclear as to how one can choose the design parameter ϵ in order to achieve the minimal guaranteed cost of the closed-loop system. Denote

$$
\epsilon_a = \sup\{\epsilon > 0 : S_a(P, \epsilon) = 0 \text{ has a stabilizing solution } P \geq 0
$$
$$
\text{and (3.8) holds}\} \tag{3.20}
$$

Then, the design parameter ϵ for achieving the suboptimal guaranteed cost of the closed-loop system falls in the range of $0 < \epsilon < \epsilon_a$. The next theorem shows that the optimal guaranteed cost control (i.e., the control law that yields the minimal cost as defined in (3.2)) is obtained at the boundary value of $\epsilon = \epsilon_a$.

Theorem 3.2 *Consider the system (3.1) with the cost function (3.2). Suppose that the pair (A, B) is stabilizable. If there exists a state feedback gain K such that the control law (3.3) with additive uncertainty (3.4) is a quadratic guaranteed cost control with a cost matrix P_0, then the following Riccati equation with ϵ_a defined by (3.20) has a unique stabilizing solution $P_{opt} > 0$ satisfying $P_{opt} \leq P_0$ and*

$$R_2(P_{opt}, \epsilon_a) \geq 0, \tag{3.21}$$

$$S_a(P_{opt}, \epsilon_a) = 0 \tag{3.22}$$

and the control law (3.3) with

$$K = -(B^T P_{opt} B + R)^{-1} B^T P_{opt} A \tag{3.23}$$

is such that the resulting closed-loop system (3.6) is quadratically stable, and $J \leq x_0^T P_{opt} x_0$ for all uncertainties ΔK of the form (3.4).

Proof 3.3 *By Theorem 3.1, there exists a constant $\epsilon' > 0$ such that the inequalities (3.8) and (3.9) hold for $\epsilon = \epsilon'$ and $P = P_0$. Let $P_{01} \geq 0$ be a stabilizing solution to $S_a(P, \epsilon') = 0$. By the comparison theorem (Theorem 13.3.1 in Lancaster and Rodman [83]), we have $P_{01} \leq P_0$ and $R_2(P_{01}, \epsilon') > 0$. Thus, ϵ_a in (3.20) is well defined. Choose sequences $\{\epsilon_n\}_{n=1}^\infty$ and $\{P_n\}_{n=1}^\infty$ such that $0 < \epsilon_n \leq \epsilon_{n+1}$, $\epsilon_n \to \epsilon_a$ $(n \to \infty)$, P_n is a stabilizing solution to $S_a(P, \epsilon_n) = 0$ and $R_2(P_n, \epsilon_n) > 0$. By the definition of $S_a(P, \epsilon)$ in (3.9) and the comparison theorem, we have $P_n \geq P_{n+1} > 0$ $(n = 1, 2, \cdots)$. Thus, $\lim_{n \to \infty} P_n = P_\infty \geq 0$ exists, and P_∞ satisfies $S_a(P_\infty, \epsilon_a) = 0$ and $R_2(P_\infty, \epsilon_a) \geq 0$. By Theorem 16.6.4 in Lancaster and Rodman [83], it follows that P_∞ is a stabilizing solution to $S_a(P_\infty, \epsilon_a) = 0$ and $P_\infty > 0$. Consider a sequence $\{\sigma_n\}_{n=1}^\infty$ with $\sigma_n > 0$, $\sigma_n \to 0$ $(n \to \infty)$, then there exists a sequence $\{\epsilon_{0n}\}_{n=1}^\infty$ with $0 < \epsilon_{0n} < \epsilon_a$, $\epsilon_{0n} \to \epsilon_a$ $(n \to \infty)$ such that*

$$S_a(P_\infty, \epsilon_{0n}) - \sigma_n I < 0, \ n = 1, 2, \cdots.$$

By the proof of Theorem 3.1, it follows that

$$[A + B\mathcal{K}]^T P_\infty [A + B\mathcal{K}] - P_\infty + \mathcal{K}^T R \mathcal{K} + Q - \sigma_n I < 0, \ n = 1, 2, \cdots,$$

where K is given by (3.23) with $P_{opt} = P_\infty$ and $\mathcal{K} = K + \Delta K$, and ΔK is given by (3.4). Let $n \to \infty$, and we have

$$[A + B\mathcal{K}]^T P_\infty [A + B\mathcal{K}] - P_\infty + \mathcal{K}^T R \mathcal{K} + Q \leq 0.$$

Let $P_\omega = \omega P_\infty$ with $\omega > 1$. Then, from $Q > 0$ and the above inequality, it follows that

$$\begin{aligned}
&[A + B\mathcal{K}]^T P_\omega [A + B\mathcal{K}] - P_\omega + \mathcal{K}^T R \mathcal{K} + Q \\
&= \omega\{[A + B\mathcal{K}]^T P_\omega [A + B\mathcal{K}] - P_\omega + \mathcal{K}^T R \mathcal{K} + Q\} - (\omega - 1)[\mathcal{K}^T R \mathcal{K} + Q] \\
&\leq -(\omega - 1)[\mathcal{K}^T R \mathcal{K} + Q] \\
&< 0.
\end{aligned}$$

Thus, $u_k = (K + \Delta K)x_k$ is a guaranteed cost control with P_w. By Lemma 3.1 and letting $\omega \to 1$, we have $J \leq \lim_{\omega \to 1} x_0^T P_w x_0 = x_0^T P_\infty x_0$. Since $\epsilon' \leq \epsilon_a$, it follows that $P_\infty \leq P_{01} \leq P_0$. The proof is completed by letting $P_{opt} = P_\infty$.

Remark 3.2 *Theorem 3.2 presents a design procedure for optimal guaranteed cost control, and the closed-loop value of the cost function J is bounded by the minimal value $x_0^T P_{opt} x_0$. From (3.8), the parameter ϵ_a in Theorem 3.2 lies in the range of $0 < \epsilon_a \leq \lambda_a$ where*

$$\lambda_a = \begin{cases} [\lambda_{max}\{H_1^T(B^T P_a B + R)H_1\}]^{-1} & \text{if } H_1 \neq 0 \\ \infty & \text{if } H_1 = 0 \end{cases} \qquad (3.24)$$

and $P_a > 0$ is the stabilizing solution to the Riccati equation

$$S_{a,\infty}(P) = A^T PA - P + Q - A^T PB(B^T PB + R)^{-1} B^T PA = 0. \qquad (3.25)$$

Then, from (3.20), (3.21), and (3.22), it follows that

$$\epsilon_a = \max \{0 < \epsilon \leq \lambda_a : S_a(P, \epsilon) = 0 \text{ has a stabilizing solution } P \geq 0$$
$$\text{and } R_2(P, \epsilon_a) \geq 0\} \qquad (3.26)$$

Thus, ϵ_a can be searched by solving the Riccati equation $S_a(P, \epsilon) = 0$ to find a stabilizing solution $P \geq 0$, and checking if P satisfies $R_2(P, \epsilon_a) \geq 0$ for $\epsilon \in (0, \lambda_a]$ in an increasing sequence. However, the search for ϵ_a may be difficult if the interval $(0, \lambda]$ is very large. Moreover, since the optimal parameter ϵ_a is on the boundary of the interval $(0, \epsilon_a)$, yielding a family of guaranteed cost controls, it will be safe to choose an ϵ slightly smaller than ϵ_a for achieving a suboptimal guaranteed cost control in a practical design. It should be noted that Equation (3.25) turns out to be that of the standard quadratic optimal control [82] for the system (3.1) with the cost function (3.2). Also, the Riccati Equation (3.22) corresponds to that of the standard quadratic optimal control for the system (3.1) with a cost function

$$J_a = \sum_{k=0}^{\infty} (x_k^T \bar{Q} x_k + u_k^T R u_k)$$

where $\bar{Q} = \frac{\rho}{\epsilon_a} E_1^T E_1 + Q$.

3.3.2   Multiplicative Controller Gain Uncertainty Case

In this section, we consider the guaranteed cost control problem under the multiplicative gain uncertainties (3.5).

Theorem 3.3 *Consider the system (3.1) with the cost function (3.2). There exists a state feedback gain matrix K such that the control law (3.3) with* *multiplicative uncertainty (3.5) is a quadratic guaranteed cost control with a cost matrix P if and only if there exists a constant $\epsilon > 0$ such that*

$$R_{20} = R_{20}(P, \epsilon) \triangleq I - \epsilon H_2^T (B^T P B + R) H_2 > 0 \tag{3.27}$$

and

$$S_m(P, \epsilon) \triangleq A^T P A - P + Q - A^T P B \Delta_0 B^T P A < 0 \tag{3.28}$$

where

$$\Delta_0 \triangleq \left(I - \rho H_2 H_2^T E_2^T E_2 \right) \left[\left(B^T P B + R \right) \left(I - \rho H_2 H_2^T E_2^T E_2 \right) + \frac{\rho}{\epsilon} E_2^T E_2 \right]^{-1}. \tag{3.29}$$

Furthermore, if both (3.27) and (3.28) hold, then a guaranteed cost control law with the cost matrix P is given by (3.3) with

$$K = - \left\{ I - \rho \left(X^{-1} - \epsilon H_2 H_2^T \right) E_2^T \left[\epsilon \left(I - \rho E_2 H_2 H_2^T E_2^T \right) + \rho E_2 X^{-1} E_2^T \right]^{-1} \right. $$
$$\left. E_2 \right\} \times X^{-1} B^T P A, \tag{3.30}$$

where $X = B^T P B + R$.

Proof 3.4 *From the proof of Theorem 3.1, it is easy to see that*

$$[A + B(K + \Delta K)]^T P [A + B(K + \Delta K)] - P + (K + \Delta K)^T R (K + \Delta K) + Q < 0$$

for all uncertainties ΔK satisfying (3.5) if and only if there exists a constant $\epsilon > 0$ such that inequality (3.27) holds and

$$
\begin{aligned}
\Delta_2 &= \begin{bmatrix} (A + BK)^T & K^T \end{bmatrix} M_0 \begin{bmatrix} A + BK \\ K \end{bmatrix} - P + \frac{\rho}{\epsilon} K^T E_2^T E_2 K + Q \\
&= A^T P A - P + Q - A^T P B \left[R_4 R_3^{-1} R_4^T - \epsilon H_2 R_{20}^{-1} H_2^T \right] B^T P A \\
&\quad + \left[K^T + A^T P B R_4 R_3^{-1} \right] R_3^{-1} \times \left[K^T + A^T P B R_4 R_3^{-1} \right]^T < 0 \quad (3.31)
\end{aligned}
$$

where

$$M_0 = \left(\begin{bmatrix} P^{-1} & 0 \\ * & R^{-1} \end{bmatrix} - \epsilon \begin{bmatrix} BH_2 \\ H_2 \end{bmatrix} \begin{bmatrix} BH_2 \\ H_2 \end{bmatrix}^T \right)^{-1} > 0, \tag{3.32}$$

$$R_3 = X + \epsilon X H_2 R_{20}^{-1} H_2^T X + \frac{\rho}{\epsilon} E_2^T E_2, \tag{3.33}$$

$$R_4 = I + \epsilon H_2 R_{20}^{-1} H_2^T X, \tag{3.34}$$

$$X = B^T P B + R. \tag{3.35}$$

Denote

$$R_{10} = X + \epsilon X H_2 R_{20}^{-1} H_2^T X. \tag{3.36}$$

Then, from (3.27), (3.33), and (3.36), it follows that

$$R_3^{-1} = R_{10}^{-1} - \frac{\rho}{\epsilon} R_{10}^{-1} E_2^T \left(I + \frac{\rho}{\epsilon} E_2 R_{10}^{-1} E_2^T \right)^{-1} E_2 R_{10}^{-1}, \quad (3.37)$$

$$R_{10}^{-1} = \left[X \left(I + \epsilon H_2 R_{20}^{-1} H_2^T X \right) \right]^{-1}$$

$$= \left\{ X \left[I + \epsilon H_2 H_2^T \left(I - \epsilon X H_2 H_2^T \right)^{-1} X \right] \right\}^{-1} \quad (3.38)$$

$$= X^{-1} - \epsilon H_2 H_2^T, \quad (3.39)$$

$$R_4 = X^{-1} \left(X^{-1} - \epsilon H_2 H_2^T \right)^{-1}, \quad (3.40)$$

$$\epsilon H_2 R_{20}^{-1} H_2^T = \epsilon H_2 H_2^T \left(I - \epsilon X H_2 H_2^T \right)^{-1}. \quad (3.41)$$

By combining (3.29) and (3.37)–(3.41), it follows that

$$R_4 R_3^{-1} R_4^T - \epsilon H_2 R_{20}^{-1} H_2^T$$

$$= X^{-1} \left(X^{-1} - \epsilon H_2 H_2^T \right)^{-1} X^{-1} - \epsilon H_2 H_2^T \left(I - \epsilon X H_2 H_2^T \right)^{-1}$$

$$\quad - \frac{\rho}{\epsilon} X^{-1} E_2^T \left(I + \frac{\rho}{\epsilon} E_2 X^{-1} E_2^T - \rho E_2 H_2 H_2^T E_2^T \right)^{-1} E_2 X^{-1}$$

$$= X^{-1} - \frac{\rho}{\epsilon} X^{-1} E_2^T \left(I - \rho E_2 H_2 H_2^T E_2^T + \frac{\rho}{\epsilon} E_2 X^{-1} E_2^T \right)^{-1} E_2 X^{-1}$$

$$= \Delta_0(P, \epsilon) \quad (3.42)$$

and

$$R_3^{-1} R_4^T B^T P A \Phi \Phi \Phi = \left\{ I - \rho \left(X^{-1} - \epsilon H_2 H_2^T \right) E_2^T \times \left[\epsilon \left(I - \rho E_2 H_2 H_2^T E_2^T \right) \right. \right.$$

$$\left. \left. + \rho E_2 X^{-1} E_2^T \right]^{-1} E_2 \right\} X^{-1} B^T P A. \quad (3.43)$$

Thus, the proof is completed by (3.31), (3.42), and (3.43).

Although it is not obvious at first glance, the inequality (3.28) is actually equivalent to a standard algebraic Riccati inequality, as shown in the next lemma. Suppose that the matrix $I - \rho E_2 H_2 H_2^T E_2^T$ is singular. Then there exists an orthonormal matrix T_1 such that

$$T_1 E_2 H_2 H_2^T E_2^T T_1^T = diag \left\{ \Psi_{s_1}, \frac{1}{\rho} I_{s_0} \right\}, \quad (3.44)$$

where $\Psi_{s_1} \geq 0$ is a diagonal with eigenvalues not including $\frac{1}{\rho}$, and I_{s_0} is an $s_0 \times s_0$ identity matrix with $s_0 > 0$. Denote

$$\bar{E}_2 = T_1 E_2 = \left[\begin{array}{c} \bar{E}_{2s_1} \\ \bar{E}_{2s_0} \end{array} \right], \quad \bar{E}_{2s_0} \in R^{s_0 \times m}. \quad (3.45)$$

If $\bar{E}_{2s_0} \neq 0$, then let T_0 be an orthonormal matrix such that

$$T_0 \bar{E}_{2s_0}^T \bar{E}_{2s_0} T_0^T = diag\{0, \Phi_s\}, \quad \Phi_s > 0, \quad (3.46)$$

where $\Phi_s \in R^{s \times s}$ is diagonal and $0 < s \le s_0$. If $\bar{E}_{2s_0} = 0$, then let $T_0 = I$ and $s = 0$. Denote

$$\bar{B} = BT_0^T, \quad \bar{R} = T_0 R T_0^T, \quad \bar{N} = \bar{N}(\epsilon) = \frac{\rho}{\epsilon} T_0 \bar{E}_{2s_1}^T (I_{s_1} - \rho \Psi_{s_1})^{-1} \bar{E}_{2s_1} T_0^T \tag{3.47}$$

and decompose $\bar{B}$, $\bar{R}$, and $\bar{N}$ as follows,

$$\bar{B} = \begin{bmatrix} \bar{B}_{m-s} & \bar{B}_s \end{bmatrix}, \quad \bar{R} = \begin{bmatrix} \bar{R}_{m-s} & \bar{R}_{s_1} \\ * & \bar{R}_s \end{bmatrix}, \quad \bar{N} = \frac{\rho}{\epsilon} \begin{bmatrix} \bar{N}_{m-s} & \bar{N}_{s_1} \\ * & \bar{N}_s \end{bmatrix}, \tag{3.48}$$

where $\bar{B}_s \in R^{n \times s}$, $\bar{R}_s \in R^{s \times s}$, and $\bar{N}_s \in R^{s \times s}$. Then, we have the following lemma.

Lemma 3.2 *(i) If the matrix $I - \rho E_2 H_2 H_2^T E_2^T$ is nonsingular, then*

$$B \Delta_0 B^T = B \left[B^T P B + R + \frac{\rho}{\epsilon} E_2^T (I - \rho E_2 H_2 H_2^T E_2^T)^{-1} E_2 \right]^{-1} B^T. \tag{3.49}$$

(ii) If the matrix $I - \rho E_2 H_2 H_2^T E_2^T$ is singular, then

$$B \Delta_0 B^T = \bar{B}_{m-s} \left(\bar{B}_{m-s}^T P \bar{B}_{m-s} + \bar{R}_{m-s} + \frac{\rho}{\epsilon} \bar{N}_{m-s} \right)^{-1} \bar{B}_{m-s}^T, \tag{3.50}$$

where $\bar{B}_{m-s}$, $\bar{R}_{m-s}$, and $\bar{N}_{m-s}$ are defined by (3.48), and Δ_0 is given by (3.29).

Proof 3.5 *(i) The proof of (3.49) is immediate from (3.29).*

(ii) Choose a sequence $\{\sigma_n\}_{n=1}^{\infty}$ with $\sigma_n \to 1$ $(n \to \infty)$ such that $I - \sigma_n \rho E_2 H_2 H_2^T E_2^T$ is nonsingular for $n = 1, 2, \cdots$. Denote

$$\Delta_{0n} = \mathcal{H} \left[(B^T P B + R) \mathcal{H} + \frac{\rho}{\epsilon} E_2^T E_2 \right]^{-1}, \tag{3.51}$$

$$\bar{N}_n = \frac{\rho}{\epsilon} T_0 \bar{E}_{2s_1}^T (I_{s_1} - \sigma_n \rho \Psi_{s_1})^{-1} \bar{E}_{2s_1} T_0^T, \tag{3.52}$$

where $\mathcal{H} = I - \sigma_n \rho H_2 H_2^T E_2^T E_2$.
Then, from (3.44)-(3.47), (3.51), and (3.52), we have

$$\begin{aligned}
\Delta_{0n} &= \left[B^T P B + R + \frac{\rho}{\epsilon} E_2^T (I - \sigma_n \rho E_2 H_2 H_2^T E_2^T)^{-1} E_2 \right]^{-1} \\
&= \left[B^T P B + R + \frac{\rho}{\epsilon} \bar{E}_2^T diag \left\{ (I_{s_1} - \sigma_n \rho \Psi_{s_1})^{-1}, \frac{1}{1 - \sigma_n} I_{s_0} \right\} \bar{E}_2 \right]^{-1} \\
&= \left[B^T P B + R + \frac{\rho}{\epsilon} \bar{E}_{2s_1}^T (I_{s_1} - \sigma_n \rho \Psi_{s_1})^{-1} \bar{E}_{2s_1} + \frac{\rho}{\epsilon(1 - \sigma_n)} \bar{E}_{2s_0}^T \bar{E}_{2s_0} \right]^{-1} \\
&= T_0^T \left[\bar{B}^T P \bar{B} + \bar{R} + \bar{N}_n + diag \left\{ 0, \frac{\rho}{\epsilon(1 - \sigma_n)} \Phi_s \right\} \right]^{-1} T_0. \tag{3.53}
\end{aligned}$$

Decompose the matrix $\bar{N}_n$ as follows,

$$\bar{N}_n = \frac{\rho}{\epsilon} \begin{bmatrix} \bar{N}_{n \ m-s} & \bar{N}_{ns_1} \\ * & \bar{N}_{ns} \end{bmatrix}, \tag{3.54}$$

where $\bar{N}_{ns} \in R^{s \times s}$ for $n = 1, 2, \cdots$. By (3.48), (3.53), and (3.54), it follows that

$$
\begin{aligned}
B\Delta_{0n}B^T &= \bar{B} \begin{bmatrix} \bar{Y}_{11n} & \bar{Y}_{12n} \\ * & \bar{Y}_{22n} + \frac{\rho}{\epsilon(1-\sigma_n)}\Phi_s \end{bmatrix}^{-1} \bar{B}^T \\
&= \bar{B} \begin{bmatrix} \Delta_{an} & -\Delta_{an}\bar{Y}_{12n}\Delta_{bn} \\ * & \Delta_{bn} + \Delta_{bn}\bar{Y}_{12n}^T\Delta_{an}\bar{Y}_{12n}\Delta_{bn} \end{bmatrix} \bar{B}^T, \quad (3.55)
\end{aligned}
$$

where

$$\Delta_{an} = \left(\bar{Y}_{11n} - \bar{Y}_{12n}\Delta_{bn}\bar{Y}_{12n}^T\right)^{-1}, \quad \Delta_{bn} = \left[\bar{Y}_{22n} + \frac{\rho}{\epsilon(1-\sigma_n)}\Phi_s\right]^{-1} \tag{3.56}$$

with

$$\bar{Y}_{11n} = \bar{B}_{m-s}^T P \bar{B}_{m-s} + \bar{R}_{m-s} + \frac{\rho}{\epsilon}\bar{N}_{n \ m-s},$$

$$\bar{Y}_{12n} = \bar{B}_{m-s}^T P \bar{B}_s + \bar{R}_{s_1} + \frac{\rho}{\epsilon}\bar{N}_{ns_1},$$

$$\bar{Y}_{22n} = \bar{B}_s^T P \bar{B}_s + \bar{R}_s + \frac{\rho}{\epsilon}\bar{N}_{ns}.$$

Since $\Phi_s > 0$, it follows from (3.47), (3.48), (3.52), (3.54), and (3.56) that

$$\lim_{n\to\infty} \Delta_{bn} = 0, \quad \lim_{n\to\infty} \Delta_{an} = \left(\bar{B}_{m-s}^T P \bar{B}_{m-s} + \bar{R}_{m-s} + \frac{\rho}{\epsilon}\bar{N}_{m-s}\right)^{-1}. \tag{3.57}$$

By (3.55) and (3.57), we have

$$
\begin{aligned}
B\Delta_0 B^T &= \lim_{n\to\infty} B\Delta_{0n}B^T \\
&= \bar{B} \begin{bmatrix} (\bar{B}_{m-s}^T P \bar{B}_{m-s} + \bar{R}_{m-s} + \frac{\rho}{\epsilon}\bar{N}_{m-s})^{-1} & 0 \\ * & 0 \end{bmatrix} \bar{B}^T \\
&= \bar{B}_{m-s} \left(\bar{B}_{m-s}^T P \bar{B}_{m-s} + \bar{R}_{m-s} + \frac{\rho}{\epsilon}\bar{N}_{m-s}\right)^{-1} \bar{B}_{m-s}^T.
\end{aligned}
$$

Thus, the proof is complete.

From Lemma 3.2, $S_m(P, \epsilon) = 0$ is a standard Riccati equation. Theorem 3.3 provides a necessary and sufficient condition for the solution to the quadratic guaranteed cost control problem with multiplicative uncertainty. But, similar to the case of additive controller *gain uncertainty*, it remains unclear as to how one can choose the design parameter ϵ in order to achieve the minimal guaranteed cost of the closed-loop system. Denote

$$\epsilon_m = \sup\{\epsilon > 0: \ S_m(P, \epsilon) = 0 \text{ has a stabilizing solution} \tag{3.58}$$

$$P \geq 0 \text{ and (3.27) holds}\}. \tag{3.59}$$

Then, the design parameter ϵ for achieving suboptimal guaranteed cost of the closed-loop system falls in the range of $0 < \epsilon < \epsilon_m$. The next theorem shows that the optimal guaranteed cost control (i.e., the control law that yields the minimal cost as defined in (3.2)) is obtained at the boundary value of $\epsilon = \epsilon_m$.

Theorem 3.4 *Consider the system (3.1) with the cost function (3.2) and the multiplicative gain uncertainty (3.5). Suppose that the pair (A, B) is stabilizable if $I - \rho E_2 H_2 H_2^T E_2^T$ is nonsingular; or the pair $(A, \bar{B}_{m-s})$ with $\bar{B}_{m-s}$ given by (3.47) and (3.48) is stabilizable if $I - \rho E_2 H_2 H_2^T E_2^T$ is singular, and*

$$I - \rho E_2 H_2 H_2^T E_2^T \geq 0. \tag{3.60}$$

If there exists a state feedback gain K such that the control (3.3) with multiplicative uncertainty (3.5) is a quadratic guaranteed cost control with a cost matrix P_0, then the following Riccati equation with ϵ_m defined by (3.58) has a unique stabilizing solution $P_{opt} > 0$ satisfying $P_{opt} \leq P_0$ and

$$R_{20}(P_{opt}, \epsilon_m) \geq 0, \tag{3.61}$$

$$S_m(P_{opt}, \epsilon_m) = 0, \tag{3.62}$$

and the control law (3.3) with

$$K = -\left\{I - \rho(X^{-1} \epsilon_m H_2 H_2^T) E_2^T \left[\epsilon_m(I - \rho E_2 H_2 H_2^T E_2^T) + \rho E_2 X^{-1} E_2^T\right]^{-1} E_2\right\}$$
$$\times X^{-1} B^T P_{opt} A \tag{3.63}$$

where $X = B^T P_{opt} B + R$ is such that the resulting closed-loop system (3.6) is quadratically stable, and $J \leq x_0^T P_{opt} x_0$ for all uncertainties ΔK of the form (3.5).

Proof 3.6 *By Lemma 3.2 and Equation (3.28), it follows that if $I - \rho E_2 H_2 H_2^T E_2^T$ is nonsingular, then*

$$S_m(P, \epsilon) = A^T P A - P + Q - A^T P B \left[B^T P B + R\right.$$
$$\left. + \frac{\rho}{\epsilon} E_2^T (I - \rho E_2 H_2 H_2^T E_2^T)^{-1} E_2\right]^{-1} B^T P A \tag{3.64}$$

and if $I - \rho E_2 H_2 H_2^T E_2^T$ is singular, then

$$S_m(P, \epsilon) = A^T P A - P + Q - A^T P \bar{B}_{m-s} \left(\bar{B}_{m-s}^T P \bar{B}_{m-s} + \bar{R}_{m-s}\right.$$
$$\left. + \frac{\rho}{\epsilon} \bar{N}_{m-s}\right)^{-1} \bar{B}_{m-s}^T P A. \tag{3.65}$$

From (3.60), (3.47), and (3.48), it follows that

$$S_m(P, \epsilon_1) \leq S_m(P, \epsilon_2) \quad \text{if } \epsilon_1 \geq \epsilon_2 > 0. \tag{3.66}$$

By using Theorem 3.3 and (3.64)–(3.66), the rest of the proof is similar to that of Theorem 3.2, and is omitted.

Remark 3.3 *Theorem 3.4 presents a design of an optimal guaranteed cost control under the multiplicative gain uncertainties, and the closed-loop value of the cost function J is bounded by the minimal value of $x_0^T P_{opt} x_0$. It is easy to show that the condition (3.60) can be satisfied by the bound condition $\|H_2 F_2 E_2\|_2 \leq 1$ on the gain uncertainties for all F_2 satisfying $F_2^T F_2 \leq \rho I$. This implies that the control effort is at most permitted to degrade to zero. From (3.27), we have that the design parameter ϵ_m in Theorem 3.4 satisfies $0 < \epsilon_m \leq \lambda_m$ with*

$$\lambda_m = \begin{cases} [\lambda_{max}\{H_2^T(B^T P_m B + R)H_2\}]^{-1} & \text{if } H_2 \neq 0 \\ \infty & \text{if } H_2 = 0 \end{cases} \qquad (3.67)$$

where $P_m = P_a$ is either the stabilizing solution to the Riccati equation (3.25) if $I - \rho E_2 H_2 H_2^T E_2^T > 0$, or the stabilizing solution to the Riccati equation $S_m(P, \infty) = 0$ with $S_m(P, \epsilon)$ given by (3.65) if $I - \rho E_2 H_2 H_2^T E_2^T$ is singular. From (3.58), (3.61), and (3.62) it follows that

$$\epsilon_m = \max\{0 < \epsilon \leq \lambda_m : S_m(P, \epsilon) = 0 \text{ has a stabilizing solution } P \geq 0$$
$$\text{and } R_{20}(P, \epsilon_m) \geq 0\}. \qquad (3.68)$$

Thus, similar to search ϵ_a for the additive case in Remark 3.2, ϵ_m can be obtained by solving the Riccati equation $S_m(P, \epsilon) = 0$ to find a stabilizing solution $P \geq 0$, and checking if P satisfies $R_{20}(P, \epsilon) \geq 0$ for $\epsilon \in (0, \lambda_m]$ in an increasing sequence. When the condition (3.60) is not satisfied, the suboptimal guaranteed cost control can be searched by solving (3.28). If $I - \rho E_2 H_2 H_2^T E_2^T > 0$ in Theorem 3.4, then from (3.64) it follows that the Riccati equation 3.62 corresponds to that of the standard quadratic optimal control [82] for the system (3.1) with a cost function

$$J_m = \sum_{k=0}^{\infty} (x_k^T Q x_k + u_k^T \bar{R} u_k)$$

where $\bar{R} = R + \frac{\rho}{\epsilon_m} E_2^T (I - \rho E_2 H_2 H_2^T E_2^T)^{-1} E_2$.

For extending the reliable control design in Veillette [126] for continuous-time systems to discrete-time systems, we consider a special case in which $\rho = 1$, $H_2 = E_2 = diag\{0, I_s\}$ with an $s \times s$ identity matrix I_s ($s < m$), which corresponds to permitting the partial control effort $diag\{0, I_s\}u$ (the last s actuators) to degrade to zero. It covers the case of the last s actuator outages considered in Veillette [126]. We decompose the matrices B and R as follows,

$$B = [B_{m-s} \quad B_s], \quad R = \begin{bmatrix} R_{m-s} & R_{s_1} \\ * & R_s \end{bmatrix} \qquad (3.69)$$

with $B_s \in R^{n \times s}$ and $R_s \in R^{s \times s}$. Then the following result presents an optimal guaranteed cost control.

Theorem 3.5 *Consider the system (3.1) with the cost function (3.2) and multiplicative gain uncertainty (3.5). Suppose that the pair (A, B_{m-s}) is stabilizable and the controller uncertainty in (3.5) is given by*

$$\rho = 1, \quad H_2 = E_2 = diag\{0, I_s\}. \tag{3.70}$$

Let $P > 0$ be the stabilizing solution to the following Riccati equation:

$$S_0(P) \triangleq A^T PA - P + Q - A^T PB_{m-s}(B_{m-s}^T PB_{m-s} + R_{m-s})^{-1} B_{m-s}^T PA = 0. \tag{3.71}$$

Then the control law $u_k = (K + \Delta K)x_k$ with

$$K = -\begin{bmatrix} I_{m-s} & X_{11}^{-1} X_{12} \\ 0 & \lambda_0(X_{22} - X_{12}^T X_{11}^{-1} X_{12}) \end{bmatrix} X^{-1} B^T PA \tag{3.72}$$

is such that the resulting closed-loop system (3.6) is quadratically stable, and $J \le x_0^T P x_0$ for all uncertainties ΔK with (3.5) and (3.70), where X_{11}, X_{12}, $X_{22} \in R^{s \times s}$, and λ_0 are defined as follows:

$$\lambda_0 = (\lambda_{max}[B_s^T PB_s + R_s])^{-1}, \tag{3.73}$$

$$\begin{bmatrix} X_{11} & X_{12} \\ * & X_{22} \end{bmatrix} \triangleq \begin{bmatrix} B_{m-s}^T PB_{m-s} + R_{m-s} & B_{m-s}^T PB_s + R_{s_1} \\ * & B_s^T PB_s + R_s \end{bmatrix}$$

$$= X$$

$$= B^T PB + R. \tag{3.74}$$

Furthermore, if any other feedback gain K_0 is such that $u_k = (K_0 + \Delta K)x_k$ (with ΔK given by (3.5) and (3.70)) is a guaranteed cost control with cost matrix P_0, then $P \le P_0$.

Proof 3.7 *Let $T_1 = I$ and $T_0 = I$. Then, from (3.69), (3.70), and (3.44)-(3.48), we have $\bar{B}_{m-s} = B_{m-s}$, $\bar{R}_{m-s} = R_{m-s}$, and $\bar{N}_{m-s} = 0$. By (3.65), it follows that*

$$S_m(P, \epsilon) = S_0(P), \tag{3.75}$$

which is independent of ϵ. By (3.63) and $\epsilon_m = \lambda_0$, we have

$$K = -\left\{I - \rho\left(X^{-1} - \lambda_0 H_2 H_2^T\right) E_2^T \left[\lambda_0 \left(I - \rho H_2 H_2^T E_2^T E_2\right)\right.\right.$$
$$\left.\left. + \rho E_2 X^{-1} E_2^T\right]^{-1} E_2\right\} X^{-1} B^T PA$$

$$= -\left\{I - \left(X^{-1} - \lambda_0 diag\{0, I_s\}\right) X \left[\lambda_0 [I_{m-s}, 0]X\right.\right.$$
$$\left.\left. + diag\{0, I_s\}\right]^{-1} diag\{0, I_s\}\right\} X^{-1} B^T PA$$

$$= -\left\{I - (I - \lambda_0 \begin{bmatrix} 0 & 0 \\ X_{12}^T & X_{22} \end{bmatrix}) \begin{bmatrix} \lambda_0 X_{11} & \lambda_0 X_{12} \\ 0 & I_s \end{bmatrix}^{-1}\right.$$
$$\left. \times diag\{0, I_s\}\right\} X^{-1} B^T PA$$

$$= -\begin{bmatrix} I_{m-s} & X_{11}^{-1} X_{12} \\ 0 & \lambda_0(X_{22} - X_{12}^T X_{11}^{-1} X_{12}) \end{bmatrix} X^{-1} B^T PA.$$

By Theorem 3.4, the conclusion follows.

Remark 3.4 *Theorem 3.5 presents an optimal guaranteed cost control for the special case of $H_2 = E_2 = diag\{0, I_s\}$ and $\rho = 1$, which covers the case of the outages of the last s actuators. The design Equation (3.71) corresponds to that of the standard quadratic optimal control for the system $x_{k+1} = Ax_k + B_{m-s}u_k^{m-s}$ with a cost function*

$$J_m = \sum_{k=0}^{\infty}(x_k^T Q x_k + [u_k^{m-s}]^T R_{m-s} u_k^{m-s}).$$

The result also is an extension of the reliable control design in Veillette [126] for continuous-time systems to discrete-time systems.

3.4   Example

Example 3.1 *Consider the uncertain system (3.1), the performance index (3.2), and the state feedback controller (3.3) with*

$$A = \begin{bmatrix} -1 & 0.5 \\ 1 & 1.5 \end{bmatrix}, \quad B = \begin{bmatrix} 0 \\ 1 \end{bmatrix}, \quad Q = \begin{bmatrix} 1 & 0 \\ 0 & 1 \end{bmatrix}, \quad R = 1.$$

Obviously, (A, B) is a controllable pair (hence, stabilizable), and the eigenvalues are at -1.1861 and 1.6761, both unstable.
Case 1: *For additive controller uncertainties of the form (3.4) with*

$$H_1 = \begin{bmatrix} 1 & 1 \end{bmatrix}, \quad E_1 = \begin{bmatrix} 1 & 0 \\ 0 & 1 \end{bmatrix}, \quad \text{and} \quad \rho = 0.2,$$

we can use the result in Theorem 3.2 to design an optimal guaranteed cost control law. First, we compute the bound λ_a in (3.24) for the optimal parameter value ϵ_a in (3.20). By solving (3.25) and using (3.24) we have $\lambda_a = 0.0836$. Then, by the method given in Remark 3.2, we have that the optimal value of ϵ as given in (3.26) is $\epsilon_a = 0.0266$. The corresponding smallest performance matrix P_{opt} and optimal feedback gain K are given by

$$P_{opt} = \begin{bmatrix} 44.1062 & -14.3698 \\ -14.3698 & 17.7787 \end{bmatrix}, \quad K = \begin{bmatrix} -1.7120 & -1.0375 \end{bmatrix}$$

and the closed-loop eigenvalues are at -0.6915 and 0.1540, both stable.
Case 2: *For multiplicative controller uncertainties of the form (3.6) with*

$$H_2 = 1, \quad E_2 = 1, \quad \text{and} \quad \rho = 0.2,$$

we use Theorem 3.4 to design an optimal guaranteed cost control law. First, by using Remark 3.3, we have that $\lambda_m = 0.0836$. Then, by the method given in Remark 3.3, we have that the optimal value of ϵ as given in (3.68) is $\epsilon_m = 0.04$. The corresponding smallest performance matrix P_{opt} and optimal feedback gain K are given by

$$P_{opt} = \begin{bmatrix} 91.4905 & -23.2850 \\ -23.2850 & 23.9163 \end{bmatrix}, \quad K = [-1.8931, -0.9719]$$

and the closed-loop eigenvalues are at -0.6064 and 0.1345, both stable.

3.5   Conclusion

In this chapter, we have investigated the problem of guaranteed cost control of discrete-time *linear systems* under two classes of controller gain uncertainties. For additive controller gain uncertainties, an optimal guaranteed cost control design is presented in terms of an algebraic Riccati equation, which corresponds to the standard optimal control design for the same system with a modified cost function. Under a bound condition for the gain uncertainties, an optimal guaranteed cost control design is also given for the case of the multiplicative gain uncertainties. A numerical example is given to illustrate the design procedures.

4

Non-Fragile Dynamic Output Feedback Control with Norm-Bounded Gain Uncertainty

4.1 Introduction

In Chapter 3, based on algebraic Riccati inequality techniques, the problem of *non-fragile* guaranteed cost control for discrete-time *linear systems* under state feedback controller gain uncertainties was investigated. However, in many systems, all of the system states are often not fully available, not measurable, or too expensive to measure, in particular for the case that only the system output is available and therefore this condition limits the practical applicability of state-feedback control schemes [14, 117]. Correspondingly, *dynamic output feedback* controllers are desired and every important both in theories and applications, and also are a very challenging problem [98]. On the other hand, it is well known that the controller *coefficient variations* caused by the imprecision inherent in analog systems and the need for additional tuning of parameters in the final controller implementation have significant influence on the performance of the control system [66] due to the fact that small perturbations on the controller coefficients may cause the designed *closed-loop system* to go unstable. Therefore, the problem of designing non-fragile dynamic output feedback H_∞ *controllers* with respect to controller coefficient variations is a comparable worthy research issue and is exceptionally challenging and much more difficult.

This chapter is concerned with the non-fragile H_∞ controller design problem for linear time-invariant systems. The controller to be designed is assumed to have time-varying, but *additive* or *multiplicative* norm-bounded uncertainties. Design methods are presented for dynamic output (measurement) feedback. The designed controllers with uncertainty (i.e., non-fragile controllers) are such that the closed-loop system is quadratically stable and has an H_∞ disturbance attenuation bound. Furthermore, these non-fragile controllers degenerate to the standard H_∞ output feedback control designs, when the controller uncertainties are set to zero. It is also shown that the existence of solutions to the standard H_∞ *control* problem (without controller uncertainty) guarantees the existence of such non-fragile H_∞ *controllers* with a certain level of controller uncertainties.

4.2 Problem Statement

Consider a linear time-invariant system Σ described by equations of the form

$$\Sigma: \quad \dot{x}(t) = Ax(t) + B_1 w(t) + B_2 u(t) \tag{4.1}$$
$$z(t) = C_1 x(t) + D_{12} u(t) \tag{4.2}$$
$$y(t) = C_2 x(t) + D_{21} w(t) \tag{4.3}$$

where $x(t) \in \mathbf{R}^n$ is the state, $u(t) \in \mathbf{R}^m$ is the control input, $w(t) \in \mathbf{R}^q$ is the disturbance input, $z(t) \in \mathbf{R}^p$ the regulated output, $y(t) \in \mathbf{R}^r$ the measured output, and A, B_1, B_2, C_1, C_2, D_{12}, and D_{21} are real constant matrices of appropriate dimensions.

The problem under consideration in this chapter is as follows.

Non-fragile H_∞ controller design problem: Given the system Σ described by Equations (4.1)–(4.3) and a constant $\gamma > 0$, find a controller of the following form,

$$\dot{\xi}(t) = (A_c + \Delta A_c)\,\xi(t) + (L + \Delta L)y(t) \tag{4.4}$$
$$u(t) = (K + \Delta K)\xi(t) \tag{4.5}$$

where $\xi(t) \in \mathbf{R}^n$, and ΔA_c, ΔL, and ΔK represent the controller *gain uncertainty* of the following forms.

(a) ΔA_c, ΔL, and ΔK are of the additive form

$$\Delta K = K_{11} F_1(t) K_{22}, \quad \Delta L = L_{11} F_2(t) L_{22}, \quad \Delta A_c = A_{11} F_3(t) A_{22} \tag{4.6}$$

with K_{11}, K_{22}, L_{11}, L_{22}, A_{11}, and A_{22} being known constant matrices of appropriate dimensions, and

$$F_1^T(t) F_1(t) \le \rho_1 I, \quad F_2^T(t) F_2(t) \le \rho_2 I, \quad F_3^T(t) F_3(t) \le \rho_3 I \tag{4.7}$$

such that the resulting closed-loop uncertain system is quadratically stable, and the H_∞ norm of the closed-loop transfer function matrix from w to z is bounded by γ for any uncertainties $F_1(t), F_2(t)$, and $F_3(t)$ satisfying (4.7), where the positive constants ρ_1, ρ_2, and ρ_3 are given.

(b) ΔA_c, ΔL, and ΔK are of the multiplicative form

$$\Delta K = \left[\, k_{ij}(\theta_{K1i} + \theta_{K2ij})\,\right]_{m \times n} \tag{4.8}$$
$$\Delta A_c = \left[\, a_{ij}(\theta_{A1j} + \theta_{A2ij})\,\right]_{n \times n} \tag{4.9}$$
$$\Delta L = \left[\, l_{ij}(\theta_{L1j} + \theta_{L2ij})\,\right]_{n \times r} \tag{4.10}$$

with

$$K = \begin{bmatrix} K_1^T & K_2^T & \cdots & K_m^T \end{bmatrix}^T = \begin{bmatrix} k_{ij} \end{bmatrix}_{m \times n}$$
$$A_c = \begin{bmatrix} a_{ij} \end{bmatrix}_{n \times n}, \quad L = \begin{bmatrix} l_{ij} \end{bmatrix}_{n \times r} \tag{4.11}$$

where θ_{K1i}, θ_{K2ij}, θ_{A1i}, θ_{A2ji}, θ_{L1i}, and θ_{L2ji} are uncertain parameters satisfying the following bound conditions:

$$|\theta_{K1i}| \leq \bar{\theta}_{K1i} \leq \theta_{K1} \leq 1, \quad |\theta_{K2ij}| \leq \bar{\theta}_{K2i} \leq \theta_{K2} \leq 1,$$
$$i = 1, \cdots, m; \ j = 1, \cdots, n \tag{4.12}$$
$$|\theta_{A1i}| \leq \bar{\theta}_{A1i} \leq \theta_{A1} \leq 1, \quad |\theta_{A2ji}| \leq \bar{\theta}_{A2i} \leq \theta_{K2} \leq 1,$$
$$i, j = 1, \cdots, n \tag{4.13}$$
$$|\theta_{L1i}| \leq \bar{\theta}_{L1i} \leq \theta_{L1} \leq 1, \quad |\theta_{L2ji}| \leq \bar{\theta}_{L2i} \leq \theta_{L2} \leq 1,$$
$$i = 1, \cdots, r; \ j = 1, \cdots, n \tag{4.14}$$

Remark 4.1 *In (4.6) and (4.7), $F_1(t)$, $F_2(t)$, and $F_3(t)$ are uncertain matrices of appropriate dimensions. For a state feedback gain K with the dimension $m \times n$ to be designed, the matrices K_{11} and K_{22} can be specified. For example, a 2×2 uncertain matrix* $\begin{bmatrix} \bar{F}_1(t) & \bar{F}_2(t) \\ \bar{F}_3(t) & \bar{F}_4(t) \end{bmatrix}$ *with $(\bar{F}_i(t))^2 \leq \rho_i > 0$, $i = 1, \cdots, 4$ can be written as* $\begin{bmatrix} \bar{F}_1(t) & \bar{F}_2(t) \\ \bar{F}_3(t) & \bar{F}_4(t) \end{bmatrix} = \sum_{i=1}^{4} \bar{F}_i \bar{E}_i$ *where* $\bar{E}_1 = \begin{bmatrix} 1 & 0 \\ 0 & 0 \end{bmatrix}$, $\bar{E}_2 = \begin{bmatrix} 0 & 1 \\ 0 & 0 \end{bmatrix}$, $\bar{E}_3 = \begin{bmatrix} 0 & 0 \\ 1 & 0 \end{bmatrix}$, *and* $\bar{E}_4 = \begin{bmatrix} 0 & 0 \\ 0 & 1 \end{bmatrix}$, *which can be over-bounded by an uncertain matrix $K_{11}F(t)K_{22}$ of form (4.6) and (4.7) with* $K_{11} = \begin{bmatrix} \frac{\rho_1}{\rho} & \frac{\rho_2}{\rho} & 0 & 0 \\ 0 & 0 & \frac{\rho_3}{\rho} & \frac{\rho_4}{\rho} \end{bmatrix}$, $K_{22} = \begin{bmatrix} 1 & 0 & 1 & 0 \\ 0 & 1 & 0 & 1 \end{bmatrix}^T$, *and $F^T(t)F(t) \leq \rho I$. By using a finite word length (FWL) implementation, each element of the state feedback gain K can be implemented only up to a finite accuracy. Thus, in implementation, each element of K may be perturbed by $\pm \rho$ [43]. The uncertainty can be described by the above model of the additive controller uncertainty given by (4.6) and (4.7). The uncertainty of the form (4.6) and (4.7) was first introduced in Petersen [107] and has been widely used in robust control for uncertain linear systems.*

Remark 4.2 *The multiplicative gain variation model ΔK (ΔA_c and ΔL) of form (4.8) ((4.9) and (4.10)) means that the identical relative percentage drift with errors θ_{K2ij} (θ_{A2ij} and θ_{L2ij}) from the nominal entries of every row of the matrix K (A_c and L) is allowed. When $\theta_{K2} = 0$, that is, $\theta_{K2ij} = 0$ ($i = 1, \cdots, m; \ j = 1, \cdots, n$), the model ΔK of form (4.8) corresponds to the degradation of actuators [112]. Similarly, when $\theta_{L2} = 0$, the model ΔL of form (4.10) corresponds to the degradation of sensors [112]. As a more special case in which $\theta_{K1i} = \theta_{A1i} = \theta_{L1i} = \theta$ and $\theta_{K2ij} = \theta_{A2ji} = \theta_{L2ji} = 0$, then the models of (4.8)–(4.10) are reduced to the case considered in Haddad and Corrado [52]. If $\theta_{K1} = 0$, that is, $\theta_{K1i} = 0$ ($i = 1, \cdots, m$), the model ΔK*

of (4.8) represents more general multiplicative gain variations introduced in Famularo et al. [34].

The multiplicative gain variation model can be used to represent the errors due to implementation of continuous controllers in digital form. Examples of the errors include roundoff error, quantization errors, controller realization errors, and so forth. In practice, with finite precision in the calculations, different choices of the state-space representation shall introduce some uncertainties into the controller dynamic A_c matrix [5]. Under some circumstances, the model ΔA_C of the form (4.9) can also be used to represent the errors due to controller state errors. As the controller is a dynamic system, the controller state errors could occur due to the switching of controllers (and hence their parameters) at different operating points, and improper initialization when the controller is switched on or switched from manual to automatic control [5].

Under the control law given by (4.4) and (4.5) for the system Σ, we obtain the following closed-loop system Σ_c,

$$\Sigma_c: \quad \dot{x}_e(t) = A_{cl} x_e(t) + B_{cl} w(t) \qquad (4.15)$$

$$z(t) = C_{cl} x_e(t) \qquad (4.16)$$

where $x_e(t) = \begin{bmatrix} x^T(t) & \xi^T(t) \end{bmatrix}^T$, $A_{cl} = A_e + \Delta A_e$, $B_{cl} = B_e + \Delta B_e$, $C_{cl} = H_e + \Delta H_e$, and

$$A_e = \begin{bmatrix} A & B_2 K \\ LC_2 & A_c \end{bmatrix}, \quad \Delta A_e = \begin{bmatrix} 0 & B_2 \Delta K \\ \Delta LC_2 & \Delta A_c \end{bmatrix} \qquad (4.17)$$

$$B_e = \begin{bmatrix} B_1 \\ LD_{21} \end{bmatrix}, \quad \Delta B_e = \begin{bmatrix} 0 \\ \Delta LD_{21} \end{bmatrix}$$

$$H_e = \begin{bmatrix} C_1 & D_{12} K \end{bmatrix}, \quad \Delta H_e = \begin{bmatrix} 0 & D_{12} \Delta K \end{bmatrix}. \qquad (4.18)$$

Then, the following lemma can be obtained.

Lemma 4.1 *Consider the closed-loop system Σ_c described by (4.15) and (4.16). If there exists a symmetric positive-definite matrix P and a constant $\epsilon > 0$ such that*

$$He\{XA_{cl}\} + \frac{1}{\gamma^2} XB_{cl}B_{cl}^T X + C_{cl}^T C_{cl} + \epsilon I < 0 \qquad (4.19)$$

holds for all uncertainties satisfying (4.6) and (4.7). Then the uncertain closed-loop system is quadratically stable (the notion of quadratic stability can refer to Barmish [6]) and the H_∞ norm of the closed-loop transfer-function matrix from y to u is bounded by γ.

Proof 4.1 *It is immediate by the definition of quadratic stability [6] and Lemma 2.12.*

We also make the following assumptions throughout this chapter:

Assumption 4.1 *The matrix $R_2 = D_{12}^T D_{12}$ is nonsingular.*

Assumption 4.2 *The matrix $R_1 = D_{21} D_{21}^T$ is nonsingular.*

Assumption 4.3 *Rank $\begin{bmatrix} A - j\omega I & B_2 \\ C_1 & D_{12} \end{bmatrix} = n + m$ for each $\omega \in R$.*

4.3 Non-Fragile H_∞ Dynamic Output Feedback Controller Design

4.3.1 Additive Controller Gain Uncertainty Case

In this section, we consider the non-fragile H_∞ controller design problem via dynamic measurement feedback with respect to additive controller gain uncertainties. Denote

$$\Delta_{os}(P) \triangleq He\{PA\} + C_1^T C_1 + \frac{1}{\gamma^2} P B_1 B_1^T P - \Xi_1 R_2^{-1} \Xi_1^T$$
$$+ \frac{\rho_2}{\sigma_2} \Xi_2 (I - \frac{\rho_2}{\gamma^2 \sigma_2} L_{22} R_1 L_{22}^T)^{-1} \Xi_2^T \tag{4.20}$$

where σ_2 is a constant, $\Xi_1 = PB_2 + C_1^T D_{12}$, $\Xi_2 = (\frac{1}{\gamma^2} P B_1 D_{21}^T + C_2^T) L_{22}^T$, and denote

$$\Delta_{oo}(Q) \triangleq He\{Q\bar{A}_0\} + Q\left[B_1 R_{01}^{-1} B_1^T + \sigma_3 A_{11} A_{11}^T + \sigma_1 B_2 K_{11}(I + \sigma_1 K_{11}^T R_{20} \right.$$
$$\left. K_{11}) K_{11}^T B_2^T + \sigma_2 L_{11} L_{11}^T\right] Q + K^T R_{20} K - \Xi_3 R_{10}^{-1} \Xi_3^T \tag{4.21}$$

where σ_1, σ_2, and σ_3 are positive constants, and

$$\Xi_3 = QB_1 R_{01}^{-1} D_{21}^T + \bar{C}_{20}^T$$
$$R_{10} = R_{10}(\sigma_2) = D_{21} R_{01}^{-1} D_{21}^T \tag{4.22}$$
$$R_{20} = R_{20}(\sigma_1) = D_{12}^T R_0^{-1}(\sigma_1) D_{12} \tag{4.23}$$
$$\bar{A}_0 = \bar{A} + B_1 R_{01}^{-1} B_1^T P + \sigma_1 B_2 K_{11}(I + \sigma_1 K_{11}^T R_{20} K_{11}) K_{11}^T B_2^T P \tag{4.24}$$
$$\bar{C}_{20} = \bar{C}_2 + D_{21} R_{01}^{-1} B_1^T P$$
$$= (I + \frac{\rho_2}{\sigma_2} D_{21} R_{01}^{-1} D_{21}^T L_{22}^T L_{22}) C_2 + D_{21} R_{01}^{-1} B_1^T P \tag{4.25}$$

with

$$R_0 = R_0(\sigma_1) = I - \sigma_1 D_{12} H_1 H_1^T D_{12}^T > 0 \tag{4.26}$$
$$R_{01} = R_{01}(\sigma_2, \rho_2) = \gamma^2 I - \frac{\rho_2}{\sigma_2} D_{21}^T L_{22}^T L_{22} D_{21} > 0 \tag{4.27}$$
$$K = -R_2^{-1}(B_2^T P + D_{12}^T C_1) \tag{4.28}$$
$$\bar{A} = A + \sigma_1 B_2 K_{11} K_{11}^T D_{12}^T R_0^{-1}(\sigma_1) C_1$$
$$+ \frac{\rho_2}{\sigma_2} B_1 R_{01}^{-1}(\sigma_2) D_{21}^T L_{22}^T L_{22} C_2. \tag{4.29}$$

Then, we have the following theorem for the non-fragile H_∞ controller design problem with dynamic measurement feedback.

Theorem 4.1 *Consider the system Σ described by Equations (4.1)–(4.3) and positive constants γ, ρ_1, ρ_2, and ρ_3. Suppose that*
(i) Assumptions 4.1 and 4.2 hold;
(ii) there exist positive constants σ_1, σ_2, and σ_3, and symmetric positive-definite matrices P and Q such that
    (a) the inequalities (4.26) and (4.27) hold;
    (b) the following inequality

$$\begin{bmatrix} \Delta_{os}(P) & 0 \\ 0 & \Delta_{oo}(Q) \end{bmatrix} + T_0 < 0 \qquad (4.30)$$

    holds, where

$$T_0 = \begin{bmatrix} T_{01} & T_{01} \\ T_{01} & T_{01} \end{bmatrix} \qquad (4.31)$$

with $T_{01} = \frac{\rho_1}{\sigma_1} K_{22}^T K_{22} + \frac{\rho_3}{\sigma_3} A_{22}^T A_{22}$.
    Then the controller given by (4.4) and (4.5) with (4.28) and

$$A_c = \bar{A}_0 + \bar{B}_2 K - L\bar{C}_{20}, \quad L = Q^{-1}\bar{C}_{20}^T R_{10}^{-1} + B_1 R_{01}^{-1} D_{21}^T R_{10}^{-1} \qquad (4.32)$$

with $\bar{A}_0$ and $\bar{C}_{20}$ as given by (4.24) and (4.25), and $\bar{B}_2$ defined by

$$\bar{B}_2 = B_2(I + \sigma_1 K_{11} K_{11}^T D_{12}^T R_0^{-1}(\sigma_1) D_{12}) \qquad (4.33)$$

is a solution of the non-fragile H_∞ controller design problem for the system Σ.

The following preliminaries will be used in the proof of Theorem 4.1. Denote

$$\bar{\Delta}_{os}(P) = He\{P\bar{A}\} + C_1^T R_0^{-1} C_1 + \frac{\rho_2}{\sigma_2} C_2^T L_{22}^T \bar{R}_{10} L_{22} C_2 + P\left[B_1 R_{01}^{-1} B_1^T\right.$$
$$\left. + \sigma_1 B_2 K_{11} \bar{R}_{20} K_{11}^T B_2^T\right] P - \Omega R_{20}^{-1}(\sigma_1) \Omega^T \qquad (4.34)$$

$$\bar{K} = -R_{20}^{-1}[B_2^T P + D_{12}^T R_0^{-1}(C_1 + \epsilon D_{12} H_1 H_1^T B_2^T P)] \qquad (4.35)$$

where

$$\bar{R}_{20} = I + \sigma_1 K_{11}^T R_{20} K_{11}, \bar{R}_{10} = I + \frac{\rho_2}{\sigma_2} L_{22} R_{10} L_{22}^T, \qquad (4.36)$$

$$\Omega = \left[P\bar{B}_2 + D_{12}^T R_0^{-1}(\sigma_1) C_1\right]$$

with $\bar{A}$, $\bar{B}_2$, $R_0(\sigma_1)$, $R_{01}(\sigma_2)$, $R_{10}(\sigma_2)$, and $R_{20}(\sigma_1)$ defined by (4.29), (4.33), (4.26), (4.27), (4.22), and (4.23), respectively. Then we have the following lemma.

Lemma 4.2 *Suppose that the inequalities (4.26) and (4.27) hold for $\sigma_1 > 0$ and $\sigma_2 > 0$, respectively. Then*

$$\Delta_{os}(P) = \bar{\Delta}_{os}(P), \quad \bar{K} = K \tag{4.37}$$

where $\Delta_{os}(P)$ and K are defined by (4.20) and (4.28), respectively.

Proof 4.2 *From (4.23) and (4.26), we have*

$$R_0^{-1} = I + D_{12}(I - \epsilon H_1 H_1^T R_2)^{-1}\epsilon H_1 H_1^T D_{12}^T \tag{4.38}$$
$$R_{20}^{-1} = (I - \epsilon H_1 H_1^T R_2)R_2^{-1}. \tag{4.39}$$

Combining (4.38), (4.39), (4.34), (4.35), (4.20), and (4.28), and the routine algebraic manipulations, (4.37) follows. The details are omitted.

Lemma 4.3 *Consider the closed-loop system Σ_c described by Equations. (4.15) and (4.16) with (4.28) and (4.32). Suppose that the inequalities (4.26) and (4.27) hold for $\sigma_1 > 0$ and $\sigma_2 > 0$, respectively. If there exists a symmetric matrix $X \geq 0$ and a constant $\sigma_3 > 0$ such that*

$$He\{X\bar{A}_e\} + X\left\{B_e R_{01}^{-1} B_e^T + diag\{0, \sigma_2 L_{11}L_{11}^T + \sigma_3 A_{11}A_{11}^T\}\right.$$
$$+ \left.diag\{\sigma_1 B_2 K_{11}\bar{R}_{20}K_{11}^T B_2^T, 0\}\right\}X + diag\left\{\frac{\rho_2}{\sigma_2}C_2^T L_{22}^T \bar{R}_{10}L_{22}C_2,\right.$$
$$\left.\frac{\rho_1}{\sigma_1}K_{22}^T K_{22} + \frac{\rho_3}{\sigma_3}A_{22}^T A_{22}\right\} + \bar{C}_e^T R_0^{-1}\bar{C}_e < 0 \tag{4.40}$$

where

$$\bar{A}_e = \begin{bmatrix} \bar{A} & \bar{B}_2 K \\ L\bar{C}_2 & A_c \end{bmatrix}, \bar{C}_e = \begin{bmatrix} C_1 & D_{12}K \end{bmatrix} \tag{4.41}$$

with $\bar{A}$, $\bar{B}_2$, $\bar{C}_2$, $\bar{R}_{10}$, and $\bar{R}_{20}$, defined by (4.29), (4.33), (4.25), and (4.36), respectively, then there exists a constant $\epsilon > 0$ such that (4.19) holds for all uncertainties satisfying Equations. (4.6) and (4.7).

Proof 4.3 *By (4.6), (4.7), (4.26), (4.27), (4.17), and (4.18), we have*

$$
Y = \begin{bmatrix} He\{XA_{cl}\} & * & * \\ B_{cl}^T X & -\gamma^2 I & * \\ C_{cl} & 0 & -I \end{bmatrix}
$$

$$
= \Psi_1 + He\left\{ \begin{bmatrix} X\begin{bmatrix} B_2 K_{11} \\ 0 \end{bmatrix} \\ 0 \\ D_{12}K_{11} \end{bmatrix} F_1\Psi_2 + \begin{bmatrix} X\begin{bmatrix} 0 \\ A_{11} \end{bmatrix} \\ 0 \\ 0 \end{bmatrix} F_3\Psi_3 \right.
$$

$$
\left. + \begin{bmatrix} X\begin{bmatrix} 0 \\ L_{11} \end{bmatrix} \\ 0 \\ 0 \end{bmatrix} F_2\Psi_4 \right\} \tag{4.42}
$$

$$
\le \begin{bmatrix} M_3 & * & * \\ (XB_e + \frac{\rho_2}{\sigma_2}[L_{22}C_2\ 0]^T L_{22}D_{21})^T & -R_{01}(\sigma_2,\rho_2) & * \\ M_{31} & 0 & -R_0(\sigma_1) \end{bmatrix}
$$

where

$$
\Psi_1 = \begin{bmatrix} He\{XA_e\} & * & * \\ B_e^T X & -\gamma^2 I & * \\ \tilde{C}_e & 0 & -I \end{bmatrix}, \Psi_2 = \begin{bmatrix} [\ 0 & K_{22}\ ] & 0 & 0 \end{bmatrix}
$$

$$
\Psi_3 = \begin{bmatrix} [\ 0 & A_{22}\ ] & 0 & 0 \end{bmatrix}, \Psi_4 = \begin{bmatrix} [\ L_{22}C_2 & 0\ ] & L_{22}D_{21} & 0 \end{bmatrix}
$$

$$
M_3 = He\{XA_e\} + X\left\{ \sigma_1 \begin{bmatrix} B_2 K_{11} \\ 0 \end{bmatrix}\begin{bmatrix} B_2 K_{11} \\ 0 \end{bmatrix}^T + \sigma_2 \begin{bmatrix} 0 \\ L_{11} \end{bmatrix}\begin{bmatrix} 0 \\ L_{11} \end{bmatrix}^T \right.
$$

$$
\left. +\sigma_3 \begin{bmatrix} 0 \\ A_{11} \end{bmatrix}\begin{bmatrix} 0 \\ A_{11} \end{bmatrix}^T \right\} X + diag\left\{ \frac{\rho_2}{\sigma_2}C_2^T L_{22}^T L_{22}C_2, \right.
$$

$$
\left. \frac{\rho_1}{\sigma_1}K_{22}^T K_{22} + \frac{\rho_3}{\sigma_3}A_{22}^T A_{22} \right\}
$$

$$
M_{31} = [C_1\ D_{12}K] + \sigma_1[D_{12}K_{11}K_{11}^T B_2\ 0]X
$$

By (4.40) and Lemma 2.1, it follows that

$$
\begin{bmatrix} M_3 & * & * \\ (XB_e + \frac{\rho_2}{\sigma_2}[L_{22}C_2\ 0]^T L_{22}D_{21})^T & -R_{01}(\sigma_2,\rho_2) & * \\ M_{31} & 0 & -R_0(\sigma_1) \end{bmatrix} < 0.
$$

Thus, from (4.42), there exists a constant $\epsilon > 0$ such that

$$
Y + diag\{\epsilon I, 0, 0\} < 0
$$

which further implies from Lemma 2.1 that the inequality (4.19) holds.

Consider an auxiliary closed-loop system Σ_{ca} is given by

$$\Sigma_{ca}: \quad \dot{x}_e(t) = \bar{A}_e x_e(t) + \bar{B}_e \begin{bmatrix} w(t) \\ w_{11}(t) \\ w_{12}(t) \end{bmatrix} \tag{4.43}$$

$$\bar{z}(t) = \bar{H}_e x_e(t) \tag{4.44}$$

where $w_{11}(t)$ and $w_{12}(t)$ are new disturbance inputs, and $\bar{A}_e$ is given by (4.41):

$$\bar{B}_e = \begin{bmatrix} B_1 R_{01}^{-\frac{1}{2}} & \sigma_1^{\frac{1}{2}} B_2 K_{11}(I + \sigma_1 K_{11}^T R_{20} K_{11})^{\frac{1}{2}} & 0 \\ L D_{21} R_{01}^{-\frac{1}{2}} & 0 & \begin{bmatrix} \sigma_2^{\frac{1}{2}} L_{11} & \sigma_3^{\frac{1}{2}} A_{11} \end{bmatrix} \end{bmatrix}$$

$$\bar{H}_e = \begin{bmatrix} R_0^{-\frac{1}{2}} \bar{C}_e \\ \begin{bmatrix} (\frac{\rho_2}{\sigma_2})^{\frac{1}{2}}(I + \frac{\rho_2}{\sigma_2} L_{22} R_{10} L_{22}^T)^{\frac{1}{2}} L_{22} C_2 & 0 \end{bmatrix} \\ \begin{bmatrix} 0 & (\frac{\rho_1}{\sigma_1})^{\frac{1}{2}} K_{22} \end{bmatrix} \\ \begin{bmatrix} 0 & (\frac{\rho_3}{\sigma_3})^{\frac{1}{2}} A_{22} \end{bmatrix} \end{bmatrix}.$$

Proof 4.4 *Let* $X = \begin{bmatrix} P + Q & -Q \\ -Q & Q \end{bmatrix}$. *Then, from (4.43), (4.44), (4.20), (4.21), (4.28), (4.32), (4.34), and (4.37), and routine algebraic manipulations, it follows that*

$$\bar{A}_e^T X + X \bar{A}_e + X \bar{B}_e \bar{B}_e^T X + \bar{H}_e^T \bar{H}_e$$

$$= \begin{bmatrix} \bar{H}_{os}(P) + \Delta_{oo}(Q) - \Delta_{oo}(Q) \\ -\Delta_{oo}(Q)\Delta_{oo}(Q) + \frac{\rho_1}{\sigma_1} K_{22}^T K_{22} + \frac{\rho_3}{\sigma_3} A_{22}^T A_{22} \end{bmatrix}$$

$$= \begin{bmatrix} I & -I \\ 0 & I \end{bmatrix} \{ \begin{bmatrix} \Delta_{os}(P) & 0 \\ 0 & \Delta_{oo}(P) \end{bmatrix} + T_0 \} \begin{bmatrix} I & 0 \\ -I & I \end{bmatrix}.$$

By (4.30), we have

$$He\{X\bar{A}_e\} + X\bar{B}_e \bar{B}_e^T X + \bar{H}_e^T \bar{H}_e < 0.$$

Thus, the proof is completed by Lemma 4.3 and Lemma 4.1.

Remark 4.3 *Theorem 4.1 presents a solution of the non-fragile H_∞ controller design problem based on the existence of solutions to a matrix inequality (4.30) under the constraints (4.26) and (4.27) on the design parameters σ_1 and σ_2. In the case of no controller uncertainty, Theorem 4.1 recovers exactly the standard H_∞ control design. In fact, in this case, $R_0 = I$, $R_{20} = R_2$, $R_{01} = \gamma^2 I$, and $R_{10} = \frac{1}{\gamma^2} R_1$, the inequality (4.30) is reduced to*

$$H_{os}(P) = He\{PA\} + \frac{1}{\gamma^2} P B_1 B_1^T P + C_1^T C_1$$

$$- (PB_2 + C_1^T D_{12}) R_2^{-1} (PB_2 + C_1^T D_{12})^T < 0 \tag{4.45}$$

and

$$H_{oo}(Q) = He \left\{ Q \left[A + \frac{1}{\gamma^2} B_1 B_1^T P - B_1 D_{21}^T R_1^{-1} (C_2 + \frac{1}{\gamma^2} D_{21} B_1^T P) \right] \right\}$$

$$+ K^T R_2 K - \gamma^2 (C_2 + \frac{1}{\gamma^2} D_{21} B_1^T P)^T R_1^{-1} (C_2 + \frac{1}{\gamma^2} D_{21} B_1^T P)$$

$$+ \frac{1}{\gamma^2} Q B_1 (I - D_{21}^T R_1^{-1} D_{21}) B_1^T Q \ < \ 0. \tag{4.46}$$

The Riccati inequalities (4.45) and (4.46) are precisely the inequalities corresponding to the standard H_∞ design in Tadmor [120].

Theorem 4.1 presents a sufficient condition for the solvability of the problem of non-fragile output feedback H_∞ control with controller uncertainty in terms of a matrix inequality. However, to solve the inequality directly may be very difficult. Next, we present an alternative sufficient condition in terms of two Riccati inequalities, which is immediate from Theorem 4.1 and Lemma 2.1.

Theorem 4.2 *Consider the system Σ described by Equations (4.1)–(4.3) and positive constants γ and ρ. Suppose that*
(i) Assumptions 4.1 and 4.2 hold;
(ii) there exist positive constants σ_1, σ_2, and σ_3, and symmetric positive-definite matrices P and Q such that
(a) the inequalities (4.26) and (4.27) hold;
(b) the following inequalities

$$\Delta_{os}(P) + T_{01} < 0 \tag{4.47}$$

$$\Delta_{oo}(Q) + T_{01} - T_{01}(\Delta_{os}(P) + T_{01})^{-1} T_{01} < 0 \tag{4.48}$$

hold, where the matrix T_{011} is defined by (4.31).
Then the controller given by (4.4) and (4.5) with (4.28) and (4.32) is a solution of the non-fragile H_∞ controller design problem for the system Σ.

Remark 4.4 *Theorem 4.2 gives an alternative sufficient condition for the non-fragile H_∞ control problem in terms of two Riccati inequalities (4.47) and (4.48), and the inequality constraints (4.26) and (4.27) on the design parameters σ_1 and σ_2. The solutions of the Riccati inequalities (4.47) and (4.48) can be obtained by choosing matrices $T_a > 0$ and $T_b > 0$, and solving the following Riccati equations:*

$$\Delta_{os}(P) + T_{01} + T_a = 0 \tag{4.49}$$

$$\Delta_{oo}(Q) + T_{01} + T_{01} T_a^{-1} T_{01} + T_b = 0 \tag{4.50}$$

For the choices of the design parameters σ_1, σ_2, and σ_3, σ_1 and σ_2 can be searched in the range given by the constraints (4.26) and (4.27). However, σ_3 may be any positive constant. Moreover, from Theorem 4.1 and Theorem 4.2, it is easy to see that reducing the uncertainty bounds ρ_1, ρ_2, and ρ_3 will enhance the existence of a non-fragile controller.

Remark 4.5 *For the more general gain variations described by*

$$\Delta K = \sum_{i=1}^{n_k} K_{1i} F_{1i}(t) K_{2i}, \quad F_{1i}^T(t) F_{1i}(t) \leq \rho_{1i} I, \quad i = 1, \cdots, n_k$$

$$\Delta L = \sum_{j=1}^{n_l} L_{1j} F_{2j} L_{2j}, \quad F_{2j}^T(t) F_{2j}(t) \leq \rho_{2j} I, \quad j = 1, \cdots, n_l$$

and

$$\Delta A = \sum_{s=1}^{n_a} A_{1s} F_{3s}(t) A_{3s}, \quad F_{3s}^T(t) F_{3s}(t) \leq \rho_{3s} I, \quad s = 1, \cdots, n_a$$

the corresponding designs can be derived by using Theorem 4.1 and Theorem 4.2; the details are omitted.

The following theorem establishes the relationship between the solvability of the standard H_∞ control problem and the solvability of the non-fragile H_∞ control problem.

Theorem 4.3 *Consider the system Σ described by Equations (4.1)–(4.3), and suppose that Assumptions 4.1 and 4.2 hold. If there exists a dynamic output feedback control*

$$\dot{\xi}(t) = A_c \xi(t) + L y(t)$$
$$u(t) = K \xi(t)$$

which asymptotically stabilizes the system Σ and renders the closed-loop system having an H_∞ norm bounded by $\gamma > 0$ (with $\gamma > \gamma_{opt}$, where γ_{opt} is the infimum of such γs). Then for the same γ, there exist uncertainty bounds $\rho_1 > 0$, $\rho_2 > 0$, and $\rho_3 > 0$ and positive constants σ_1, σ_2, and σ_3 such that the inequalities (4.26) and (4.27) hold, and (4.30) has symmetric positive-definite solutions P and Q. Furthermore, the non-fragile H_∞ controller design problem is solvable for the uncertainty bounds $\rho_1 > 0$, $\rho_2 > 0$, and $\rho_3 > 0$.

Proof 4.5 *By the result in Tadmor [120], the Riccati inequalities (4.49) and (4.50) have symmetric positive definite solutions P and Q, respectively. Let $\sigma_1 = \sigma_2 = \sigma_3$, and $\rho_1 = \rho_2 = \rho_3 = \sigma_1^2$. Then, from (4.45), (4.46), (4.20), (4.21), (4.31), (4.26), and (4.27), it follows that*

$$\lim_{\sigma_1 \to 0} \left(\begin{bmatrix} \Delta_{os}(P) & 0 \\ 0 & \Delta_{oo}(Q) \end{bmatrix} + T_0 \right) = \begin{bmatrix} H_{os}(P) & 0 \\ 0 & H_{oo}(Q) \end{bmatrix} < 0$$

$$\lim_{\sigma_1 \to 0} R_0(\sigma_1) = I > 0$$

$$\lim_{\sigma_1 \to 0} R_{01}(\sigma_1, \sigma_1^2) = \gamma^2 I > 0$$

which further implies that, for a sufficient small $\sigma_1 > 0$, the inequalities (4.26), (4.27), and (4.30) hold. Thus, by Theorem 4.1, the conclusion follows.

Remark 4.6 *Theorem 4.3 shows that the solvability of the non-fragile H_∞ control problem is guaranteed for some level of controller uncertainties, provided that the corresponding standard H_∞ control problem has a solution and that the achieved disturbance attenuation bound γ is strictly greater than the optimal value γ_{opt} in the standard H_∞ control problem. The conclusion can also be derived by using the small-gain theorem [154]. However, Theorem 4.1 and Theorem 4.2 provide a non-fragile H_∞ controller design for a given level of controller uncertainties.*

4.3.2   Multiplicative Controller Gain Uncertainty Case

The following preliminaries will be used in the sequel.

For a symmetric matrix $E = [e_{ij}] \in R^{n \times n}$, $\lambda_{min}(E)$ and $\lambda_{max}(E)$ denote the minimal eigenvalue and the maximal eigenvalue of E, respectively, and E_d is defined as the diagonal matrix $diag\{e_{11}, e_{22}, \cdots, e_{nn}\}$ with e_{ii}, $i = 1, 2, \cdots, n$ the diagonal entries of E.

$$T_{K1} = diag\{\bar{\theta}^2_{K11}, \bar{\theta}^2_{K12}, \cdots, \bar{\theta}^2_{K1m}\}$$
$$T_{K2} = diag\{\bar{\theta}^2_{K21}, \bar{\theta}^2_{K22}, \cdots, \bar{\theta}^2_{K2m}\} \tag{4.51}$$
$$T_{A1} = diag\{\bar{\theta}^2_{A11}, \bar{\theta}^2_{A12}, \cdots, \bar{\theta}^2_{A1n}\}$$
$$T_{A2} = diag\{\bar{\theta}^2_{A21}, \bar{\theta}^2_{A22}, \cdots, \bar{\theta}^2_{A2n}\} \tag{4.52}$$
$$T_{L1} = diag\{\bar{\theta}^2_{L11}, \bar{\theta}^2_{L12}, \cdots, \bar{\theta}^2_{L1r}\}$$
$$T_{L2} = diag\{\bar{\theta}^2_{L21}, \bar{\theta}^2_{L22}, \cdots, \bar{\theta}^2_{L2r}\} \tag{4.53}$$

Lemma 4.4 *Let ΔK, ΔA_c, and ΔL be described by (4.8)–(4.10) with (4.12)–(4.14). $R_K = diag\{R_{K1}, R_{K2}, \cdots, R_{Km}\} > 0$, $R_A = diag\{R_{A1}, R_{A2}, \cdots, R_{An}\} > 0$, and $R_L = diag\{R_{L1}, R_{L2}, \cdots, R_{Lr}\} > 0$ are constant matrices. Then for any constants $\beta_K > 0$, $\beta_A > 0$, and $\beta_L > 0$, the following inequalities hold:*

$$(\Delta K)^T R_K^{-1} \Delta K \leq (1 + \beta_K) K^T T_{K1} R_K^{-1} K$$
$$+ (1 + \frac{1}{\beta_K}) J_1(K, R_K^{-1}, T_{K2}) \tag{4.54}$$

$$\Delta A_c R_A^{-1} (\Delta A_c)^T \leq (1 + \beta_A) A_c T_{A1} R_A^{-1} A_c^T$$
$$+ (1 + \frac{1}{\beta_A}) J_2(A_c, R_A^{-1}, T_{A2}) \tag{4.55}$$

$$\Delta L R_A^{-1} (\Delta L)^T \leq (1 + \beta_L) L T_{L1} R_L^{-1} L^T$$
$$+ (1 + \frac{1}{\beta_L}) J_3(L, R_L^{-1}, T_{L2}) \tag{4.56}$$

$$J_1(K, R_K^{-1}, T_{K2}) \leq \theta^2_{K2} n [K^T R_K^{-1} K]_d \tag{4.57}$$
$$J_2(A_c, R_A^{-1}, T_{A2}) \leq \theta^2_{A2} n [A_c R_A^{-1} A_c^T]_d \tag{4.58}$$
$$J_3(L, R_L^{-1}, T_{L2}) \leq \theta^2_{L2} n [L R_L^{-1} L^T]_d \tag{4.59}$$

where

$$J_1(K, R_K^{-1}, T_{K2}) = diag \left\{ \sum_{i=1}^{m} nR_{Ki}\bar{\theta}_{K2i}^2 k_{i1}^2, \sum_{i=1}^{m} nR_{Ki}\bar{\theta}_{K2i}^2 k_{i2}^2, \right.$$
$$\left. \cdots, \sum_{i=1}^{m} nR_{Ki}\bar{\theta}_{K2i}^2 k_{in}^2 \right\} \tag{4.60}$$

$$J_2(A_c, R_A^{-1}, T_{A2}) \leq diag \left\{ \sum_{i=1}^{n} nR_{Ai}\bar{\theta}_{A2i}^2 a_{1i}^2, \sum_{i=1}^{n} nR_{Ai}\bar{\theta}_{A2i}^2 a_{2i}^2, \right.$$
$$\left. \cdots, \sum_{i=1}^{n} nR_{Ai}\bar{\theta}_{A2i}^2 a_{ni}^2 \right\} \tag{4.61}$$

$$J_3(L, R_L^{-1}, T_{L2}) \leq diag \left\{ \sum_{i=1}^{r} nR_{Li}\bar{\theta}_{L2i}^2 l_{1i}^2, \sum_{i=1}^{r} nR_{Li}\bar{\theta}_{L2i}^2 l_{2i}^2, \right.$$
$$\left. \cdots, \sum_{i=1}^{r} nR_{Li}\bar{\theta}_{L2i}^2 l_{ni}^2 \right\} \tag{4.62}$$

Proof 4.6 *From (4.8), we have* $\Delta K = \Delta_1(K) + \Delta_2(K)$, *where*

$$\begin{aligned} \Delta_1(K) &= diag\{\theta_{K11}, \theta_{K12}, \cdots, \theta_{K1m}\}K, \\ \Delta_2(K) &= [k_{ij}\theta_{K2ij}]_{m \times n}. \end{aligned} \tag{4.63}$$

Thus,

$$\begin{aligned} (\Delta K)^T R_K^{-1} \Delta K &\leq (1 + \beta_K)[\Delta_1(K)]^T R_K^{-1} \Delta_1(K) \\ &\quad + (1 + \frac{1}{\beta_K})[\Delta_2(K)]^T R_K^{-1} \Delta_2(K). \end{aligned} \tag{4.64}$$

By (4.63), (4.11), (4.12), (4.51), and (4.60), it follows that

$$\begin{aligned} [\Delta_1(K)]^T R_K^{-1} \Delta_1(K) &= K^T R_K^{-1} diag\{\theta_{K11}^2, \theta_{K12}^2, \cdots, \theta_{K1m}^2\}K \\ &\leq K^T T_{K1} R_K^{-1} K, \end{aligned} \tag{4.65}$$

$$\begin{aligned} [\Delta_2(K)]^T R_K^{-1} \Delta_2(K) &= \Lambda_{11}^T R_K^{-1} \Lambda_{11} \\ &= \sum_{i=1}^{m} \Lambda_{12} K_i^T R_{K_i}^{-1} K_i \Lambda_{12} \\ &= \sum_{i=1}^{m} \Lambda_{13} R_{K_i}^{-1} \Lambda_{14}^T \Lambda_{14} \Lambda_{13} \\ &\leq J_1(K, R_K^{-1}, T_{K2}) \end{aligned} \tag{4.66}$$

where

$$\Lambda_{11} = \begin{bmatrix} K_1 diag\{\theta_{K211}, \theta_{K212}, \cdots, \theta_{K21n}\} \\ K_2 diag\{\theta_{K221}, \theta_{K222}, \cdots, \theta_{K22n}\} \\ \cdots \\ K_m diag\{\theta_{K2m1}, \theta_{K2m2}, \cdots, \theta_{K2mn}\} \end{bmatrix},$$

$$\Lambda_{12} = diag\{\theta_{K2i1}, \theta_{K2i2}, \cdots, \theta_{K2in}\},$$
$$\Lambda_{13} = diag\{k_{i1}, k_{i2}, \cdots, k_{in}\},$$
$$\Lambda_{14} = \begin{bmatrix} \theta_{K2i1} & \theta_{K2i2} & \cdots & \theta_{K2in} \end{bmatrix}.$$

Combining (4.64)–(4.66), inequality (4.54) follows. The proof of (4.55) and (4.56) are similar, and omitted. Inequalities (4.57), (4.58), and (4.59) are immediate from (4.11), (4.60)–(4.62), and (4.12)–(4.14). This completes the proof.

Remark 4.7 *The inequalities (4.54)–(4.56) give the bounds of* $(\Delta K)^T R_K^{-1} \Delta K$, $\Delta A_c R_A^{-1}(\Delta A_c)^T$, *and* $\Delta L R_L^{-1}(\Delta L)^T$ *in terms of the bounds of the uncertain parameters and the entries of matrices* K, A_c, *and* L, *respectively. If* $\theta_{K2} = 0$, $\theta_{A2} = 0$, *and* $\theta_{L2} = 0$, *then (4.54)–(4.56) can be simplified as*

$$(\Delta K)^T R_K^{-1} \Delta K \le K^T T_{K1} R_K^{-1} K, \Delta A_c R_A^{-1}(\Delta A_c)^T \le A_c T_{A1} R_A^{-1} A_c^T$$
$$\Delta L R_A^{-1}(\Delta L)^T \le L T_{L1} R_L^{-1} L^T$$

respectively. If $\theta_{K1} = 0$, $\theta_{A1} = 0$, *and* $\theta_{L1} = 0$, *then we have*

$$(\Delta K)^T R_K^{-1} \Delta K \le J_1(K, R_K^{-1}, T_{K2}), \Delta A_c R_A^{-1}(\Delta A_c)^T \le J_2(A_c, R_A^{-1}, T_{A2})$$
$$\Delta L R_A^{-1}(\Delta L)^T \le J_3(L, R_L^{-1}, T_{L2}),$$

respectively.

In the following, sufficient conditions for the solvability of the non-fragile H_∞ control problem with respect to multiplicative controller gain uncertainties are presented.

Consider an auxiliary closed-loop system Σ_{ac} given by

$$\Sigma_{ac} : \quad \dot{x}_e(t) = \bar{A}_e x_e(t) + \bar{B}_e \bar{w}(t) \tag{4.67}$$
$$\bar{z}(t) = \bar{H}_e x_e(t) \tag{4.68}$$

where

$$\bar{A}_e = \begin{bmatrix} \bar{A} & \bar{B}_2 K \\ L\bar{C}_2 & A_c \end{bmatrix}$$

$$\bar{B}_e = \begin{bmatrix} \bar{B}_1 & 0 & 0 \\ L\bar{D}_{21} & \gamma[(1 + \beta_A)A_c T_{A1} R_A^{-1} A_c^T]^{\frac{1}{2}} & \gamma \pi_1^{\frac{1}{2}} \end{bmatrix}$$

$$\bar{H}_e = \begin{bmatrix} \bar{C}_1 & \bar{D}_{12} K \\ 0 & (R_A + \pi_2)^{\frac{1}{2}} \end{bmatrix}$$

with

$$\bar{A} = A + \frac{1}{\gamma^2} B_1 R_{L0}^{-1} D_{21}^T R_L C_2 + B_2 R_K D_{12}^T R_{K0}^{-1} C_1 \qquad (4.69)$$

$$\bar{B}_2 = B_2 (I + R_K D_{12}^T R_{K0}^{-1} D_{12}) \qquad (4.70)$$

$$\bar{C}_2 = (I + \frac{1}{\gamma^2} D_{21} R_{L0}^{-1} D_{21}^T R_L) C_2 \qquad (4.71)$$

$$\bar{B}_1 = \begin{bmatrix} B_1 R_{L0}^{-\frac{1}{2}} & \gamma [B_2(R_K + R_K D_{12}^T R_{K0}^{-1} D_{12} R_K) B_2^T]^{\frac{1}{2}} & 0 \end{bmatrix} \qquad (4.72)$$

$$\bar{D}_{21} = \begin{bmatrix} D_{21} R_{L0}^{-\frac{1}{2}} & 0 & \gamma (1 + \beta_L)^{\frac{1}{2}} T_{L1}^{\frac{1}{2}} R_L^{-\frac{1}{2}} \end{bmatrix} \qquad (4.73)$$

$$\bar{C}_1 = \begin{bmatrix} R_{K0}^{-\frac{1}{2}} C_1 \\ 0 \\ [C_2^T (R_L + \frac{1}{\gamma^2} R_L D_{21} R_{L0}^{-1} D_{21}^T R_L) C_2]^{\frac{1}{2}} \end{bmatrix} \qquad (4.74)$$

$$\bar{D}_{12} = \begin{bmatrix} R_{K0}^{-\frac{1}{2}} D_{12} \\ (1 + \beta_K)^{\frac{1}{2}} T_{K1}^{\frac{1}{2}} R_K^{-\frac{1}{2}} \\ 0 \end{bmatrix} \qquad (4.75)$$

$$\pi_1 = (1 + \frac{1}{\beta_A}) J_2(A_c, R_A^{-1}, T_{A2}) + (1 + \frac{1}{\beta_L}) J_3(L, R_L^{-1}, T_{L2}) \qquad (4.76)$$

$$\pi_2 = (1 + \frac{1}{\beta_K}) J_1(K, R_K^{-1}, T_{K2}) \qquad (4.77)$$

where $\gamma > 0$, $\beta_K > 0$, $\beta_A > 0$, and $\beta_L > 0$ are constants, $R_K > 0$, $R_A > 0$, and $R_L > 0$ are diagonal matrices, $T_{K1}, T_{K2}, T_{A1}, T_{A2}, T_{L1}, T_{L2}, J_1(K, R_K^{-1}, T_{K2})$, $J_2(A_c, R_A^{-1}, T_{A2})$, and $J_3(L, R_L^{-1}, T_{L2})$ are defined by (4.51)–(4.53) and (4.60)–(4.62), respectively, and

$$R_{K0} = I - D_{12} R_K D_{12}^T, \quad R_{L0} = I - \frac{1}{\gamma^2} D_{21}^T R_L D_{21}. \qquad (4.78)$$

Then, we have the following lemma.

Lemma 4.5 *Let $\gamma > 0$ be a constant. If there exists a symmetric positive-definite matrix X, constants $\beta_K > 0$, $\beta_A > 0$, and $\beta_L > 0$, diagonal matrices $R_K > 0$, $R_A > 0$, and $R_L > 0$ such that*

$$R_K < R_2^{-1} \qquad (4.79)$$
$$R_L < \gamma^2 R_1^{-1} \qquad (4.80)$$

and

$$He\{X\bar{A}_e\} + \frac{1}{\gamma^2} X \bar{B}_e \bar{B}_e^T X + \bar{H}_e^T \bar{H}_e < 0 \qquad (4.81)$$

then the inequality

$$He\{X(A_e + \Delta A_e)\} + \frac{1}{\gamma^2} X(B_e + \Delta B_e)(B_e + \Delta B_e)^T X$$

$$+ (H_e + \Delta H_e)^T (H_e + \Delta H_e) < 0 \qquad (4.82)$$

holds for any θ_{K1i}, θ_{K2ij}, θ_{A1i}, θ_{A2ji}, θ_{L1i}, *and* θ_{L2ji} *satisfying (4.12)–(4.14). Furthermore, the controller given by (4.4) and (4.5) solves the non-fragile* H_∞ *control problem for the system* Σ *described by (4.1)–(4.3).*

Proof 4.7 *By Lemma 2.1, it is sufficient to show that*

$$
M_1 \;=\; \begin{bmatrix} He\{X(A_e + \Delta A_e)\} & * & * \\ \frac{1}{\gamma}(B_e + \Delta B_e)^T X & -I & * \\ H_e + \Delta H_e & 0 & -I \end{bmatrix} < 0.
$$

By (4.8)–(4.14), (4.54)–(4.56), (4.15), and (4.16), it follows that

$$
M_1 \;\leq\; \begin{bmatrix} He\{X(A_e)\} & * & * \\ \frac{1}{\gamma}B_e^T X & -I & * \\ H_e & 0 & -I \end{bmatrix} + \begin{bmatrix} X\begin{bmatrix} B_2 \\ 0 \end{bmatrix} \\ 0 \\ D_12 \end{bmatrix}\begin{bmatrix} X\begin{bmatrix} B_2 \\ 0 \end{bmatrix} \\ 0 \\ D_12 \end{bmatrix}^T
$$

$$
+ \begin{bmatrix} \begin{bmatrix} 0 & I \end{bmatrix} & 0 & 0 \end{bmatrix}^T R_A \begin{bmatrix} \begin{bmatrix} 0 & I \end{bmatrix} & 0 & 0 \end{bmatrix}
$$

$$
+ \begin{bmatrix} \begin{bmatrix} C_2 & 0 \end{bmatrix} & \frac{1}{\gamma}D_{21} & 0 \end{bmatrix}^T R_L \begin{bmatrix} \begin{bmatrix} C_2 & 0 \end{bmatrix} & \frac{1}{\gamma}D_{21} & 0 \end{bmatrix}
$$

$$
+ \begin{bmatrix} \begin{bmatrix} 0 & 0 \\ 0 & \Upsilon_K \end{bmatrix} & * & * \\ 0 & 0 & * \\ 0 & 0 & 0 \end{bmatrix} + \begin{bmatrix} X\begin{bmatrix} 0 & 0 \\ 0 & \Upsilon_A \end{bmatrix}X & * & * \\ 0 & 0 & * \\ 0 & 0 & 0 \end{bmatrix}
$$

$$
+ \begin{bmatrix} X\begin{bmatrix} 0 & 0 \\ 0 & \Upsilon_L \end{bmatrix}X & * & * \\ 0 & 0 & * \\ 0 & 0 & 0 \end{bmatrix}
$$

$$
= \; M_2 < 0 \tag{4.83}
$$

where

$$
M_2 \;=\; \begin{bmatrix} M_{11} & * & * \\ M_{12}^T & -I + \frac{1}{\gamma^2}D_{21}^T R_L D_{21} & * \\ M_{13}^T & 0 & -I + D_{12}^T R_K D_{12} \end{bmatrix}, \tag{4.84}
$$

$$
\Upsilon_K \;=\; (1 + \beta_K)K^T T_{K1} R_K^{-1} K + (1 + \frac{1}{\beta_K})J_1(K, R_K^{-1}, T_{K2}),
$$

$$
\Upsilon_A \;=\; (1 + \beta_A)A_c T_{A1} R_A^{-1} A_c^T + (1 + \frac{1}{\beta_A})J_2(A_c, R_A^{-1}, T_{A2}),
$$

$$
\Upsilon_L \;=\; (1 + \beta_L)L T_{L1} R_L^{-1} L^T + (1 + \frac{1}{\beta_L})J_3(L, R_L^{-1}, T_{L2})
$$

with

$$
\begin{aligned}
M_{11} &= He\{XA_e\} + X\left\{\begin{bmatrix} B_2 R_K B_2^T & * \\ 0 & \Upsilon_A + \Upsilon_L \end{bmatrix}\right\}X \\
&\quad + \begin{bmatrix} C_2 R_L C_2^T & * \\ 0 & R_A + \Upsilon_K \end{bmatrix}, \\
M_{12} &= \frac{1}{\gamma}(XB_e + \begin{bmatrix} C_2^T \\ 0 \end{bmatrix}R_L D_{21}), M_{13} = H_e + X\begin{bmatrix} B_2 \\ 0 \end{bmatrix}R_K D_{21}^{12}.
\end{aligned}
$$

From (4.78), (4.79), and (4.80), we have $R_{K0} > 0$ and $R_{L0} > 0$. Then, $M_2 < 0$ holds if and only if $M_3 < 0$ holds, where

$$
M_3 = M_{11} + M_{12}R_{L0}^{-1}M_{12}^T + M_{13}R_{K0}^{-1}M_{13}^T. \tag{4.85}
$$

Combining (4.67), (4.68), (4.84), and (4.85), it is easy to check that

$$
M_3 = He\{X\bar{A}_e\} + \frac{1}{\gamma^2}X\bar{B}_e\bar{B}_e^T X + \bar{H}_e^T \bar{H}_e.
$$

Thus, the conclusion follows from (4.81), (4.83), and Lemma 2.13.

Lemma 4.4 shows that the controller given by (4.4) and (4.5) will solve the non-fragile H_∞ control problem for the system Σ if the controller renders the auxiliary closed-loop system Σ_{ac} with an H_∞ disturbance attenuation γ.

Denote

$$
\begin{aligned}
H_{os}(P) &= H_{os}(P, R_K, R_L, \beta_K) \\
&= He\{PA\} + \frac{1}{\gamma^2}PB_1 B_1^T P + C_1^T C_1 - (PB_2 + C_1^T D_{12})R_{2K} \\
&\quad \times (PB_2 + C_1^T D_{12})^T + (\frac{1}{\gamma^2}PB_1 D_{21}^T + C_2^T)(R_L^{-1} - \frac{1}{\gamma^2}R_1)^{-1} \\
&\quad \times (\frac{1}{\gamma^2}PB_1 D_{21}^T + C_2^T)^T
\end{aligned} \tag{4.86}
$$

where

$$
\begin{aligned}
R_{2K} &= R_{2K}(R_K, \beta_K) \\
&= (I - (1 + \beta_K)T_{K1})[R_2 - (R_2 - R_K^{-1})(1 + \beta_K)T_{K1}]^{-1}
\end{aligned} \tag{4.87}
$$

and denote

$$
\begin{aligned}
H_{oo}(Q) &= H_{oo}(Q, R_K, R_A, R_L, \beta_A, \beta_L) \\
&= He\{Q\bar{A}_0\} + K^T \bar{R}_2 K + Q[\frac{1}{\gamma^2}\bar{B}_1\bar{B}_1^T + (1 + \beta_A)\bar{A}_c T_{A1} R_A^{-1}\bar{A}_c^T]Q \\
&\quad - \frac{1}{\gamma^2}\{\gamma^2\bar{C}_{20} + [\bar{D}_{21}\bar{B}_1^T + \gamma^2(1 + \beta_A)\bar{C}_{20}T_{A1}R_A^{-1}\bar{A}_c^T]Q\}^T\bar{R}_{10}^{-1} \\
&\quad \times \{\gamma^2\bar{C}_{20} + [\bar{D}_{21}\bar{B}_1^T + \gamma^2(1 + \beta_A)\bar{C}_{20}T_{A1}R_A^{-1}\bar{A}_c^T]Q\}
\end{aligned} \tag{4.88}
$$

where

$$\bar{A}_0 = \bar{A} + \frac{1}{\gamma^2}\bar{B}_1\bar{B}_1^T P, \quad \bar{C}_{20} = \bar{C}_2 + \frac{1}{\gamma^2}\bar{D}_{21}\bar{B}_1^T P \tag{4.89}$$

$$\bar{R}_{10} = \bar{D}_{21}\bar{D}_{21}^T + \gamma^2(1 + \beta_A)\bar{C}_{20}T_{A1}R_A^{-1}\bar{C}_{20}^T, \quad \bar{R}_2 = \bar{D}_{12}^T\bar{D}_{12} \tag{4.90}$$

$$K = K(R_K, \beta_K) = -[R_2 - (R_2 - R_K^{-1})(1 + \beta_K)T_{K1}]^{-1}$$
$$\times(B_2^T P + D_{12}^T C_1) \tag{4.91}$$

$$\bar{A}_c = \bar{A}_0 + \bar{B}_2 K, A_c = A_c(\beta_K, \beta_A, \beta_L) = \bar{A}_c - L\bar{C}_{20} \tag{4.92}$$

$$L = L(\beta_L, \beta_A) = [\gamma^2 Q^{-1}\bar{C}_{20}^T + \bar{B}_1\bar{D}_{21}^T + \gamma^2(1 + \beta_A)$$
$$\times \bar{A}_c T_{A1} R_A^{-1}\bar{C}_{20}^T]\bar{R}_{10}^{-1} \tag{4.93}$$

with $\bar{A}$, $\bar{B}_2$, $\bar{C}_2$, $\bar{B}_1$, $\bar{D}_{21}$, $\bar{C}_1$, and $\bar{D}_{12}$ defined by (4.69)–(4.75), respectively.

Then the following theorem presents a sufficient condition of solving the non-fragile H_∞ control problem.

Theorem 4.4 *Consider the system Σ described by Equations (4.1)–(4.3), $\gamma > 0$ is a constant. Suppose the following:*
(i) Assumptions 4.1 and 4.2 hold;
(ii) there exist positive constants β_K, β_A, and β_L, diagonal matrices $R_K > 0$, $R_A > 0$, and $R_L > 0$, and symmetric positive-definite matrices P and Q such that
(a) the inequalities (4.79) and (4.80) hold; and
(b)

$$\begin{bmatrix} H_{os}(P) & 0 \\ 0 & H_{oo}(Q) + Q\pi_1 Q \end{bmatrix} + \begin{bmatrix} R_A + \pi_2 & R_A + \pi_2 \\ R_A + \pi_2 & R_A + \pi_2 \end{bmatrix} < 0 \tag{4.94}$$

where π_1 and π_2 are defined by (4.76) and (4.77), respectively. Then the controller given by (4.4) and (4.5) with (4.8)–(4.10) and (4.91)–(4.93) solves the non-fragile H_∞ control problem for the system Σ.

Proof 4.8 *Denote $X = \begin{bmatrix} P+Q & -Q \\ -Q & Q \end{bmatrix}$.*

Then, from $P > 0$ and $Q > 0$, we have $X > 0$, and routine algebraic manipulations show

$$He\{X\bar{A}_e\} + \frac{1}{\gamma^2}X\bar{B}_e\bar{B}_e^T X + \bar{H}_e^T\bar{H}_e$$

$$= \begin{bmatrix} H_{os}(P) + H_{oo}(Q) + Q\pi_1 Q & -H_{oo}(Q) - Q\pi_1 Q \\ -H_{oo}(Q) - Q\pi_1 Q & H_{oo}(Q) + Q\pi_1 Q \end{bmatrix} + \begin{bmatrix} 0 & 0 \\ 0 & R_A + \pi_2 \end{bmatrix}$$

$$= \begin{bmatrix} I & -I \\ 0 & I \end{bmatrix}\left\{ \begin{bmatrix} H_{os}(P) & 0 \\ 0 & H_{oo}(Q) + Q\pi_1 Q \end{bmatrix} \right.$$
$$\left. + \begin{bmatrix} R_A + \pi_2 & R_A + \pi_2 \\ R_A + \pi_2 & R_A + \pi_2 \end{bmatrix} \right\} \begin{bmatrix} I & -I \\ 0 & I \end{bmatrix}^T < 0 \tag{4.95}$$

From assumption (ii), (4.95), and Lemma 4.5, the conclusion follows.

Remark 4.8 *Theorem 4.4 presents a sufficient condition for the solvability of the non-fragile H_∞ control problem. When the controller given by (4.4) and (4.5) contains no gain variations, that is, $\Delta K = 0$, $\Delta A_c = 0$, and $\Delta L = 0$, then from (4.76), (4.77), (4.86), and (4.88), it is easy to show that condition (ii) of Theorem 4.4 holds if and only if the following two Riccati inequalities hold:*

$$He\{PA\} + \frac{1}{\gamma^2}PB_1B_1^T P + C_1^T C_1$$

$$-(PB_2 + C_1^T D_{12})R_2^{-1}(PB_2 + C_1^T D_{12})^T < 0$$

$$He\{QA_0\} + K_0 R_2 K_0 + \frac{1}{\gamma^2}QB_1B_1^T Q$$

$$-\frac{1}{\gamma^2}(\gamma^2 C_{20} + D_{21}B_1^T Q)^T R_1^{-1}(\gamma^2 C_{20} + D_{21}B_1^T Q) < 0$$

where

$$A_0 = A + \frac{1}{\gamma^2}B_1 B_1^T P, \quad C_{20} = C_2 + \frac{1}{\gamma^2}D_{21}B_1^T P$$

$$K_0 = -R_2^{-1}(B_2^T P + D_{12}^T C_1)$$

The above two Riccati inequalities are those for the standard H_∞ control problem in Tadmor [120]. From (4.76) and (4.77), we know that π_1 and π_2 is a function of A_c and L, and π_2 is a function of K, respectively, while A_c, L, and K are to be designed. So, it may be very difficult to solve the inequality (4.94) directly.

The following theorem presents another sufficient condition in terms of two Riccati inequalities, which can be solved directly.

Theorem 4.5 *Consider the system Σ described by Equations (4.1)–(4.3), where $\gamma > 0$ is a constant. Suppose the following:*
(i) Assumptions 4.1 and 4.2 hold;
(ii) there exist positive constants β_K, β_A, and β_L, diagonal matrices $R_K > 0$, $R_A > 0$, $R_L > 0$, $R_{\pi_1} > 0$, and $R_{\pi_2} > 0$, and symmetric positive-definite matrices P and Q such that
(a) the inequalities (4.79) and (4.80) hold; and
(b)

$$\Delta(H) = H_{os}(P) + R_A + R_{\pi_2} < 0 \tag{4.96}$$

and

$$H_{oo}(Q) + QR_{\pi_1}Q + R_A + R_{\pi_2} - (R_A + R_{\pi_2})\Delta^{-1}(H)(R_A + R_{\pi_2}) < 0. \tag{4.97}$$

Then the controller given by (4.4) and (4.5) with (4.8)–(4.10) and (4.91)–(4.93), and θ_{K2}, θ_{A2}, and θ_{L2} satisfying

$$\theta_{A2}^2\{(1+\frac{1}{\beta_A})[A_c R_A^{-1}A_c^T]_d\} + \theta_{L2}^2\{(1+\frac{1}{\beta_L})[LR_L^{-1}L^T]_d\} \leq \frac{R_{\pi_1}}{n} \tag{4.98}$$

$$\theta_{K2}^2\{(1+\frac{1}{\beta_K})[K^T R_K^{-1}K]_d\} \leq \frac{R_{\pi_2}}{n} \tag{4.99}$$

solves the non-fragile H_∞ control problem for the system Σ.

Proof 4.9 *From (4.57)–(4.59), (4.76), (4.77), (4.98), and (4.99), it follows that $\pi_1 \le R_{\pi_1}$ and $\pi_2 \le R_{\pi_2}$, which further implies that*

$$\begin{bmatrix} H_{os}(P) & 0 \\ 0 & H_{oo}(Q) + Q\pi_1 Q \end{bmatrix} + \begin{bmatrix} R_A + \pi_2 & R_A + \pi_2 \\ R_A + \pi_2 & R_A + \pi_2 \end{bmatrix}$$

$$\le \begin{bmatrix} H_{os}(P) + R_A + R_{\pi_2} & R_A + R_{\pi_2} \\ R_A + R_{\pi_2} & H_{oo}(Q) + Q\pi_1 Q + R_A + R_{\pi_2} \end{bmatrix}.$$

By Lemma 2.1 and Theorem 4.4, the conclusion follows.

Remark 4.9 *Based on the positive-definite solutions to the Riccati inequalities (4.96) and (4.97), Theorem 4.5 presents a non-fragile H_∞ controller design for the given bounds $\bar\theta_{K1i}$, $\bar\theta_{A1i}$, and $\bar\theta_{L1i}$ of the uncertain parameters θ_{K1i}, θ_{A1i}, and θ_{L1i}, and the uncertain parameters θ_{K2ij}, θ_{A2ji}, and θ_{L2ji} with the bounds θ_{K2}, θ_{A2}, and θ_{L2} determined by (4.98) and (4.99). From (4.99), we have*

$$\theta_{K2} = [\lambda_{max}\{n(1 + \frac{1}{\beta_K})R_{\pi_2}^{-\frac{1}{2}}[K^T R_K^{-1} K]_d R_{\pi_2}^{-\frac{1}{2}}\}]^{-\frac{1}{2}} \tag{4.100}$$

and if $\theta_{A2} = \theta_{L2}$, then from (4.98),

$$\theta_{A2} = \sqrt{\lambda_{max}\{nR_{\pi_1}^{-\frac{1}{2}}((1 + \frac{1}{\beta_A})[A_c R_A^{-1} A_c^T]_d + (1 + \frac{1}{\beta_L})[LR_L^{-1} L^T]_d)R_{\pi_1}^{-\frac{1}{2}}\}}. \tag{4.101}$$

As an important special case in which $\theta_{K2} = \theta_{A2} = \theta_{L2} = 0$, we have the following theorem.

Theorem 4.6 *Consider the system Σ described by Equations (4.1)–(4.3), where $\gamma > 0$ is a constant. Suppose the following:*
(i) Assumptions 4.1 and 4.2 hold;
(ii) there exist diagonal matrices $R_K > 0$, $R_A > 0$, and $R_L > 0$, and symmetric positive-definite matrices P and Q such that
(a) the inequalities (4.79) and (4.80) hold; and
(b)

$$\Delta_0(H) = H_{os}(P, R_K, R_L, 0) + R_A < 0 \tag{4.102}$$

and

$$H_{oo}(Q, R_K, R_A, R_L, 0, 0) + R_A - R_A \Delta_0^{-1}(H)R_A < 0 \tag{4.103}$$

where $H_{os}(P, R_K, R_L, 0)$ and $H_{oo}(Q, R_K, R_A, R_L, 0, 0)$ are defined by (4.86) and (4.88), respectively. Then the controller given by (4.4) and (4.5) with (4.8)–(4.10), and $\theta_{K2} = \theta_{A2} = \theta_{L2} = 0$, and

$$K = K(R_K, 0), \quad A_c = A_c(0, 0, 0), \quad L = L(0, 0) \tag{4.104}$$

where $K(R_K, \beta_K)$, $A_c(\beta_K, \beta_A, \beta_L)$, and $L(\beta_A, \beta_L)$ are defined by (4.91)–(4.93), solves the non-fragile H_∞ control problem for the system Σ.

Proof 4.10 *It is immediate from Remark 4.7 and the proof of Theorem 4.4.*

Remark 4.10 *For the special case of $\theta_{K2} = \theta_{A2} = \theta_{L2} = 0$, the design Equations (4.102)–(4.104) in Theorem 4.6 are simpler. Similarly, for the case of $\theta_{K1} = \theta_{A1} = \theta_{L1} = 0$, the corresponding design can be derived, and the details are omitted. It should be noted that the Riccati inequalities (4.102) and (4.103), choose a positive-definite matrix S and a small constant $\varepsilon > 0$, and then solve the Riccati equations*

$$\Delta_0(H) + S = 0 \tag{4.105}$$

and

$$H_{00}(Q, R_K, R_A, R_L, 0, 0) + Q_A - R_A S^{-1} R_A + \varepsilon I = 0 \tag{4.106}$$

to obtain positive-definite solutions P and Q. Thus, P and Q will be solutions of the Riccati inequalities (4.102) and (4.103).

Remark 4.11 *Theorems 4.4, 4.5, and 4.6 provide a method of designing H_∞ controllers that can tolerate the given amount of controller uncertainty, which are derived based on considering the auxiliary closed-loop system Σ_{ac} given by (4.67) and (4.68) corresponding to inequality (4.81) in Lemma 4.5. Since the matrices $\bar{B}_e$ and $\bar{H}_e$ in (4.67) and (4.68) contain the submatrices $\gamma\sqrt{(1 + \beta_A)}A_c T_{A1} R_A^{-1} A_c^T$, $\gamma\sqrt{\pi_1}$, and $\sqrt{R_A + \pi_2}$, the auxiliary closed-loop system Σ_{ac} cannot be generated by using the standard H_∞ control design [28, 154]. Theorem 4.4 is given essentially by applying the differential game approach [7, 79] to the auxiliary closed-loop system Σ_{ac}. In comparison with the results in Corrado and Haddad [22, 52] for designing robust non-fragile H_2 controllers under multiplicative controller gain uncertainty and plant parameter uncertainties, a more general class of multiplicative controller gain variations given by (4.8)–(4.10) is dealt with in this chapter, and the design inequalities (4.96) and (4.97) in Theorem 4.5 ((4.102) and (4.103) in Theorem 4.6) are independent of the to-be-designed controller parameters, which can be effectively solved by using MATLAB™ as indicated in Remark 4.10, while Corrado and Haddad [22, 52] considered the multiplicative controller gain variation as a single parameter, and the design equations are dependent on the to-be-designed controller parameters.*

4.4 Example

To illustrate the effectiveness of our results, two examples are given. Example 4.1 is for the additive controller gain uncertainty case and Example 4.2 is for the multiplicative controller gain uncertainty case.

Example 4.1 *Given a system of the form (4.1)–(4.3) with*

$$A = \begin{bmatrix} -8 & 0.5 \\ 0.5 & 1 \end{bmatrix}, \quad B_1 = \begin{bmatrix} 0 & 0 \\ 0.5 & 0 \end{bmatrix}, \quad B_2 = \begin{bmatrix} 0 \\ 6 \end{bmatrix},$$

$$C_1 = \begin{bmatrix} 0.2 & 1 \\ 0 & 0 \end{bmatrix}, \quad C_2 = [0.2 \ \ 2], \quad D_{12} = \begin{bmatrix} 0 \\ 1 \end{bmatrix}, \quad D_{21} = [0 \ \ 0.5].$$

We consider a prescribed performance level of $\gamma = 1.6$, and controller uncertainties of the form (4.7)–(4.15) with $\rho_1 = 0.9$, $\rho_2 = 0.1$, $\rho_3 = 0.5$, and

$$K_{11} = 1, \qquad L_{11} = \begin{bmatrix} 0 \\ 1 \end{bmatrix}, \qquad A_{11} = \begin{bmatrix} 1 & 0 \\ 0 & 1 \end{bmatrix},$$

$$K_{22} = \begin{bmatrix} 1 & 0 \\ 0 & 1 \end{bmatrix}, \qquad L_{22} = 1, \qquad A_{22} = \begin{bmatrix} 1 & 0 \\ 0 & 1 \end{bmatrix}.$$

It is easy to see that (4.17) and (4.27) hold for $0 < \sigma_1 < 1$ and $\sigma_2 > 1/102.4$. Let $\sigma_1 = 0.2$, $\sigma_2 = 2$, and $\sigma_3 = 4$, $T_a = \begin{bmatrix} 4.6570 & 0 \\ 0 & 4.8750 \end{bmatrix}$ and $T_b = \begin{bmatrix} 0.0995 & 0 \\ 0 & 0.2875 \end{bmatrix}$. Then, the Riccati equations (4.30) and (4.31) have symmetric positive-definite solutions

$$P = \begin{bmatrix} 0.5837 & 0.0283 \\ 0.0283 & 0.5701 \end{bmatrix}, \quad Q = \begin{bmatrix} 3.3186 & 0.0522 \\ 0.0522 & 0.7327 \end{bmatrix}.$$

So, from Theorem 4.2, the problem of H_∞ control with controller uncertainty for this system has a solution. By using Equations (4.31) and (4.37), the controller parameters are computed as follows,

$$K = [-0.1697 \ \ -3.4206], L = \begin{bmatrix} 0.1775 \\ 27.9369 \end{bmatrix}, A_c = \begin{bmatrix} -8.0357 & 0.1432 \\ -6.1303 & -75.6156 \end{bmatrix}$$

with $\gamma = 0.4697$.

For comparison, a standard H_∞ controller is also designed. The optimal γ that can be reached is $\gamma_{opt} = 0.41$ and the corresponding feedback matrices are

$$K_{opt} = \begin{bmatrix} -0.1297 \\ -1.2207 \end{bmatrix}^T, L_{opt} = \begin{bmatrix} 0.1817 \\ 3.5929 \end{bmatrix}, A_{opt} = \begin{bmatrix} -8.0363 & 0.1366 \\ -0.9647 & -13.2073 \end{bmatrix}.$$

However, the closed-loop system by using the standard controller becomes unstable with an unstable eigenvalue at 0.6223 when there are controller uncertainties of the form (4.6) with $F_1 = [0 \ \ 0.94]$, $F_2 = -0.31$, and $F_3 = diag\{-0.7, -0.7\}$, while, by using our non-fragile controller subject to the same controller uncertainties (which satisfy the bounds in (4.7)), the closed-loop system is stable with eigenvalues at -63.5683, -11.6816, -8.1267, and -8.6747, which shows the superiority of the proposed method.

Example 4.2 *Consider a linear system of the form (4.1)–(4.3) with*

$$A = \begin{bmatrix} 0.4 & 0.5 \\ 1 & -0.1 \end{bmatrix}, B_1 = \begin{bmatrix} 0 & 0 \\ 0.5 & 0 \end{bmatrix},$$

$$B_2 = \begin{bmatrix} 0 \\ 6 \end{bmatrix}, C_1 = \begin{bmatrix} 0.2 & 1 \\ 0 & 0 \end{bmatrix}, D_{12} = \begin{bmatrix} 0 \\ 1 \end{bmatrix},$$

$$C_2 = \begin{bmatrix} 3 & 2 \end{bmatrix}, D_{21} = \begin{bmatrix} 0 & 0.5 \end{bmatrix}$$

Suppose that the controller with gain variations is described by (4.4) and (4.5), and the controller gain variations are given by

$$\Delta K = 0, \quad \Delta L = 0, \quad \Delta A_c = \begin{bmatrix} a_{11}\theta_{A11} & a_{12}\theta_{A12} \\ a_{21}\theta_{A11} & a_{22}\theta_{A12} \end{bmatrix}$$

with $|\theta_{A11}| \le \theta_{A1}$, $|\theta_{A12}| \le \theta_{A1}$ *and* $|\theta_{A1}| = 0.1$. *By using MATLAB, the inequalities (4.105) and (4.106) have positive definite solutions*

$$P = \begin{bmatrix} 4.1378 & 0.3610 \\ 0.3610 & 0.2068 \end{bmatrix}$$

and

$$Q = \begin{bmatrix} 2.1272 & -0.2579 \\ -0.2579 & 0.0326 \end{bmatrix},$$

respectively, for $R_K = R_L = 0.000002$, $R_A = diag\{0.22, 0.08\}$, *and* $\gamma = 0.425$. *By using Equations (4.91)–(4.93) and (4.104), the controller parameters are computed as follows:*

$$K = \begin{bmatrix} -2.1663 & -1.2405 \end{bmatrix}, L = \begin{bmatrix} 92.6371 \\ 751.7789 \end{bmatrix}, A_c = \begin{bmatrix} -277.5 & -184.8 \\ -2281.8 & -1520.8 \end{bmatrix}.$$

By Theorem 4.6, the controller renders the closed-loop system with the gain variations to be stable with disturbance attenuation $\gamma < 0.425$ *for* $|\theta_{A11}| \le 0.1$ *and* $|\theta_{A12}| \le 0.1$. *It should be noted that 0.425 is a guaranteed* H_∞ *performance of the closed-loop system against to the controller uncertainty* ΔA_c, *which means that the actual achieved performance may be better. In fact, the worst* H_∞ *performance and the best* H_∞ *performance of the closed-loop system with the controller gain variations* ΔA *are* $\gamma = 0.3373$ *and* $\gamma = 0.3032$.

In the following, we make some comparisons with the standard H_∞ *designs. By using the standard* H_∞ *control design method (see Tadmor [120] and Zhou, Doyle, and Glover [154]), the optimal* H_∞ *control law is given by*

$$K = \begin{bmatrix} -1.7866 & -1.1696 \end{bmatrix}, L = \begin{bmatrix} 234.1116 \\ 453.5845 \end{bmatrix},$$

$$A_c = \begin{bmatrix} -701.9 & -467.7 \\ -136.96 & -913.7 \end{bmatrix}$$

which renders the closed-loop system stable with disturbance attenuation $\gamma_{opt} = 0.295$, *provided that there are no controller gain variations. However, with the controller uncertainty* $\Delta A_c = A_c diag\{-0.07, 0.07\}$, *which are in the percentage range of* $|\theta_{A11}| \le 0.1$ *and* $|\theta_{A12}| \le 0.1$, *the closed-loop system is unstable.*

4.5   Conclusion

In this chapter, we have investigated the non-fragile H_∞ dynamic output feedback   controller
design problem for linear continuous-time systems. The procedures of designing non-fragile controllers that can tolerate some additive/multiplicative controller gain uncertainties are presented. The design methods are presented in terms of symmetric positive-definite solutions of algebraic Riccati inequalities. The resulting design is such that the closed-loop system is quadratically stable and satisfies an H_∞ disturbance attenuation bound. When the controller contains no gain variations, the results are reduced to those for the standard H_∞ control.

5

Robust Non-Fragile Kalman Filtering with Norm-Bounded Gain Uncertainty

5.1 Introduction

The *Kalman filtering* is a very popular approach for estimating the states of a nominal system by using past measurements due to its simplicity, optimality, tractability, and robustness [4,73,82]. A great number of results on the Kalman filter have been reported, and different approaches have been proposed [see 10, 51, 69, 110, 115, 139, and the references therein]. Noting that in the above-mentioned works on the filter designs, an implicit assumption is that the filter will be implemented exactly. However, inaccuracies or uncertainties do occur in the implementation of a designed filter or controller due to, among other things, round-off errors in numerical computation during the filter or controller implementation and the need of providing practicing engineers with safe-tuning margins. Filter or controller uncertainties can also be generated from the degradation of sensors and actuators [1]. This finding has received some attention from the control systems community, and some preliminary results have appeared recently to tackle the problem of designing controllers that are insensitive to variations in controller coefficients, as well as plant uncertainties [see 8, 27, 34, 52–54, 68, 74, 77, 78].

This chapter is concerned with the problem of robust *non-fragile* Kalman filter design for *linear systems* with norm-bounded uncertainties. Such robust non-fragile Kalman filters are required to be robust with respect to uncertainties in both the plant and the Kalman filter gains. Two classes of gain uncertainties are considered, namely, *additive* and *multiplicative*, and design methods are given for such robust non-fragile Kalman filters in terms of solutions to algebraic Riccati equations.

5.2   Problem Statement

Consider a class of uncertain linear systems described by the state equations

$$\dot{x}(t) \;=\; [A + D_1\Delta(t)E_1]x(t) + w_1(t), \quad x(t_0) = x_0 \tag{5.1}$$

$$y(t) \;=\; [C + D_2\Delta(t)E_1]x(t) + w_2(t) \tag{5.2}$$

where $x(t) \in R^n$ is the state, $y(t) \in R^p$ is the measured output, x_0 is the initial condition which is assumed to be a zero mean Gaussian random vector, $\Delta(t)$ is a time-varying matrix of uncertain parameters satisfying

$$\Delta^T(t)\Delta(t) \leq I \tag{5.3}$$

and $w_1(t)$ and $w_2(t)$ are zero mean white Gaussian noise processes with the joint covariance matrix

$$E\{ \begin{bmatrix} w_1 \\ w_2 \end{bmatrix} [w_1^T \; w_2^T] \} = \begin{bmatrix} V_1 & 0 \\ 0 & V_2 \end{bmatrix} > 0. \tag{5.4}$$

The state estimators under consideration are of form

$$\dot{\xi} = A_0\xi(t) + \tilde{G}y(t), \quad \xi(t_0) = \xi_0 \tag{5.5}$$

where $\tilde{G} = G + \Delta G$, $\xi(t) \in R^n$ is the state estimate, ξ_0 is the estimator initial condition which is assumed to be a zero mean Gaussian random vector, and ΔG represents the estimator *gain uncertainty* of the following forms:

(a) ΔG is of the additive form

$$\Delta G = G_{11}FG_{22}, \quad F^TF \leq \rho_a I, \quad \rho_a \geq 0 \tag{5.6}$$

with G_{11}, G_{22} constant matrices, and F the uncertain parameter matrix; and

(b) ΔG is of the multiplicative form

$$\Delta G = G\bar{G}_{11}F\bar{G}_{22}, \quad F^TF \leq \rho_m I, \quad \rho_m \geq 0 \tag{5.7}$$

with $\bar{G}_{11}$, $\bar{G}_{22}$ constant matrices, and F the uncertain parameter matrix.

Remark 5.1 *The above additive and multiplicative uncertainty models are used to describe the controller gain variations in Haddad and Corrado [53] and Famularo et al. [34], respectively. The multiplicative uncertainty model can also be used to describe degradations of sensors [1]. For example, if some percentage of degradation of a sensor is permitted in the design, then the characterization of the corresponding multiplicative gain uncertainty prior to the design can be given. A characterization of the additive gain uncertainty prior to the design corresponds to provide a set of gain variations for the designed estimator gain.*

When the state estimator (5.5) is applied to the system described by (5.1) and (5.2), we obtain the following augmented system,

$$\dot{x}_e(t) = A_e x_e(t) + D_e \Delta(t) E_e x_e(t) + B_e w_e(t), \qquad x_e(t_0) = x_{e0} \qquad (5.8)$$

where $x_e(t) = \begin{bmatrix} e^T(t) & \xi^T(t) \end{bmatrix}^T$, $e(t)$ is the estimation error defined by $e(t) = x(t) - \xi(t)$, and

$$x_{e0} = \begin{bmatrix} x_0 - \xi_0 \\ \xi_0 \end{bmatrix}, \qquad w_e = \begin{bmatrix} V_1^{-\frac{1}{2}} w_1 \\ V_2^{-\frac{1}{2}} w_2 \end{bmatrix}$$

$$A_e = \begin{bmatrix} A - \tilde{G}C & A - A_0 - \tilde{G}C \\ \tilde{G}C & A_0 + \tilde{G}C \end{bmatrix}, \qquad E_e = [E_1 \;\; E_1] \qquad (5.9)$$

$$B_e = \begin{bmatrix} V_1^{\frac{1}{2}} & -\tilde{G}V_2^{\frac{1}{2}} \\ 0 & \tilde{G}V_2^{\frac{1}{2}} \end{bmatrix}, \qquad D_e = \begin{bmatrix} D_1 - \tilde{G}D_2 \\ \tilde{G}D_2 \end{bmatrix} \qquad (5.10)$$

with Δ satisfying (5.3), and $E\{w_e w_e^T\} = I$. The following definition, similar to that in Petersen and McFarlane [110] will be used in the sequel.

Definition 5.1 *The state estimator (5.5) with the gain uncertainty ΔG of the form (5.6) (or of the form (5.7)) is said to be a non-fragile quadratic guaranteed cost state estimator with associated cost matrix $Q > 0$ for the system described by (5.1) and (5.2) if there exists a constant $\delta > 0$ and a symmetric matrix*

$$Q_e = \begin{bmatrix} Q & Q_{12} \\ Q_{12}^T & Q_{22} \end{bmatrix} \qquad (5.11)$$

such that

$$He\{(A_e + D_e \Delta E_e)Q_e\} + B_e B_e^T + \delta I \leq 0 \qquad (5.12)$$

for all matrices Δ satisfying (5.3) and all ΔG of the form (5.6) (or of the form (5.7)), where A_e, D_e, E_e, and B_e are as defined by (5.9) and (5.10).

For each uncertainty matrix Δ, the steady-state error covariance $Q_\Delta(t)$ at time t is defined by $Q_\Delta(t) = \lim_{t_0 \to -\infty} E\{e(t)e^T(t)\}$. The following lemma shows that Q in (5.11) defines an upper bound on $Q_\Delta(t)$ over all Δ satisfying (5.3), and the corresponding augmented system (5.8) is quadratically stable [6] if (5.5) is a non-fragile quadratic guaranteed cost state estimator for the system described by (5.1) and (5.2) with cost matrix Q.

Lemma 5.1 *Suppose that (5.5) is a non-fragile quadratic guaranteed cost state estimator with cost matrix $Q > 0$ for the system described by (5.1) and (5.2). Then the corresponding augmented system (5.8) is quadratically stable and the steady-state error covariance at time t satisfies $Q_\Delta(t) \leq Q$ for all admissible uncertainties $\Delta(t)$ and ΔG of the form (5.6) (or of the form (5.7)).*

Proof 5.1 *It is similar to that of Theorem 4.1 in Petersen and McFarlane [110], and is, hence, omitted here.*

In this chapter, the problem under consideration is to design non-fragile quadratic guaranteed cost state estimators for the system described by (5.1) and (5.2). The following preliminaries will be used in the sequel.

Lemma 5.2 *For symmetric matrices Q_1, R_1, Q_2, and R_2 with $Q_2 > 0$, let*

$$J_1(P) \triangleq He\{PA_1\} + PR_1P + Q_1$$
$$J_2(P) \triangleq He\{PA_2\} + PR_2P + Q_2$$

If $J_2(P) \leq J_1(P)$ for any symmetric matrix P, and $J_1(P) = 0$ has a stabilizing solution $P_1 > 0$, then the Riccati equation $J_2(P) = 0$ has a stabilizing solution $P_2 > 0$ and $P_2 \leq P_1$.

Proof 5.2 *By Lemma A.2.9 in Knobloch, Isidori, and Flockerzi [79], it follows that there exists a matrix sequence $\{P_n\}_{n=1}^{\infty}$ such that $P_n > P_1$, $P_n \to P_1$ $(n \to \infty)$, and $J_1(P_n) < 0$, $n = 1, 2, \cdots$. By $Q_2 > 0$, and Lemmas A.2.6 and A.2.7 in Knobloch, Isidori, and Flockerzi [79], it follows that $J_2(P) = 0$ has a stabilizing solution $P_2 > 0$ and $P_2 < P_n$ $(n = 1, 2, \cdots)$. Let $n \to \infty$, and we have $P_2 \leq P_1$. Thus, the proof is complete.*

5.3   Robust Non-Fragile Filter Design

5.3.1   Additive Gain Uncertainty Case

In this section, we consider the non-fragile quadratic guaranteed cost estimator design with additive gain uncertainty of the form (5.6). Denote

$$H_1(P, \epsilon, \sigma) \triangleq He\{(A + \frac{\sigma}{\epsilon}D_1 D_2^T G_{22}^T R_0^{-1} G_{22}C)P\} + \sigma PC^T G_{22}^T R_0^{-1} G_{22}CP$$

$$\frac{1}{\epsilon}D_1(I + \frac{\sigma}{\epsilon}D_2^T G_{22}^T R_0^{-1} G_{22}D_2)D_1^T + \epsilon PE_1^T E_1 P + V_1 \quad (5.13)$$

$$H_2(Q, \epsilon) \triangleq He\{(A - \frac{1}{\epsilon}D_1 D_2^T R_a^{-1}C)Q\} + \frac{1}{\epsilon}D_1(I - \frac{1}{\epsilon}D_2^T R_a^{-1}D_2)D_1^T$$

$$+V_1 - Q(C^T R_a^{-1}C - \epsilon E_1^T E_1)Q \quad (5.14)$$

where $\epsilon > 0$ and $\sigma > 0$ are constants, and

$$R_a = R_a(\epsilon) \triangleq \frac{1}{\epsilon}D_2 D_2^T + V_2 \quad (5.15)$$

$$R_0 = R_0(\epsilon, \sigma) \triangleq I - \sigma G_{22}R_a(\epsilon)G_{22}^T \quad (5.16)$$

For G_{11} in Equation (5.6), choose symmetric matrices $N_1 \geq 0$ and $N_2 \geq 0$ such that

$$\begin{bmatrix} N_2 & 0 \\ 0 & N_1 - N_2 \end{bmatrix} \geq \begin{bmatrix} G_{11}G_{11}^T & -G_{11}G_{11}^T \\ -G_{11}G_{11}^T & G_{11}G_{11}^T \end{bmatrix} \tag{5.17}$$

Lemma 5.3 *Consider the augmented system (5.8) with ΔG given by (5.6). Assume that there exist constants $\epsilon > 0$ and $\sigma > 0$, and a symmetric matrix $Q_{e0} > 0$ such that the inequality (5.55) holds and*

$$
\begin{aligned}
T_0(Q_{e0}, \epsilon, \sigma) &= He\left\{ \begin{bmatrix} A_a & 0 \\ GC_a & A_0 \end{bmatrix} Q_{e0} \right\} + \frac{\rho_a}{\sigma} \begin{bmatrix} 0 \\ G_{11} \end{bmatrix} \begin{bmatrix} 0 \\ G_{11} \end{bmatrix}^T \\
&+ Q_{e0} \begin{bmatrix} \epsilon E_1^T E_1 + \sigma C^T G_{22}^T R_0^{-1} G_{22} C & 0 \\ 0 & 0 \end{bmatrix} Q_{e0} \\
&+ \begin{bmatrix} V_1 & 0 \\ 0 & GV_2 G^T \end{bmatrix} + \frac{1}{\epsilon} \begin{bmatrix} D_1 \\ GD_2 \end{bmatrix} \begin{bmatrix} D_1 \\ GD_2 \end{bmatrix}^T \\
&+ \sigma \begin{bmatrix} \frac{1}{\epsilon} D_1 D_2^T \\ GR_a \end{bmatrix} G_{22}^T R_0^{-1} G_{22} \begin{bmatrix} \frac{1}{\epsilon} D_1 D_2^T \\ GR_a \end{bmatrix}^T < 0 \tag{5.18}
\end{aligned}
$$

where R_a and R_0 are defined by (5.15) and (5.16), respectively, and

$$A_a = A + \frac{\sigma}{\epsilon} D_1 D_2^T G_{22}^T R_0^{-1} G_{22} C \tag{5.19}$$

$$C_a = (I + \sigma R_a G_{22}^T R_0^{-1} G_{22}) C. \tag{5.20}$$

Then there exists a constant $\delta_0 > 0$ such that

$$Q_e = \begin{bmatrix} I & -I \\ 0 & I \end{bmatrix} Q_{e0} \begin{bmatrix} I & -I \\ 0 & I \end{bmatrix}^T \tag{5.21}$$

satisfies the inequality (5.12) with $\delta = \delta_0$.

Proof 5.3 *Notice that*

$$
\begin{aligned}
&He\left\{ \begin{bmatrix} A + D_1 \Delta(t) E_1 & 0 \\ \tilde{G}(C + D_2 \Delta(t) E_1) & A_0 \end{bmatrix} Q_{e0} \right\} + \begin{bmatrix} V_1 & 0 \\ 0 & \tilde{G}V_2\tilde{G}^T \end{bmatrix} \\
&= He\left\{ \begin{bmatrix} A & 0 \\ \tilde{G}C & A_0 \end{bmatrix} Q_{e0} \right\} + \begin{bmatrix} V_1 & 0 \\ 0 & \tilde{G}V_2\tilde{G}^T \end{bmatrix} \\
&\quad + He\left\{ \begin{bmatrix} D_1 \\ \tilde{G}D_2 \end{bmatrix} \Delta(t)[E_1 \ 0] Q_{e0} \right\} \\
&\leq T_1(\epsilon) \tag{5.22}
\end{aligned}
$$

where

$$
\begin{aligned}
T_1(\epsilon) &= He\left\{ \begin{bmatrix} A & 0 \\ \tilde{G}C & A_0 \end{bmatrix} Q_{e0} \right\} + \begin{bmatrix} V_1 & 0 \\ 0 & \tilde{G}V_2\tilde{G}^T \end{bmatrix} \\
&\quad + \frac{1}{\epsilon} \begin{bmatrix} D_1 \\ \tilde{G}D_2 \end{bmatrix} \begin{bmatrix} D_1 \\ \tilde{G}D_2 \end{bmatrix}^T + \epsilon Q_{e0} \begin{bmatrix} E_1^T E_1 & 0 \\ 0 & 0 \end{bmatrix} Q_{e0} \tag{5.23}
\end{aligned}
$$

In the following, we show that $T_1(\epsilon) < 0$ for any ΔG of the form (5.6). By Lemma 2.1 and (5.6), $T_1(\epsilon) < 0$ holds if and only if the following inequality holds for any F satisfying $F^T F \leq \rho_a I$,

$$
\begin{bmatrix} M_0 & * & * \\ \epsilon^{-\frac{1}{2}}\begin{bmatrix} D_1 \\ \tilde{G}D_2 \end{bmatrix}^T & -I & * \\ diag\{V_1^{\frac{1}{2}}, \tilde{G}V_2^{\frac{1}{2}}\}^T & 0 & -I \end{bmatrix} = \begin{bmatrix} M_1 & * & * \\ \epsilon^{-\frac{1}{2}}\begin{bmatrix} D_1 \\ GD_2 \end{bmatrix}^T & -I & * \\ diag\{V_1^{\frac{1}{2}}, GV_2^{\frac{1}{2}}\}^T & 0 & -I \end{bmatrix}
$$

$$
+ He\left\{ \begin{bmatrix} \begin{bmatrix} 0 \\ G_{11} \\ 0 \end{bmatrix} \\ 0 \end{bmatrix} FM_2 \right\} < 0 \quad (5.24)
$$

where

$$
M_0 = He\left\{ \begin{bmatrix} A & 0 \\ \tilde{G}C & A_0 \end{bmatrix} Q_{e0} \right\} + \epsilon Q_{e0} \begin{bmatrix} E_1^T E_1 & 0 \\ 0 & 0 \end{bmatrix} Q_{e0} \quad (5.25)
$$

$$
M_1 = He\left\{ \begin{bmatrix} A & 0 \\ GC & A_0 \end{bmatrix} Q_{e0} \right\} + \epsilon Q_{e0} \begin{bmatrix} E_1^T E_1 & 0 \\ 0 & 0 \end{bmatrix} Q_{e0} \quad (5.26)
$$

$$
M_2 = \begin{bmatrix} \begin{bmatrix} G_{22}C & 0 \end{bmatrix} Q_{e0} & \epsilon^{-\frac{1}{2}}G_{22}D_2 & \begin{bmatrix} 0 & G_{22}V_2^{\frac{1}{2}} \end{bmatrix} \end{bmatrix}. \quad (5.27)
$$

By Lemma 2.12, the inequality (5.24) holds if and only if there exists a constant $\sigma > 0$ such that

$$
\begin{bmatrix} M_1 & * & * \\ \epsilon^{-\frac{1}{2}}\begin{bmatrix} D_1 \\ GD_2 \end{bmatrix}^T & -I & * \\ diag\{V_1^{\frac{1}{2}}, GV_2^{\frac{1}{2}}\}^T & 0 & -I \end{bmatrix} + \frac{\rho_a}{\sigma}\begin{bmatrix} 0 \\ G_{11} \\ 0 \\ 0 \end{bmatrix}\begin{bmatrix} 0 \\ G_{11} \\ 0 \\ 0 \end{bmatrix}^T
$$

$$
+ \sigma M_2^T M_2
$$

$$
= \begin{bmatrix} M_3 & [M_4 \quad M_5] \\ [M_4 \quad M_5]^T & -M \end{bmatrix} < 0 \quad (5.28)
$$

where

$$
M_3 = M_1 + \frac{\rho_a}{\sigma}\begin{bmatrix} 0 \\ G_{11} \end{bmatrix}\begin{bmatrix} 0 \\ G_{11} \end{bmatrix}^T
$$

$$
+ \sigma Q_{e0}\begin{bmatrix} G_{22}C & 0 \end{bmatrix}^T\begin{bmatrix} G_{22}C & 0 \end{bmatrix} Q_{e0} \quad (5.29)
$$

$$
M_4 = \epsilon^{-\frac{1}{2}}(\begin{bmatrix} D_1 \\ GD_2 \end{bmatrix} + \sigma Q_{e0}\begin{bmatrix} G_{22}C & 0 \end{bmatrix}^T G_{22}D_2) \quad (5.30)
$$

$$
M_5 = diag\{V_1^{\frac{1}{2}}, GV_2^{\frac{1}{2}}\} + \sigma Q_{e0}[G_{22}C, \quad 0]^T\begin{bmatrix} 0 & G_{22}V_2^{\frac{1}{2}} \end{bmatrix} \quad (5.31)
$$

$$
M = \begin{bmatrix} I & 0 \\ 0 & I \end{bmatrix} - \sigma\begin{bmatrix} \epsilon^{-\frac{1}{2}}G_{22}D_2 & \begin{bmatrix} 0 & G_{22}V_2^{\frac{1}{2}} \end{bmatrix} \end{bmatrix}^T
$$

$$
\times \begin{bmatrix} \epsilon^{-\frac{1}{2}}G_{22}D_2 & \begin{bmatrix} 0 & G_{22}V_2^{\frac{1}{2}} \end{bmatrix} \end{bmatrix}. \quad (5.32)
$$

By using Lemma 2.1 again, the inequality (5.28) holds if and only if $M > 0$ and

$$T_2(\epsilon, \sigma) = M_3 + \begin{bmatrix} M_4 & M_5 \end{bmatrix} M^{-1} \begin{bmatrix} M_4 & M_5 \end{bmatrix}^T < 0. \tag{5.33}$$

From (5.16) and (5.32), it is easy to show that $M > 0$ if and only if the inequality (5.55) holds, and

$$
M^{-1} = \begin{bmatrix} I & 0 \\ 0 & I \end{bmatrix} + \sigma \begin{bmatrix} \epsilon^{-\frac{1}{2}} D_2 & \begin{bmatrix} 0 & V_2^{\frac{1}{2}} \end{bmatrix} \end{bmatrix}^T
$$
$$
\times G_{22}^T R_0^{-1} G_{22} \begin{bmatrix} \epsilon^{-\frac{1}{2}} D_2 & \begin{bmatrix} 0 & V_2^{\frac{1}{2}} \end{bmatrix} \end{bmatrix}. \tag{5.34}
$$

By using (5.15), (5.26), (5.29)–(5.31), (5.33), and (5.34), it follows that

$$
T_2(\epsilon, \sigma) = M_1 + \frac{\rho_a}{\sigma} \begin{bmatrix} 0 \\ G_{11} \end{bmatrix} \begin{bmatrix} 0 \\ G_{11} \end{bmatrix}^T + \frac{1}{\epsilon} \begin{bmatrix} D_1 \\ GD_2 \end{bmatrix} \begin{bmatrix} D_1 \\ GD_2 \end{bmatrix}^T
$$
$$
+ \begin{bmatrix} V_1 & 0 \\ 0 & GV_2 G^T \end{bmatrix} + \sigma \begin{bmatrix} \frac{1}{\epsilon} D_1 D_2^T \\ GR_a \end{bmatrix} G_{22}^T R_0^{-1} G_{22} \begin{bmatrix} \frac{1}{\epsilon} D_1 D_2^T \\ GR_a \end{bmatrix}^T
$$
$$
+ Q_{e0} \begin{bmatrix} M_6 & 0 \\ 0 & 0 \end{bmatrix} Q_{e0} + \left\{ \begin{bmatrix} M_7 & 0 \\ M_8 & 0 \end{bmatrix} Q_{e0} \right\} \tag{5.35}
$$

where

$$M_6 = C^T G_{22}^T G_{22} [(\sigma^2 R_a + \sigma^3 R_a G_{22}^T R_0^{-1} G_{22} R_a) G_{22}^T G_{22} C + \sigma C] \tag{5.36}$$

$$M_7 = \frac{\sigma}{\epsilon} D_1 D_2^T G_{22}^T G_{22} C + \frac{\sigma^2}{\epsilon} D_1 D_2^T G_{22}^T R_0^{-1} G_{22} R_a G_{22}^T G_{22} C \tag{5.37}$$

$$M_8 = \sigma G R_a G_{22}^T G_{22} C + \sigma^2 G R_a G_{22}^T R_0^{-1} G_{22} R_a G_{22}^T G_{22} C. \tag{5.38}$$

By (5.16) and (5.36)–(5.38), it is easy to show that

$$M_6 = \sigma C^T G_{22}^T R_0^{-1} G_{22} C \tag{5.39}$$

$$M_7 = \frac{\sigma}{\epsilon} D_1 D_2^T G_{22}^T R_0^{-1} G_{22} C \tag{5.40}$$

$$M_8 = \sigma G R_a G_{22}^T R_0^{-1} G_{22} C. \tag{5.41}$$

Combining (5.18)–(5.20), (5.26), (5.35), and (5.39)–(5.41), it follows that $T_2(\epsilon, \sigma) = T_0(Q_{e0}, \epsilon, \sigma) < 0$. Thus $T_1(\epsilon) < 0$ for any ΔG of the form (5.6). By Lemma 2.12, it follows that there exists a constant $\delta_0 > 0$ such that $T_1(\epsilon) + \delta_0 I \leq 0$. From (5.22), (5.9), (5.10), and (5.21), it further implies that Q_e satisfies (5.12) for some $\delta > 0$. The proof is complete.

Denote

$$
H_{20}(Q, \epsilon, \sigma) = He \left\{ Q(A_a^T - C_a^T R_1^{-1} D_{12}^T C_1) \right\} + Q(B_1 B_1^T - C_a^T R_1^{-1} C_a) Q
$$
$$
+ C_1^T (I - D_{12} R_1^{-1} D_{12}^T) C_1 \tag{5.42}
$$

where A_a and C_a are given by (5.19) and (5.20), respectively, and

$$B_1 = \left[\begin{array}{cc} \epsilon^{\frac{1}{2}} E_1^T & \sigma^{\frac{1}{2}} C^T G_{22}^T R_0^{-\frac{1}{2}} \end{array} \right],$$

$$C_1 = \left[\begin{array}{cccc} \epsilon^{-\frac{1}{2}} D_1 & 0 & V_1^{\frac{1}{2}} & \frac{\sigma^{\frac{1}{2}}}{\epsilon} D_1 D_2^T G_{22}^T R_0^{-\frac{1}{2}} \end{array} \right]^T \tag{5.43}$$

$$D_{12} = \left[\begin{array}{cccc} \epsilon^{-\frac{1}{2}} D_2 & V_2^{\frac{1}{2}} 70 & \sigma^{\frac{1}{2}} R_a G_{22}^T R_0^{-\frac{1}{2}} \end{array} \right]^T, R_1 = D_{12}^T D_{12}. \tag{5.44}$$

Then we have the following lemma.

Lemma 5.4 *For $H_2(Q, \epsilon)$, G, and A_0 as defined in (5.14), (5.58), and (5.59), respectively, we have that*

$$H_2(Q, \epsilon) = H_{20}(Q, \epsilon, \sigma) \tag{5.45}$$

$$G = (QC_a^T + C_1^T D_{12}) R_1^{-1} \tag{5.46}$$

$$A_0 = A_a + B_1 B_1^T - GC_a. \tag{5.47}$$

Proof 5.4 *From (5.19), (5.20), and (5.42)–(5.44), it follows that*

$$\begin{aligned} H_{20}(Q, \epsilon, \sigma) &= He\{A_a Q\} + C_1^T C_1 + Q(\epsilon E_1^T E_1 + \sigma C^T G_{22}^T R_0^{-1} G_{22} C) Q \\ &\quad - (QC_a^T + C_1^T D_{12}) R_1^{-1} (QC_a^T + C_1^T D_{12})^T \\ &= He\{AQ\} + \epsilon Q E_1^T E_1 Q + \frac{1}{\epsilon} D_1 D_1^T + V_1 \\ &\quad - (QC^T + \frac{1}{\epsilon} D_1 D_2^T) \Delta_0 (QC^T + \frac{1}{\epsilon} D_1 D_2^T)^T \end{aligned} \tag{5.48}$$

where

$$\Delta_0 = (I + \sigma G_{22}^T R_0^{-1} G_{22} R_a) R_1^{-1} (I + \sigma G_{22}^T R_0^{-1} G_{22} R_a)^T - \sigma G_{22}^T R_0^{-1} G_{22}.$$

By (5.44) and (5.16), we have

$$R_1^{-1} = (R_a + \sigma R_a G_{22}^T R_0^{-1} G_{22} R_a)^{-1} = R_a^{-1} - \sigma G_{22}^T G_{22} \tag{5.49}$$

$$I + \sigma G_{22}^T R_0^{-1} G_{22} R_a = R_a^{-1} (R_a^{-1} - \sigma G_{22}^T G_{22})^{-1} \tag{5.50}$$

$$\sigma G_{22}^T R_0^{-1} G_{22} = \sigma R_a^{-1} (R_a^{-1} - \sigma G_{22}^T G_{22})^{-1} G_{22}^T G_{22}. \tag{5.51}$$

From (5.49)–(5.51), it follows that

$$\Delta_0 = R_a^{-1}. \tag{5.52}$$

Thus, the proof of (5.45) is completed by using (5.48), (5.52), and (5.14). The proof of (5.46) and (5.47) is similar, and is omitted.

Then the following result presents a sufficient condition for the solvability of the robust non-fragile filtering problem with additive estimator gain uncertainty.

Theorem 5.1 *Consider the system described by (5.1) and (5.2), and state estimators of the form (5.5) with additive estimator gain uncertainty (5.6). Suppose that the matrices $N_1 \geq 0$ and $N_2 \geq 0$ satisfy (5.17), A is stable, and there exists a constant $\beta > 0$ such that*

$$
G(s) = G(s,\beta) = \begin{bmatrix} \rho_a^{\frac{1}{2}} N_1^{\frac{1}{2}} \\ \beta^{-\frac{1}{2}} D_1^T \end{bmatrix} (sI - A^T)^{-1} \begin{bmatrix} \beta^{\frac{1}{2}} E_1^T & C^T G_{22}^T \end{bmatrix}
$$
$$
+ \begin{bmatrix} 0 & 0 \\ 0 & \beta^{-\frac{1}{2}} D_2^T G_{22}^T \end{bmatrix} \tag{5.53}
$$

satisfies

$$
\|G(s)\|_\infty < 1. \tag{5.54}
$$

Then there exists a constant $\bar{\sigma} > 0$ such that for any $\sigma \in (0, \bar{\sigma})$, there exists a constant $\epsilon > 0$ such that

$$
R_0 = R_0(\epsilon, \sigma) > 0 \tag{5.55}
$$

and the Riccati equation

$$
H_1(P, \epsilon, \sigma) + \frac{\rho_a}{\sigma} N_1 = 0 \tag{5.56}
$$

has a stabilizing solution $P > 0$. Also, for any such σ and ϵ, the Riccati equation

$$
H_2(Q, \epsilon) + \frac{\rho_a}{\sigma} N_2 = 0 \tag{5.57}
$$

has a stabilizing solution $Q > 0$ with $Q \leq P$. Furthermore, for any $\delta > 0$, there exists a symmetric matrix $Q_\delta > 0$ with $Q \leq Q_\delta < Q + \delta I$ such that the state estimator (5.5) with ΔG given by (5.6), and

$$
G = (QC^T + \frac{1}{\epsilon} D_1 D_2^T) R_a^{-1} \tag{5.58}
$$
$$
A_0 = A + \epsilon Q E_1^T E_1 - GC \tag{5.59}
$$

is a non-fragile quadratic guaranteed cost state estimator for the system described by (5.1)–(5.3) with additive estimator gain uncertainty (5.6) with cost matrix Q_δ.

Proof 5.5 *First, denote $\Xi_{a1} = I - \frac{1}{\beta} G_{22} D_2 D_2^T G_{22}^T$, $\Xi_{a2} = P_\beta C^T G_{22}^T + \frac{1}{\beta} D_1 D_2^T G_{22}^T$.*

Then, by (5.53) and (5.54), and the bounded real lemma (Corollary 13.23 in Zhou, Doyle, and Glover [154]), it follows that $\Xi_{a1} > 0$ and there exists a positive-definite matrix P_β such that

$$
He\{AP_\beta\} + \beta P_\beta E_1^T E_1 P_\beta + \frac{1}{\beta} D_1 D_1^T + \rho_a N_1 + \Xi_{a2} \Xi_{a1}^{-1} \Xi_{a2}^T < 0
$$

which further implies that there exists a constant $\bar{\sigma} > 0$ such that for any $\sigma \in (0, \bar{\sigma})$,

$$\Xi_{a3} = I - G_{22}(\frac{1}{\beta}D_2 D_2^T + \sigma V_2)G_{22}^T > 0 \tag{5.60}$$

and

$$He\{AP_\beta\} + +\beta P_\beta E_1^T E_1 P_\beta + \frac{1}{\beta}D_1 D_1^T + \rho_a N_1 + \sigma V_1 + \Xi_{a2}\Xi_{a3}^{-1}\Xi_{a2}^T < 0.$$

By Lemma A.2.6 in Knobloch, Isidori, and Flockerzi [79], the above Riccati equation has a stabilizing solution $P_{\beta\sigma} > 0$.

Let $P = \frac{1}{\sigma}P_{\beta\sigma}$, $\epsilon = \beta\sigma$. Then from (5.60), it follows that the inequality (5.55) holds and P is a stabilizing solution to (5.56). Note that, for N_1 and N_2 satisfying (5.17), we have $N_1 \geq N_2$. For the same ϵ and σ as above, by using (5.13) and (5.14), we have

$$H_2(P, \epsilon) + \frac{\rho_a}{\sigma}N_2 \leq H_1(P, \epsilon, \sigma) + \frac{\rho_a}{\sigma}N_1 = 0.$$

It follows from $V_1 > 0$ and Lemma 5.2 that the Riccati equation (5.57) has a stabilizing solution $Q > 0$ and $Q \leq P$. Let G and A_0 be given by (5.58) and (5.59), respectively, and

$$Q_{e0} = \begin{bmatrix} P & P - Q \\ P - Q & P - Q \end{bmatrix}.$$

Then $Q_{e0} \geq 0$. From (5.45)–(5.47) in Lemma 5.4, (5.18), (5.56), and (5.57), and direct algebraic manipulations, it follows that

$$
\begin{aligned}
T_0(Q_{e0}, \epsilon, \sigma) &= \begin{bmatrix} H_1(P, \epsilon, \sigma) & H_1(P, \epsilon, \sigma) - H_{20}(Q, \epsilon, \sigma) \\ H_1(P, \epsilon, \sigma) - H_{20}(Q, \epsilon, \sigma) & H_1(P, \epsilon, \sigma) - H_{20}(Q, \epsilon, \sigma) \end{bmatrix} \\
&\quad + \frac{\rho_a}{\sigma}\begin{bmatrix} 0 & 0 \\ 0 & G_{11}G_{11}^T \end{bmatrix} \\
&= \begin{bmatrix} H_1(P, \epsilon, \sigma) & H_1(P, \epsilon, \sigma) - H_2(Q, \epsilon) \\ H_1(P, \epsilon, \sigma) - H_2(Q, \epsilon) & H_1(P, \epsilon, \sigma) - H_2(Q, \epsilon) \end{bmatrix} \\
&\quad + \frac{\rho_a}{\sigma}\begin{bmatrix} 0 & 0 \\ 0 & G_{11}G_{11}^T \end{bmatrix} \\
&= -\frac{\rho_a}{\sigma}\begin{bmatrix} N_1 & N_1 - N_2 \\ N_1 - N_2 & N_1 - N_2 - G_{11}G_{11}^T \end{bmatrix} \triangleq -M_9.
\end{aligned}
$$

Then from (5.17), $M_9 \geq 0$ and

$$T_0(Q_{e0}, \epsilon, \sigma) + M_9 = 0. \tag{5.61}$$

Since Q and P are the stabilizing solutions to (5.57) and (5.56), respectively, it is easy to show that Q_{e0} is the stabilizing solution to (5.61). By Theorem

2.1 in Petersen, Anderson, and Jonckheere [108], it follows that there exist symmetric matrices $W > 0$ and $\bar{Q}_{e0} > Q_{e0}$ such that

$$T_0(\bar{Q}_{e0}, \epsilon, \sigma) + M_9 + W = 0$$

which further implies that the Riccati equation

$$T_0(Q_{e\gamma}, \epsilon, \sigma) + M_9 + \gamma W = 0 \qquad (5.62)$$

has a stabilizing solution $Q_{e\gamma} > Q_{e0}$ for any $\gamma \in (0,1)$, and $\lim_{\gamma \to 0} Q_{e\gamma} = Q_{e0}$. For any $\delta > 0$, there exists a $\bar{\gamma} > 0$ such that

$$0 \le \begin{bmatrix} I & -I \\ 0 & I \end{bmatrix} Q_{e\bar{\gamma}} \begin{bmatrix} I & -I \\ 0 & I \end{bmatrix}^T - \begin{bmatrix} Q & 0 \\ 0 & P-Q \end{bmatrix}$$

$$= \begin{bmatrix} I & -I \\ 0 & I \end{bmatrix} (Q_{e\bar{\gamma}} - Q_{e0}) \begin{bmatrix} I & -I \\ 0 & I \end{bmatrix}^T < \delta I.$$

Let Q_δ be defined as the $(1,1)$ block of the matrix

$$\begin{bmatrix} I & -I \\ 0 & I \end{bmatrix} Q_{e\bar{\gamma}} \begin{bmatrix} I & -I \\ 0 & I \end{bmatrix}^T.$$

Then we have that $Q \le Q_\delta < Q + \delta I$, and from (5.62) and $W > 0$, we further have $T_0(Q_{e\bar{\gamma}}, \epsilon, \sigma) < 0$. Thus, by Lemma 5.3 and Definition 5.1, it follows that the state estimator (5.5) with (5.6), (5.58), and (5.59) is a non-fragile guaranteed cost state estimator with cost matrix Q_δ.

Remark 5.2 *From Lemma 5.1, it is easy to see that the non-fragile guaranteed cost state estimator designed by Theorem 5.1 guarantees that the steady-state error covariance matrix at time t satisfies $Q_\Delta(t) \le Q$ for all admissible uncertainties $\Delta(t)$ and ΔG of the form (5.6), where Q is the stabilizing solution to (5.57). It should be noted that the stabilizing solution to (5.57) has the following monotone property: If $\sigma_1 < \sigma_2$, and Q_{σ_1} and Q_{σ_2} are the stabilizing solutions to $H_2(Q, \epsilon) + \frac{\rho_a}{\sigma_1} N_2 = 0$ and $H_2(Q, \epsilon) + \frac{\rho_a}{\sigma_2} N_2 = 0$, respectively, then $Q_{\sigma_2} \le Q_{\sigma_1}$. By using this property, for the given N_1 and N_2 satisfying (5.17), and ϵ, we should choose σ as big as possible for minimizing Q of the stabilizing solution to (5.57). If we consider the corresponding bound on the steady-state mean square error,*

$$\lim_{t_0 \to -\infty} E\{e^T(t)e(t)\} = tr\{Q_\Delta(t)\} \le tr(Q)$$

then we can carry out the scalar minimization of the upper bound of the filtering error covariance with respect to the scaling parameters ϵ and σ, that is,

$$\min_{\epsilon, \sigma} \{tr(Q) : Q > 0 \text{ is the stabilizing solution to (5.57)}\}$$

where ϵ and σ are subject to the constraint that the inequality (5.55) holds and the Riccati equation (5.56) has a stabilizing solution. From the inequality (5.55), it follows that σ satisfies the condition $0 < \sigma < [\lambda_{max}(G_{22}V_2G_{22}^T)]^{-1}$ if $G_{22} \neq 0$. When $\Delta G = 0$, we can choose $N_1 = N_2 = 0$, $G_{11} = 0$, and $G_{22} = 0$, and Theorem 5.1 is reduced to the result for the optimal guaranteed cost filter design for uncertain systems in Petersen and McFarlane [110].

When the system described by (5.1) and (5.2) contains no uncertainty, that is, $E_1 = 0$, $D_1 = 0$, and $D_2 = 0$, denote

$$\bar{\sigma}_a = \sup\{\sigma : I - \sigma G_{22}V_2G_{22}^T > 0 \text{ and } H_{1a}(P, \sigma) = 0$$
$$\text{has a stabilizing solution}\} \tag{5.63}$$

where

$$H_{1a}(P, \sigma) = He\{AP\} + \sigma PC^T G_{22}^T(I - \sigma G_{22}V_2G_{22}^T)^{-1}G_{22}CP + V_1 + \frac{\rho_a}{\sigma}N_1. \tag{5.64}$$

Then, we have the following simpler sufficient condition for the solvability of the robust non-fragile filtering problem.

Theorem 5.2 *Consider the system described by (5.1) and (5.2) with $E_1 = 0$, $D_1 = 0$, and $D_2 = 0$. Suppose that the matrices N_1 and N_2 satisfy (5.17), A is stable, and the transfer function $G_0(s) = (\rho_a N_1)^{\frac{1}{2}}(sI - A^T)^{-1}C^T G_{22}^T$ satisfies*

$$\|G_0(s)\|_\infty < 1. \tag{5.65}$$

Then $\bar{\sigma}_a$ given by (5.63) is well defined, and for any $\sigma \in (0, \bar{\sigma}_a)$, the Riccati equation

$$H_{2a}(Q, \sigma) = He\{AQ\} - QC^T V_2^{-1}CQ + V_1 + \frac{\rho_a}{\sigma}N_2 = 0 \tag{5.66}$$

has a stabilizing solution $Q_\sigma > 0$, and for any $\delta > 0$, there exists a symmetric matrix $Q_\delta > 0$ with $Q_\sigma \leq Q_\delta < Q_\sigma + \delta I$ such that the state estimator (5.5) with ΔG given by (5.6), and

$$G = Q_\sigma C^T V_2^{-1}, A_0 = A - GC \tag{5.67}$$

is a non-fragile quadratic guaranteed cost state estimator for the system described by (5.1) and (5.2) with cost matrix Q_δ.

Proof 5.6 *From the condition (5.65) and A stable, and the bounded real lemma, it follows that there exists a symmetric matrix $P > 0$ such that*

$$He\{AP\} + \sigma PC^T G_{22}^T G_{22}CP + \rho_a N_1 < 0$$

which further implies that for sufficient small $\sigma > 0$, $I - \sigma G_{22}V_2G_{22}^T > 0$ and

$$He\{AP\} + \sigma PC^T G_{22}^T(I - \sigma G_{22}V_2G_{22}^T)^{-1}G_{22}CP + \rho_a N_1 + \sigma V_1 < 0.$$

Thus, the Riccati equation $H_{1a}(P, \sigma) = 0$ has a stabilizing solution, and it follows that the $\bar{\sigma}_a$ given by (5.63) is well defined. The rest of the proof is similar to that of Theorem 5.1, and is omitted here.

Remark 5.3 *From the Equation (5.66), it is easy to see that, if $0 < \sigma_1 < \sigma_2 < \bar{\sigma}_a$, then $Q_{\sigma_2} \le Q_{\sigma_1}$, where Q_{σ_1} and Q_{σ_2} are the stabilizing solutions to $H_{2a}(Q, \sigma_1) = 0$ and $H_{2a}(Q, \sigma_2) = 0$, respectively. Thus, suboptimal non-fragile guaranteed cost filtering can be obtained by choosing σ sufficiently close to $\bar{\sigma}_a$ defined by (5.63). The design Equation (5.66) is also equivalent to that of the standard optimal filtering for the system*

$$\dot{x}(t) = Ax(t) + \bar{w}_1(t), \qquad x(t_0) = x_0$$
$$y(t) = Cx(t) + w_2(t)$$

with joint covariance matrix

$$E\left\{ \begin{bmatrix} \bar{w}_1 \\ w_2 \end{bmatrix} [\bar{w}_1^T \; w_2^T] \right\} = \begin{bmatrix} V_1 + \frac{\rho_a}{\sigma} N_2 & 0 \\ 0 & V_2 \end{bmatrix}.$$

5.3.2   Multiplicative Gain Uncertainty Case

In this section, we consider the non-fragile quadratic guaranteed cost estimator design with the multiplicative gain uncertainty, that is, $\Delta G = G\bar{G}_{11}F\bar{G}_{22}$ with $F^T F \le \rho_m I$ given by (5.7). Denote

$$\bar{H}_1(P, \epsilon, \sigma) = He\{\Xi_{m1}P\} + P\Xi_{m2}P + \frac{1}{\epsilon}D_1\Xi_{m3}D_1^T + V_1 \tag{5.68}$$

$$\bar{H}_2(Q, \epsilon, \sigma) = He\{(A - \frac{1}{\epsilon}D_1D_2^T ZC)Q\} + \frac{1}{\epsilon}D_1(I - \frac{1}{\epsilon}D_2^T ZD_2)D_1^T$$
$$+ V_1 - Q(C^T ZC - \epsilon E_1^T E_1)Q \tag{5.69}$$

where $\epsilon > 0$ and $\sigma > 0$ are constants,

$$\Xi_{m1} = A + \frac{\sigma}{\epsilon}D_1D_2^T \bar{G}_{22}^T \bar{R}_0^{-1}\bar{G}_{22}C, \Xi_{m3} = I + \frac{\sigma}{\epsilon}D_2^T \bar{G}_{22}^T \bar{R}_0^{-1}\bar{G}_{22}D_2$$
$$\Xi_{m2} = \epsilon E_1^T E_1 + \sigma C^T \bar{G}_{22}^T \bar{R}_0^{-1}\bar{G}_{22}C \tag{5.70}$$

$$\bar{R}_0 = \bar{R}_0(\epsilon, \sigma) \overset{\triangle}{=} I - \sigma\bar{G}_{22}R_a(\epsilon)\bar{G}_{22}^T \tag{5.71}$$

$$Z = \Xi_{m4}\left[R_a\Xi_{m4} + \frac{\rho_m}{\sigma}\bar{G}_{11}\bar{G}_{11}^T\right]^{-1} \tag{5.72}$$

with R_a given by (5.15), and $\Xi_{m4} = I - \rho_m\bar{G}_{22}^T\bar{G}_{22}\bar{G}_{11}\bar{G}_{11}^T$.

In addition, we define

$$G = (QC^T + \frac{1}{\epsilon}D_1D_2^T)R_a^{-1}(I + \frac{\rho_m}{\sigma}\bar{G}_{11}\bar{G}_{11}^T R_b)^{-1} \tag{5.73}$$

$$A_0 = A_m + Q\Xi_{m2} - GC_m \tag{5.74}$$

with Ξ_{m2} being defined by (5.70), where

$$R_b = R_b(\epsilon, \sigma) \overset{\triangle}{=} R_a^{-1} - \sigma\bar{G}_{22}^T\bar{G}_{22} \tag{5.75}$$

$$A_m = A + \frac{\sigma}{\epsilon}D_1D_2^T \bar{G}_{22}^T \bar{R}_0^{-1}\bar{G}_{22}C \tag{5.76}$$

$$C_m = (I + \sigma R_a\bar{G}_{22}^T \bar{R}_0^{-1}\bar{G}_{22})C. \tag{5.77}$$

The following lemmas are required in the derivative process of the main result.

Lemma 5.5 *Consider the augmented system (5.8) with ΔG given by (5.7). If there exist constants $\epsilon > 0$ and $\sigma > 0$, and a symmetric matrix $Q_{e0} > 0$ such that the inequality*

$$\bar{R}_0 = \bar{R}_0(\epsilon, \sigma) > 0 \tag{5.78}$$

holds and

$$
\begin{aligned}
\bar{T}_0(Q_{e0}, \epsilon, \sigma) = He &\left\{ \begin{bmatrix} A_m & 0 \\ GC_m & A_0 \end{bmatrix} Q_{e0} \right\} + \frac{\rho_m}{\sigma} \begin{bmatrix} 0 \\ G\bar{G}_{11} \end{bmatrix} \begin{bmatrix} 0 \\ G\bar{G}_{11} \end{bmatrix}^T \\
&+ \begin{bmatrix} V_1 & 0 \\ 0 & GV_2G^T \end{bmatrix} + Q_{e0} \begin{bmatrix} \Xi_{m2} & 0 \\ 0 & 0 \end{bmatrix} Q_{e0} \\
&+ \sigma \begin{bmatrix} \frac{1}{\epsilon} D_1 D_2^T \\ GR_a \end{bmatrix} \bar{G}_{22}^T \bar{R}_0^{-1} \bar{G}_{22} \begin{bmatrix} \frac{1}{\epsilon} D_1 D_2^T \\ GR_a \end{bmatrix}^T \\
&+ \frac{1}{\epsilon} \begin{bmatrix} D_1 \\ GD_2 \end{bmatrix} \begin{bmatrix} D_1 \\ GD_2 \end{bmatrix}^T < 0
\end{aligned} \tag{5.79}
$$

where R_a, $\bar{R}_0$, A_m, and C_m are as defined in (5.15), (5.71), (5.73), and (5.74), respectively. Then there exists a constant $\delta_0 > 0$ such that Q_e as defined by (5.21) satisfies the inequality (5.12) with $\delta = \delta_0$.

Proof 5.7 *It is similar to that of Lemma 5.3, and is omitted here.*

Denote

$$
\begin{aligned}
\bar{H}_{20}(Q, \epsilon, \sigma) = He&\{(A_m^T - C_m^T \bar{R}_1^{-1} \bar{D}_{12}^T \bar{C}_1)^T Q\} + Q(\bar{B}_1 \bar{B}_1^T - C_m^T \bar{R}_1^{-1} C_m)Q \\
&+ \bar{C}_1^T (I - \bar{D}_{12} \bar{R}_1^{-1} \bar{D}_{12}^T)\bar{C}_1
\end{aligned} \tag{5.80}
$$

where A_m and C_m are given by (5.73) and (5.74), respectively,

$$
\begin{aligned}
\bar{B}_1 &= \begin{bmatrix} \epsilon^{\frac{1}{2}} E_1^T & \sigma^{\frac{1}{2}} C^T \bar{G}_{22}^T \bar{R}_0^{-\frac{1}{2}} \end{bmatrix} \\
\bar{C}_1 &= \begin{bmatrix} \epsilon^{-\frac{1}{2}} D_1 & 0 & V_1^{\frac{1}{2}} & \frac{\sigma^{\frac{1}{2}}}{\epsilon} D_1 D_2^T \bar{G}_{22}^T \bar{R}_0^{-\frac{1}{2}} & 0 \end{bmatrix}^T
\end{aligned} \tag{5.81}
$$

$$
\begin{aligned}
\bar{D}_{12} &= \begin{bmatrix} \epsilon^{-\frac{1}{2}} D_2 & V_2^{\frac{1}{2}} & 0 & \sigma^{\frac{1}{2}} R_a \bar{G}_{22}^T \bar{R}_0^{-\frac{1}{2}} & (\frac{\rho_m}{\sigma})^{\frac{1}{2}} \bar{G}_{11} \end{bmatrix}^T \\
\bar{R}_1 &= \bar{D}_{12}^T \bar{D}_{12}.
\end{aligned} \tag{5.82}
$$

Lemma 5.6 *For $\bar{H}_2(Q, \epsilon, \sigma)$ and G as defined by (5.69) and (5.73), respectively, we have that*

$$\bar{H}_2(Q, \epsilon, \sigma) = \bar{H}_{20}(Q, \epsilon, \sigma) \tag{5.83}$$

$$G = (QC_m^T + \bar{C}_1^T \bar{D}_{12})\bar{R}_1^{-1}. \tag{5.84}$$

Proof 5.8 *From (5.76), (5.77), and (5.80)–(5.82), it follows that*

$$
\begin{aligned}
\bar{H}_{20}(Q, \epsilon, \sigma) &= He\{A_m Q\} + \bar{C}_1^T \bar{C}_1 + Q\Xi_{m2} Q \\
&\quad -(QC_m^T + \bar{C}_1^T \bar{D}_{12})\bar{R}_1^{-1}(QC_m^T + \bar{C}_1^T \bar{D}_{12})^T \\
&= He\{AQ\} + \epsilon Q E_1^T E_1 Q + \frac{1}{\epsilon} D_1 D_1^T + V_1 \\
&\quad -(QC^T + \frac{1}{\epsilon} D_1 D_2^T)\Delta_1(QC^T + \frac{1}{\epsilon} D_1 D_2^T)^T
\end{aligned}
\tag{5.85}
$$

where

$$
\Delta_1 = (I + \sigma \bar{G}_{22}^T \bar{R}_0^{-1} \bar{G}_{22} R_a)\bar{R}_1^{-1}(I + \sigma \bar{G}_{22}^T \bar{R}_0^{-1} \bar{G}_{22} R_a)^T - \sigma \bar{G}_{22}^T \bar{R}_0^{-1} \bar{G}_{22}
$$

and Ξ_{m2} is defined by (5.70).

By (5.71), (5.75), and (5.82), we have that

$$
\begin{aligned}
\bar{R}_1^{-1} &= (R_a + \sigma R_a \bar{G}_{22}^T \bar{R}_0^{-1} \bar{G}_{22} R_a + \frac{\rho_m}{\sigma} \bar{G}_{11} \bar{G}_{11}^T)^{-1} \\
&= (R_b^{-1} + \frac{\rho_m}{\sigma} \bar{G}_{11} \bar{G}_{11}^T)^{-1} \\
&= R_b - \frac{\rho_m}{\sigma} R_b \bar{G}_{11}(I + \frac{\rho_m}{\sigma} \bar{G}_{11}^T R_b \bar{G}_{11})^{-1} \bar{G}_{11}^T R_b
\end{aligned}
\tag{5.86}
$$

$$
I + \sigma \bar{G}_{22}^T \bar{R}_0^{-1} \bar{G}_{22} R_a = R_a^{-1} R_b^{-1}
\tag{5.87}
$$

$$
\sigma \bar{G}_{22}^T \bar{R}_0^{-1} \bar{G}_{22} = \sigma R_a^{-1} R_b^{-1} \bar{G}_{22}^T \bar{G}_{22}.
\tag{5.88}
$$

From (5.86)–(5.88) and (5.72), it follows that

$$
\begin{aligned}
\Delta_1 &= R_a^{-1}(R_b^{-1} - \frac{\rho_m}{\sigma} \bar{G}_{11}(I + \frac{\rho_m}{\sigma} \bar{G}_{11}^T R_b \bar{G}_{11})^{-1} \bar{G}_{11}^T) R_a^{-1} - \sigma R_a^{-1} R_b^{-1} \bar{G}_{22}^T \bar{G}_{22} \\
&= R_a^{-1} - \frac{\rho_m}{\sigma} R_a^{-1} \bar{G}_{11}(I + \frac{\rho_m}{\sigma} \bar{G}_{11}^T R_b \bar{G}_{11})^{-1} \bar{G}_{11}^T R_a^{-1} \\
&= R_a^{-1}\{I - \frac{\rho_m}{\sigma} \bar{G}_{11} \bar{G}_{11}^T [R_a \Xi_{m4} + \frac{\rho_m}{\sigma} \bar{G}_{11} \bar{G}_{11}^T]^{-1}\} \\
&= Z
\end{aligned}
\tag{5.89}
$$

where Ξ_{m4} is defined by (5.72).

Thus, the equality (5.83) follows from (5.69), (5.88), and (5.89). The proof of equality (5.84) is similar, and is omitted.

Lemma 5.7 *[150] The inequality (5.91) holds for all F satisfying $F^T F \leq \rho_m I$ if and only if there exists a constant $\epsilon > 0$ such that $\rho_m \|\bar{G}_{11}\|_2 \leq \epsilon$ and $\|\bar{G}_{22}\|_2 \leq \frac{1}{\epsilon}$.*

Then we have the sufficient condition for the solvability of the robust non-fragile filtering problem with multiplicative estimator gain uncertainty.

Theorem 5.3 *Consider the system described by (5.1) and (5.2). Suppose that A is stable,*

$$\|E_1(sI - A)^{-1}D_1\|_\infty < 1 \qquad (5.90)$$

and for any F satisfying $F^T F \le \rho_m I$

$$\|\bar{G}_{11} F \bar{G}_{22}\|_2 \le 1. \qquad (5.91)$$

Then there exists a constant $\bar{\epsilon} > 0$ such that for any $\epsilon \in (0, \bar{\epsilon})$, there exists a constant $\sigma > 0$ such that (5.78) and the Riccati equation

$$\bar{H}_1(P, \epsilon, \sigma) = 0 \qquad (5.92)$$

have a stabilizing solution $P > 0$. For the above ϵ and σ, the Riccati equation

$$\bar{H}_2(Q, \epsilon, \sigma) = 0 \qquad (5.93)$$

has a stabilizing solution $Q > 0$ with $Q \le P$, and for any $\delta > 0$, there exists a symmetric matrix $Q_\delta > 0$ with $Q \le Q_\delta < Q + \delta I$ such that the state estimator (5.5) with ΔG given by (5.7), (5.52) and (5.60) is a non-fragile quadratic guaranteed cost state estimator for the system described by (5.1) and (5.2) with cost matrix Q_δ.

Proof 5.9 *From the inequality (5.90) and the fact that A is stable, it follows that there exists a constant $\bar{\epsilon} > 0$ such that for any $\epsilon \in (0, \bar{\epsilon})$, and there exists a symmetric matrix $P > 0$ such that*

$$He\{AP\} + \epsilon P E_1^T E_1 P + \frac{1}{\epsilon} D_1 D_1^T + V_1 < 0.$$

By (5.71), it follows that there exists a constant σ such that the inequality (5.78) holds and

$$He\{AP\} + \epsilon P E_1^T E_1 P + \frac{1}{\epsilon} D_1 D_1^T + V_1$$
$$+ \sigma(PC^T + \frac{1}{\epsilon} D_1 D_2^T) \bar{G}_{22}^T \bar{R}_0^{-1} \bar{G}_{22}(PC^T + \frac{1}{\epsilon} D_1 D_2^T)^T < 0$$

which further implies, from Lemma A.2.6 in Knobloch, Isidori, and Flockerzi [79], that the Riccati equation (5.92) has a stabilizing solution $P > 0$. By Lemma 5.7, we have $I - \rho_m \bar{G}_{11}^T \bar{G}_{22}^T \bar{G}_{22} \bar{G}_{11} \ge 0$. From (5.68), (5.69), and (5.72), it follows that $Z \ge 0$ and

$$\bar{H}_2(P, \epsilon, \sigma) = He\{AP\} + \epsilon P E_1^T E_1 P + \frac{1}{\epsilon} D_1 D_1^T + V_1$$
$$- (PC^T + \frac{1}{\epsilon} D_1 D_2^T) Z (PC^T + \frac{1}{\epsilon} D_1 D_2^T)^T$$
$$\le \bar{H}_1(P, \epsilon, \sigma) = 0.$$

Thus, from Lemma 2.12, the Riccati equation (5.93) also has a stabilizing *solution $Q > 0$ and $Q \leq P$. Let $Q_{e0} = \begin{bmatrix} P & P-Q \\ P-Q & P-Q \end{bmatrix}$, G and A_0 be given by (5.73) and (5.74), respectively. Then $Q_{e0} \geq 0$. From (5.68), (5.69), (5.92), (5.93), (5.79), Lemma 5.6, and direct algebraic manipulations, it follows that*

$$\bar{T}_0(Q_{e0}, \epsilon, \sigma) = \begin{bmatrix} \bar{H}_1(P, \epsilon, \sigma) & \bar{H}_1(P, \epsilon, \sigma) - \bar{H}_{20}(Q, \epsilon, \sigma) \\ \bar{H}_1(P, \epsilon, \sigma) - \bar{H}_{20}(Q, \epsilon, \sigma) & \bar{H}_1(P, \epsilon, \sigma) - \bar{H}_{20}(Q, \epsilon, \sigma) \end{bmatrix}$$

$$= \begin{bmatrix} \bar{H}_1(P, \epsilon, \sigma) & \bar{H}_1(P, \epsilon, \sigma) - \bar{H}_2(Q, \epsilon, \sigma) \\ \bar{H}_1(P, \epsilon, \sigma) - \bar{H}_2(Q, \epsilon, \sigma) & \bar{H}_1(P, \epsilon, \sigma) - \bar{H}_2(Q, \epsilon, \sigma) \end{bmatrix}$$

$$= 0. \tag{5.94}$$

By using Lemma 5.5 and (5.94), the rest of the proof is similar to that of Theorem 5.1, and is omitted.

Remark 5.4 *Similar to Remark 5.2, the non-fragile guaranteed cost state estimator designed by Theorem 5.3 guarantees that the steady-state error covariance matrix at time t satisfies $Q_\Delta(t) \leq Q$ for all admissible uncertainties $\Delta(t)$ and ΔG of the form (5.7), where Q is the stabilizing solution to (5.93). From $I - \rho_m \bar{G}_{11}^T \bar{G}_{22}^T \bar{G}_{22} \bar{G}_{11} \geq 0$, it follows that the stabilizing solution to (5.93) also has the monotone property: If $\sigma_1 < \sigma_2$, then $Q_{\sigma_2} \leq Q_{\sigma_1}$ where Q_{σ_1} and Q_{σ_2} are the stabilizing solutions to $\bar{H}_2(Q, \epsilon, \sigma_1) = 0$ and $\bar{H}_2(Q, \epsilon, \sigma_2) = 0$, respectively. By using the monotone property, for a given $\epsilon > 0$, we should choose σ as big as possible for minimizing Q of the stabilizing solution to (5.93). If we consider the corresponding bound on the steady-state mean square error:*

$$\lim_{t_0 \to -\infty} E\{e^T(t)e(t)\} = tr\{Q_\Delta(t)\} \leq tr(Q)$$

then the scalar minimization of the upper bound of the filtering error covariance with respect to the scaling parameters ϵ and σ can be carried out as in Remark 5.2. Theorem 5.3 involves a bound condition (5.91) on gain uncertainty, which means from (5.7) that the measured signal is at most permitted to degrade to zero. From the proof of Theorem 5.3, it is easy to see that the condition can weaken as $I - \rho_m \bar{G}_{11}^T \bar{G}_{22}^T \bar{G}_{22} \bar{G}_{11} \geq 0$. Also, by setting $\bar{G}_{11} = 0$ and $\bar{G}_{22} = 0$, Theorem 5.3 reduces to the result in Petersen and McFarlane [110] on optimal guaranteed cost filter design for uncertain systems.

When the system described by (5.1) and (5.2) contains no uncertainty, that is, $E_1 = 0$, $D_1 = 0$, and $D_2 = 0$, denote

$$\bar{\sigma}_m = \sup \left\{ \sigma : R_m(\sigma) > 0 \text{ and } \bar{H}_{1m}(P, \sigma) = 0 \text{ has a stabilizing solution} \right\} \tag{5.95}$$

where

$$R_m = R_m(\sigma) \stackrel{\triangle}{=} I - \sigma \bar{G}_{22} V_2 \bar{G}_{22}^T > 0 \tag{5.96}$$

$$\bar{H}_{1m}(P, \sigma) \stackrel{\triangle}{=} He\{AP\} + \sigma PC^T \bar{G}_{22}^T R_m^{-1} \bar{G}_{22} CP + V_1. \tag{5.97}$$

The following lemma is required in the proof of Theorem 5.4.

Lemma 5.8 *Consider the system described by (5.1) and (5.2) with $E_1 = 0$, $D_1 = 0$, and $D_2 = 0$. Then the state estimator (5.5) with ΔG given by (5.7) is a non-fragile quadratic guaranteed cost state estimator with cost matrix Q_0, if and only if there exists a symmetric matrix $Q_e = \begin{bmatrix} Q_0 & Q_{12} \\ Q_{12}^T & Q_{22} \end{bmatrix}$ and a constant $\sigma > 0$ such that the inequality (5.103) holds and*

$$\bar{T}_m(Q_{e0}, \epsilon, \sigma) = He \left\{ \begin{bmatrix} A & 0 \\ G(I + \sigma V_2 \bar{G}_{22}^T R_m^{-1} \bar{G}_{22})C & A_0 \end{bmatrix} Q_{e0} \right\}$$
$$+ \begin{bmatrix} V_1 & 0 \\ 0 & G(V_2 + V_2 \bar{G}_{22}^T R_m^{-1} \bar{G}_{22} V_2 + \frac{\rho_m}{\sigma} \bar{G}_{11} \bar{G}_{11}^T) G^T \end{bmatrix}$$
$$+ Q_{e0} \begin{bmatrix} \sigma C^T \bar{G}_{22}^T R_m^{-1} \bar{G}_{22} C & 0 \\ 0 & 0 \end{bmatrix} Q_{e0} < 0 \qquad (5.98)$$

where

$$Q_{e0} = \begin{bmatrix} I & I \\ 0 & I \end{bmatrix} Q_e \begin{bmatrix} I & I \\ 0 & I \end{bmatrix}^T.$$

Proof 5.10 *By using Definition 5.1 and Lemma 2.12, it is similar to that of Lemma 5.3, and is omitted here.*

Then, we have the following necessary and sufficient condition for the solvability of the robust non-fragile filtering problem.

Theorem 5.4 *Consider the system described by (5.1) and (5.2) with $E_1 = 0$, $D_1 = 0$, and $D_2 = 0$. Suppose that A is stable and the condition (5.91) holds for any F satisfying $F^T F \leq \rho_m I$. Then $\bar{\sigma}_m$ given by (5.95) is well defined, and for any $\sigma \in (0, \bar{\sigma}_m)$, the Riccati equation*

$$\bar{H}_{2m}(Q, \sigma) \triangleq He\{AQ\} - QC^T Z_0 CQ + V_1 = 0 \qquad (5.99)$$

with

$$Z_0 = \Xi_{m4} \left[V_2 \Xi_{m4} + \frac{\rho_m}{\sigma} \bar{G}_{11} \bar{G}_{11}^T \right]^{-1} \qquad (5.100)$$

has a stabilizing solution $Q_\sigma > 0$, and for any $\delta > 0$, there exists a symmetric matrix $Q_\delta > 0$ with $Q_\sigma \leq Q_\delta < Q_\sigma + \delta I$ such that the state estimator (5.5) with ΔG given by (5.7) and

$$G = Q_\sigma C^T V_2^{-1} \left[I + \frac{\rho_m}{\sigma} \bar{G}_{11} \bar{G}_{11}^T (V_2^{-1} - \sigma \bar{G}_{22}^T \bar{G}_{22}) \right]^{-1} \qquad (5.101)$$

$$A_0 = A + \sigma Q_\sigma C^T \bar{G}_{22}^T R_m^{-1} \bar{G}_{22} C - G(I + \sigma V_2 \bar{G}_{22}^T R_m^{-1} \bar{G}_{22})^{-1} C \qquad (5.102)$$

is a non-fragile quadratic guaranteed cost state estimator for a system described by (5.1) and (5.2) with cost matrix Q_δ.

Conversely, suppose that the condition (5.91) holds for any F satisfying $F^T F \leq \rho_m I$, then for any given non-fragile quadratic guaranteed cost state *estimator of the form (4) with ΔG given by (5.7) for a system described by (5.1) and (5.2) with cost matrix Q_0, there exists a constant $\sigma > 0$ such that*

$$R_m = R_m(\sigma) > 0 \qquad (5.103)$$

and the Riccati equations $\bar{H}_{1m}(P,\sigma) = 0$ and (5.99) have stabilizing solutions $P > 0$ and $Q > 0$, respectively, with $Q \leq P$ and $Q < Q_0$.

Proof 5.11 *The proof of the first part of the theorem is similar to those of Theorem 5.1 and Theorem 5.3, and is omitted. Next, we prove the second part of the theorem. Suppose that the state estimator (5.5) with ΔG given by (5.7) is a non-fragile quadratic guaranteed cost estimator with cost matrix Q_0, then from Lemma 5.8, the inequalities (5.103) and (5.98) hold for some $\sigma > 0$ and $Q_e = \begin{bmatrix} Q_0 & Q_{12} \\ Q_{12}^T & Q_{22} \end{bmatrix}$. From the inequality (5.98), direct algebraic manipulations yield*

$$He\{AQ_a\} + \sigma Q_a C^T \bar{G}_{22}^T R_m^{-1} \bar{G}_{22} C Q_a + V_1 < 0$$

where $Q_a = Q_0 + Q_{12} + Q_{12}^T + Q_{22} > 0$. From Lemma A.2.6 in Knobloch, Isidori, and Flockerzi [79] and $V_1 > 0$, it follows that the Riccati equation $\bar{H}_{1m}(P,\sigma) = 0$ has a stabilizing solution $P > 0$. By (5.98), we have

$$
\begin{aligned}
[I &- (I + Q_{12}Q_{22}^{-1})]\bar{T}_m(Q_{e0},\sigma)[I - (I + Q_{12}Q_{22}^{-1})]^T \\
&= He\{[A + G_b(I + \sigma V_2 \bar{G}_{22}^T R_m^{-1} \bar{G}_{22})C]Q_b\} + \sigma Q_b C^T \bar{G}_{22}^T R_m^{-1} \bar{G}_{22} C Q_b \\
&\quad + V_1 + G_b(V_2 + V_2 \bar{G}_{22}^T R_m^{-1} \bar{G}_{22} V_2 + \frac{\rho_m}{\sigma}\bar{G}_{11}\bar{G}_{11}^T)G_b < 0 \qquad (5.104)
\end{aligned}
$$

where $G_b = -(I + Q_{12}Q_{22}^{-1})G$, $Q_b = Q_0 - Q_{12}Q_{22}^{-1}Q_{12}^T > 0$.
Denote

$$\bar{H}_{2m0}(Q,\sigma) \triangleq He\{AQ\} + V_1 - QC^T \Delta_m C Q \qquad (5.105)$$

where

$$
\begin{aligned}
\Delta_m = {}&(I + \sigma V_2 \bar{G}_{22}^T R_m^{-1} \bar{G}_{22})^T (V_2 + V_2 \bar{G}_{22}^T R_m^{-1} \bar{G}_{22} V_2 + \frac{\rho_m}{\sigma}\bar{G}_{11}\bar{G}_{11}^T)^{-1} \\
&\times (I + \sigma V_2 \bar{G}_{22}^T R_m^{-1} \bar{G}_{22}) - \sigma \bar{G}_{22}^T R_m^{-1} \bar{G}_{22}.
\end{aligned}
$$

By (5.104) and completing the square, it follows $\bar{H}_{2m0}(Q_b,\sigma) < 0$, which further implies from Lemma A.2.6 in Knobloch, Isidori, and Flockerzi [79] that $\bar{H}_{2m0}(Q,\sigma) = 0$ has a stabilizing solution $Q > 0$ and $Q < Q_b \leq Q_0$. By Lemma 5.6, (5.99), and (5.105), it follows that $\bar{H}_{2m}(Q,\sigma) = \bar{H}_{2m0}(Q,\sigma)$. $Q \leq P$ follows from the condition (5.91) and Equations (5.95) and (5.99). Thus, the proof is complete.

Remark 5.5 *Theorem 5.4 presents a necessary and sufficient condition for the solvability of the non-fragile guaranteed cost state estimator design problem with the multiplicative estimator gain uncertainty, which is different*

from Theorem 5.2 for the case of the additive estimator gain uncertainty where only a sufficient condition is provided. That is because the inequalities (5.18) and (5.98) essentially are different; (5.18) includes an additional term
$\frac{\rho_a}{\sigma} \begin{bmatrix} 0 \\ G_{11} \end{bmatrix} \begin{bmatrix} 0 \\ G_{11} \end{bmatrix}^T$ *. From* $I - \rho_m \bar{G}_{11}^T \bar{G}_{22}^T \bar{G}_{22} \bar{G}_{11} \geq 0$ *and (5.100), it follows that the stabilizing solution to Riccati equation (5.99) also has the monotone property as in Remark 5.3. So, we can obtain a suboptimal non-fragile guaranteed cost state estimator by choosing* σ *sufficiently close to* $\bar{\sigma}_m$ *as defined by (5.95). If* $I - \rho_m \bar{G}_{11}^T \bar{G}_{22}^T \bar{G}_{22} \bar{G}_{11} > 0$ *, then the design Equation (5.99) is equivalent to that of the standard optimal filtering for the system*

$$\dot{x}(t) = Ax(t) + w_1(t), \qquad x(t_0) = x_0$$
$$y(t) = Cx(t) + \bar{w}_2(t)$$

with joint covariance matrix

$$E\left\{ \begin{bmatrix} w_1 \\ \bar{w}_2 \end{bmatrix} [w_1^T, \ \bar{w}_2^T] \right\} = \begin{bmatrix} V_1 & 0 \\ 0 & V_2 + \frac{\rho_m}{\sigma} \bar{G}_{11} \bar{G}_{11}^T \Xi_{m4}^{-1} \end{bmatrix}.$$

Comparing with Remark 5.3, the above equivalence is different from that for the additive gain uncertainty case.

In the following, we present a reliable filter design that can tolerate outages within a selected subset of sensors, while maintaining stability and a known quadratic cost bound. The design problem can be covered, as a special case, of the non-fragile quadratic guaranteed cost filtering problem with multiplicative gain uncertainties (i.e., by taking $\rho_m = 1$, $\bar{G}_{11} = \bar{G}_{22} = diag\{0, I_q\}$ with I_q a $q \times q$ identity matrix ($q < p$) in (5.7)). This covers the case of permitting the partial measured signal $diag\{0, I_q\}y$ (the last q sensors) to degrade to zero. We decompose the matrices C and V_2 as follows:

$$C = \begin{bmatrix} C_{p-q} \\ C_q \end{bmatrix}, \quad V_2 = \begin{bmatrix} V_{211} & V_{212} \\ V_{212}^T & V_{222} \end{bmatrix} \tag{5.106}$$

where $C_q \in R^{q \times n}$ and $V_{222} \in R^{q \times q}$. Then we have the following theorem.

Theorem 5.5 *Consider the system described by (5.1) and (5.2) with $E_1 = 0$, $D_1 = 0$, and $D_2 = 0$. Suppose that A is stable and ΔG is given by (5.7) with*

$$\rho_m = 1, \quad \bar{G}_{11} = \bar{G}_{22} = diag\{0, I_q\}. \tag{5.107}$$

Let Q_r be the stabilizing solution to the Riccati equation

$$He\{AQ\} - QC_{p-q}^T V_{211}^{-1} C_{p-q} Q + V_1 = 0. \tag{5.108}$$

Then for any $\delta > 0$, there exists a symmetric matrix $Q_\delta > 0$ with $Q_r \leq$

$Q_\delta < Q_r + \delta I$ such that the state estimator (5.5) with ΔG given by (5.7) and (5.107),

$$G = Q_r \begin{bmatrix} C_{p-q}^T V_{211}^{-1} & \sigma C_q^T \end{bmatrix} \tag{5.109}$$

$$A_0 = A - Q_r C_{p-q}^T V_{211}^{-1} C_{p-q}$$
$$\quad - \sigma Q_r C_{p-q}^T V_{211}^{-1} V_{212} (I_q - \sigma V_{222})^{-1} C_q \tag{5.110}$$

is a non-fragile quadratic guaranteed cost state estimator with cost matrix Q_δ, where C_{p-q}, C_q, V_{211}, V_{212}, and V_{222} are defined by (5.106), and $\sigma \in (0, \sigma_r)$ with σ_r defined by

$$\sigma_r = \sup \left\{ \sigma > 0 : \begin{array}{c} I_q - \sigma V_{222} > 0 \text{ and} \\ He\{AP\} + V_1 + \sigma PC_q^T (I_q - \sigma V_{222})^{-1} C_q P = 0 \\ \text{has a stabilizing solution} \end{array} \right\} \tag{5.111}$$

which is well defined.

Proof 5.12 *By (5.107), (5.106), and (5.100)–(5.102), it follows that*

$$Z_0 = \begin{bmatrix} I_{p-q} & 0 \\ 0 & 0 \end{bmatrix} \left(\begin{bmatrix} V_{211} & 0 \\ V_{212}^T & 0 \end{bmatrix} + \frac{1}{\sigma} \begin{bmatrix} 0 & 0 \\ 0 & I_q \end{bmatrix} \right)^{-1} = \begin{bmatrix} V_{211}^{-1} & 0 \\ 0 & 0 \end{bmatrix} \tag{5.112}$$

$$G = Q_r C^T \{ V_2 + \begin{bmatrix} 0 & 0 \\ 0 & I_q \end{bmatrix} (I - \sigma \begin{bmatrix} 0 & 0 \\ 0 & I_q \end{bmatrix} V_2) \}^{-1}$$

$$= Q_r [C_{p-q}^T V_{211}^{-1}, \quad \sigma C_q^T] \tag{5.113}$$

$$A_0 = A + \sigma Q_r C_q^T (I_q - \sigma V_{222})^{-1} C_q - G(I + \sigma V_2 \begin{bmatrix} 0 & 0 \\ 0 & (I_q - \sigma V_{222})^{-1} \end{bmatrix}) C$$

$$= A - Q_r C_{p-q}^T V_{211}^{-1} C_{p-q} - \sigma Q_r C_{p-q}^T V_{211}^{-1} V_{212} (I_q - \sigma V_{222})^{-1} C_q. \tag{5.114}$$

Thus, the proof is completed by using (5.112)–(5.114) and Theorem 5.4.

Remark 5.6 *From (5.7) and (5.107), the design given by Theorem 5.5 can tolerate outages within the last q sensors, while maintaining stability and the cost bound Q_r. It should be noted that the design Equation (5.108) is independent of σ, and it corresponds to the optimal filter design for the worst-fault case: all of the last q sensors have outages. So, the system performance may be improved by including the last q sensors. A related reliable design problem for linear quadratic regulators is addressed in Veillette [126].*

Remark 5.7 *In Section 3 and this section, the robust non-fragile filter designs for the cases of both the additive estimator gain uncertainty and the multiplicative estimator gain uncertainty are presented by using the guaranteed cost approach, that is, to use a fixed quadratic Lyapunov function to establish an upper bound on the state estimation error covariance [110]. One should be aware that there could be a significant gap between the optimal guaranteed upper bound of the performance and the actual worst case performance, that* *is, minimizing the guaranteed upper bound is not the same as minimizing the actual worst case performance, which means that the guaranteed cost approach may lead to a conservative design. However, how to minimize the actual worst case performance is a very difficult problem.*

Remark 5.8 *All of the results in this chapter are presented in terms of solutions to parameter-dependent Riccati equations. Note that the optimization problems involving linear matrix inequalities (LMIs) can be solved numerically very efficiently [11], and Boyd et al. [11] has presented the LMI conditions for state mean and covariance bounds with unit energy inputs. So, by combining the method in Boyd et al. [11] and recent works [40, 67, 105], it is possible to derive the LMI versions of the results in this chapter for the robust non-fragile filtering problem, which is a further research subject.*

5.4   Example

Example 5.1 *Consider an uncertain linear system of the forms (5.1) and (5.2) with*

$$A = \begin{bmatrix} -1 & 1 \\ -0.5 & -2 \end{bmatrix}, D_1 = \begin{bmatrix} 0 \\ 0.5 \end{bmatrix}, E_1 = \begin{bmatrix} 0.1 & 0 \end{bmatrix}, C = \begin{bmatrix} 1 & 1 \end{bmatrix}, D_2 = 0$$

and uncertainty Δ satisfying (5.3). The plant and measurement noises $w_1(t)$ and $w_2(t)$ are assumed to be zero mean white Gaussian, having covariance matrix (5.4) with $V_1 = I$ and $V_2 = I$. The estimator to be designed is of the form (5.5). It is easy to check that A is stable.

As a comparison, the upper bound on the steady-state mean square error achieved by the optimal guaranteed cost filter without the estimator gain uncertainty (i.e., with $G_{11} = 0$, $G_{22} = 0$, and $\rho = 0$) is $tr(Q) = 0.6603$ achieved at $\epsilon = 9.64$.

(a) Additive Uncertainty: *First, we consider additive gain uncertainty ΔG of the form (5.6) with*

$$G_{11} = \begin{bmatrix} 1 & 0 \\ 0 & 1 \end{bmatrix}, \quad G_{22} = 1, \quad \rho_a = 0.01.$$

Then, condition (5.53) is satisfied by

$$N_1 = \begin{bmatrix} 4 & 0 \\ 0 & 4 \end{bmatrix}, \quad N_2 = \begin{bmatrix} 2 & 0 \\ 0 & 2 \end{bmatrix}$$

and condition (5.54) is satisfied for $\beta \geq 0.17$. Then, by solving (5.57)–(5.59) and by minimizing over σ and ϵ, we have solution Q with a minimum trace given by

$$Q = \begin{bmatrix} 0.4281 & -0.0183 \\ -0.0183 & 0.2579 \end{bmatrix}, \quad \text{achieved at } \sigma = 0.458 \text{ and } \epsilon = 9.3$$

with the guaranteed cost of $tr(Q) = 0.6860$. *The corresponding state estimator is given by (5.5) with*

$$A_0 = \begin{bmatrix} -1.37 & 0.5903 \\ -0.7413 & -2.2395 \end{bmatrix}, \quad G = \begin{bmatrix} 0.4097 \\ 0.2395 \end{bmatrix}.$$

It can be seen that by sacrificing the guaranteed cost by 3.9% to $tr(Q) = 0.6860$, *a non-fragile Kalman filter can be designed that can tolerate the additive gain uncertainty of* $\rho_a = 0.01$. *By considering the value of* G, *this additive uncertainty works out to be around 2.3% of* 0.4097, *the largest entry in* G. *If higher gain uncertainties are to be tolerated, the sacrifice in the guaranteed cost will have to be much higher.*

(b) Multiplicative Uncertainty: *Now we consider multiplicative gain uncertainty* ΔG *of the form (5.7) with*

$$\bar{G}_{11} = 1, \quad \bar{G}_{22} = 1, \quad \text{and} \quad \rho_m = 0.1.$$

It is easy to check that both (5.90) and (5.91) are satisfied. Then, by solving (5.78)–(5.93) and by minimizing over σ *and* ϵ, *we have solution* Q *with a minimum trace given by*

$$Q = \begin{bmatrix} 0.4256 & -0.0143 \\ -0.0143 & 0.2490 \end{bmatrix}, \quad \text{achieved at } \sigma = 0.48 \text{ and } \epsilon = 9.4$$

with the guaranteed cost of $tr(Q) = 0.6746$. *The corresponding state estimator is given by (5.5) with*

$$A_0 = \begin{bmatrix} -1.2937 & 0.6637 \\ -0.6930 & -2.1916 \end{bmatrix}, \quad G = \begin{bmatrix} 0.3736 \\ 0.2129 \end{bmatrix}.$$

Similar to the additive uncertainty case, a non-fragile Kalman filtering can be designed by sacrificing slightly (2.7%) the guaranteed cost $tr(Q)$, *in order to tolerate the multiplicative gain uncertainty of* $\rho_m = 0.1$ *(i.e., 10% degradation of the sensor).*

5.5   Conclusion

This chapter studied the problem of a robust Kalman filter design for a class of uncertain linear systems with norm-bounded uncertainty. Two classes of the state estimator gain uncertainties are investigated. The robust non-fragile state estimator designs corresponding to the state estimator gain uncertainties are given in terms of solutions to algebraic Riccati equations, which are dependent on two design parameters, one from the system uncertainty and another from the state estimator gain uncertainty. The designs guarantee known upper bounds on the steady-state error covariance.

6

Non-Fragile Dynamic Output Feedback Control with Interval-Bounded Coefficient Variations

6.1 Introduction

In Chapters 2 and 3, the *non-fragile* state feedback and *dynamic output feedback* control problems with norm-bounded uncertainties are studied, respectively. However, this kind of uncertainty cannot exactly describe the uncertain information due to the finite word length (FWL) effects. Correspondingly, the interval type of parameter uncertainty [89] can describe the uncertain information more exactly than the former type. But in the design process, due to the fact that the vertices of the set of interval uncertain parameters grow exponentially with the number of uncertain parameters, which may result in numerical problems in computation for systems with high dimensions. Moreover, similar to the case in which the problem of designing a globally optimal full-order output-feedback controller for polytopic uncertain systems is known to be a non-convex *NP-hard optimization problem* [76], the problem of designing full-order non-fragile dynamic output feedback H_∞ controllers with an interval type of *gain uncertainty* is also a non-convex NP-hard one.

The purpose of this chapter is to design the controller which is assumed to be with *additive gain variations* of the interval type. And the *full parameterized* and *sparse structured controllers* are considered, respectively. For the full parameterized controller design problem, a *two-step procedure* is adopted to solve this non-convex problem. In Step 1, we give a design method of an initial controller gain C_k. In Step 2, with the controller gain C_k designed in Step 1, a linear matrix inequality (LMI)-based sufficient condition is given for the solvability of the non-fragile H_∞ *control* problem, which requires checking all of the vertices of the set of uncertain parameters that grows exponentially with the number of uncertain parameters. It will be very difficult to apply the result to systems with high dimensions. To overcome the difficulty, a notion of a structured vertex separator is proposed to approach the problem, and is exploited to develop sufficient conditions for the non-fragile H_∞ controller design in terms of solutions to a set of LMIs. The *structured vertex separator method* can significantly reduce the number of LMI constraints involved in the design conditions. For the sparse structured controller design problem, first, a class of sparse structures is specified from a given controller, which renders the resulting *closed-loop system* to be asymptotically stable and meet an H_∞ performance requirement, but it is fully parameterized. Then, a three-step procedure for non-fragile H_∞ controller design under the restriction of the sparse structure is provided. The contribution of this method is that it not only reduces the number of nontrivial parameters but also designs the sparse structured controllers with non-fragility. The resulting designs of the two cases guarantee that the closed-loop system is *asymptotically stable* and the H_∞ performance from the disturbance to the regulated output is less than a prescribed level.

6.2 Non-Fragile H_∞ Controller Design for Discrete-Time Systems

In this section, a two-step procedure is presented for solving the non-fragile H_∞ control problem, and a comparison is made between the new proposed method and the existing method.

6.2.1 Problem Statement

Consider a linear time-invariant (LTI) discrete-time system described by

$$
\begin{aligned}
x(k+1) &= Ax(k) + B_1\omega(k) + B_2u(k) \\
z(k) &= C_1x(k) + D_{12}u(k) \\
y(k) &= C_2x(k) + D_{21}\omega(k)
\end{aligned} \tag{6.1}
$$

where $x(k) \in R^n$ is the state, $u(k) \in R^q$ is the control input, $\omega(k) \in R^r$ is the disturbance input, $y(k) \in R^p$ is the measured output, and $z(k) \in R^m$ is the regulated output, respectively, and $A, B_1, B_2, C_1, C_2, D_{12}$, and D_{21} are known constant matrices of appropriate dimensions.

To formulate the control problem, we consider a controller with gain variations of the following form:

$$
\begin{aligned}
\xi(k+1) &= (A_k + \Delta A_k)\xi(k) + (B_k + \Delta B_k)y(k) \\
u(k) &= (C_k + \Delta C_k)\xi(k).
\end{aligned} \tag{6.2}
$$

where $\xi(k) \in R^n$ is the controller state, and A_k, B_k, and C_k are controller gain matrices of appropriate dimensions to be designed. $\Delta A_k, \Delta B_k$, and ΔC_k represent the additive gain variations of the following interval types:

$$
\begin{aligned}
\Delta A_k &= [\theta_{aij}]_{n\times n}, |\theta_{aij}| \leq \theta_a, i, j = 1, \cdots, n, \\
\Delta B_k &= [\theta_{bij}]_{n\times p}, |\theta_{bij}| \leq \theta_a, i = 1, \cdots, n, j = 1, \cdots, p, \\
\Delta C_k &= [\theta_{cij}]_{q\times n}, |\theta_{cij}| \leq \theta_a, i = 1, \cdots, q, j = 1, \cdots, n.
\end{aligned} \tag{6.3}
$$

Remark 6.1 *The additive gain variations model of form (6.3) is from Li [89], which has been extensively used to describe the FWL effects.*

Let $e_k \in R^n$, $h_k \in R^p$, and $g_k \in R^q$ denote the column vectors in which the kth element equals one and the others equal zero. Then the gain variations of the form (6.3) can be described as:

$$\Delta A_k = \sum_{i=1}^{n} \sum_{j=1}^{n} \theta_{aij} e_i e_j^T, \quad \Delta B_k = \sum_{i=1}^{n} \sum_{j=1}^{p} \theta_{bij} e_i h_j^T,$$

$$\Delta C_k = \sum_{i=1}^{q} \sum_{j=1}^{n} \theta_{cij} g_i e_j^T.$$

Applying controller (6.2) to system (6.1), this yields the closed-loop system:

$$\begin{aligned} x_e(k+1) &= A_e x_e(k) + B_e \omega(k) \\ z(k) &= C_e x_e(k) \end{aligned} \tag{6.4}$$

where $x_e(k) = [x(k)^T, \xi(k)^t]^T$, and

$$A_e = \begin{bmatrix} A & B_2(C_k + \Delta C_k) \\ (B_k + \Delta B_k)C_2 & A_k + \Delta A_k \end{bmatrix},$$

$$B_e = \begin{bmatrix} B_1 \\ (B_k + \Delta B_k)D_{21} \end{bmatrix}, C_e = \begin{bmatrix} C_1 & D_{12}(C_k + \Delta C_k) \end{bmatrix}.$$

Denote the transfer function from the disturbance ω to the controlled output z, corresponding to the state-space model (6.4), as

$$G_{zw}(z) = C_e(zI - A_e)^{-1}B_e.$$

This chapter addresses the following problem.
Non-fragile H_∞ control problem with controller gain variations: Given a positive constant γ, find a dynamic output feedback controller of the form (6.2) with the gain variations (6.3) such that the resulting closed-loop system (6.4) is asymptotically stable and $\|G_{zw}(z)\| < \gamma$.

6.2.2 Non-Fragile H_∞ Controller Design Methods

In this section, the non-fragile H_∞ controller design method will be given by two steps. First, we will give the design method of the controller gain C_k. Then, with the designed controller gain C_k, the non-fragile H_∞ controller design method is presented.

In the following, we focus on the problem of finding an initial feasible solution C_k to the non-fragile H_∞ control problem.

Consider controller (6.2) with $\Delta A_k = 0$ and $\Delta B_k = 0$, which is described by

$$\dot{\xi}(k) = A_k\xi(k) + B_k y(k),$$
$$u(k) = (C_k + \Delta C_k)\xi(k) \qquad (6.5)$$

where ΔC_k is considered with the following norm-bounded form:

$$\Delta C_k = M_c F_3(t) E_c,$$

where

$$M_c = [M_{c1} \cdots M_{cnq}], \; E_c = [E_{c1}^T \cdots E_{cnq}^T]^T$$
$$M_{ck} = g_i, \; E_{ck} = e_j^T$$
$$k = n^2 + np + (i-1)n + j, \; i = 1, \cdots, q, \; j = 1, \cdots, n$$

and $F_3^T(t)F_3(t) \le \theta_a^2 I$ represent the uncertain parameters. Here θ_a is the same as before.

Combining controller (6.5) with system (6.1), we obtain the following closed-loop system:

$$\dot{x}_e(k) = A_{edc}x_e(k) + B_{edc}\omega(k),$$
$$z(k) = C_e x_e(k), \qquad (6.6)$$

where

$$A_{edc} = \begin{bmatrix} A & B_2(C_k + \Delta C_k) \\ B_k C_2 & A_k \end{bmatrix}, B_{edc} = \begin{bmatrix} B_1 \\ B_k D_{21} \end{bmatrix},$$

and C_e is the same as the one in (6.4).

Then the following theorem gives a design method of the initial controller gain C_k.

Theorem 6.1 *Consider system (6.1), where $\gamma > 0$ and $\theta_a > 0$ are constants. If there exist matrices $\hat{A}$, $\hat{B}$, $\hat{C}$, $X > 0, Y > 0$, and a constant $\varepsilon_c > 0$ such that the following LMI holds:*

$$\left[\begin{array}{ccccc} -X & -I & 0 & AX + B_2\hat{C} & A \\ * & -Y & 0 & \hat{A} & YA + \hat{B}C_2 \\ * & * & -I & C_1X + D_{12}\hat{C} & C_1 \\ * & * & * & -X & -I \\ * & * & * & * & -Y \\ * & * & * & * & * \\ * & * & * & * & * \\ * & * & * & * & * \end{array}\right.$$

$$\left.\begin{array}{ccc} B_1 & B_2M_c & 0 \\ YB_1 + \hat{B}D_{21} & YB_2M_c & 0 \\ 0 & D_{12}M_c & 0 \\ 0 & 0 & \varepsilon_c\theta_a X E_c^T \\ 0 & 0 & 0 \\ -\gamma^2 I & 0 & 0 \\ * & -\varepsilon_c I & 0 \\ * & * & -\varepsilon_c I \end{array}\right] < 0, \tag{6.7}$$

then controller (6.5) with

$$A_k = (X^{-1} - Y)^{-1}(\hat{A} - YAX - \hat{B}C_2X - YB_2\hat{C})X^{-1},$$
$$B_k = (X^{-1} - Y)^{-1}\hat{B}, \quad C_k = \hat{C}X^{-1} \tag{6.8}$$

solves the non-fragile H_∞ control problem for system (6.1).

Proof 6.1 *By Lemma 2.5, it is sufficient to show that there exists a symmetric matrix $P > 0$ with the structure (2.11), such that*

$$\begin{bmatrix} -P & 0 & PA_{edc} & PB_{edc} \\ * & -I & C_e & 0 \\ * & * & -P & 0 \\ * & * & * & -\gamma^2 I \end{bmatrix} < 0. \tag{6.9}$$

Equation (6.9) can be further written as

$$\bar{Q}_s + \Delta M_{10} + \Delta M_{10}^T < 0, \tag{6.10}$$

where

$$\bar{Q}_s = \begin{bmatrix} -P & 0 & PA_{e0} & PB_{e0} \\ * & -I & C_{e0} & 0 \\ * & * & -P & 0 \\ * & * & * & -\gamma^2 I \end{bmatrix},$$

$$\Delta M_{10} = \begin{bmatrix} 0 & 0 & P\begin{bmatrix} 0 & B_2\Delta C_k \\ 0 & 0 \end{bmatrix} & 0 \\ 0 & 0 & \begin{bmatrix} 0 & D_{12}\Delta C_k \end{bmatrix} & 0 \\ 0 & 0 & 0 & 0 \\ 0 & 0 & 0 & 0 \end{bmatrix}.$$

It is easy to see that

$$\Delta M_{10} + \Delta M_{10}^T = \Sigma_1 F_3(t)\Sigma_2 + (\Sigma_1 F_3(t)\Sigma_2)^T, \tag{6.11}$$

where $\Sigma_1 = \begin{bmatrix} P\begin{bmatrix} B_2 M_c \\ 0 \end{bmatrix} \\ D_{12} M_c \\ 0 \\ 0 \end{bmatrix}$, $\Sigma_2 = \begin{bmatrix} 0 & 0 & \begin{bmatrix} 0 & E_c \end{bmatrix} & 0 \end{bmatrix}$.

By Lemma 2.12, for any positive scalar ε_c, *(6.10) holds if and only if the following inequality holds:*

$$\bar{Q}_s + \frac{1}{\varepsilon_c}\Sigma_1\Sigma_1^T + \varepsilon_c\theta_a^2\Sigma_2^T\Sigma_2 \leq 0. \tag{6.12}$$

By the Schur complement, (6.10) is equivalent to

$$\begin{bmatrix} \bar{Q}_s & \Sigma_1 & \varepsilon_c\theta_a\Sigma_2^T \\ * & -\varepsilon_c I & 0 \\ * & * & -\varepsilon_c I \end{bmatrix} < 0. \tag{6.13}$$

As in Scherer, Gahinet, and Chilali [113], partition P^{-1} *as*

$$P^{-1} = \begin{bmatrix} X & M \\ M^T & * \end{bmatrix}. \tag{6.14}$$

Due to the fact that P *is with structure (2.11), it is easy to obtain* $X = M$ *and* $X^{-1} = Y + N$.

Similar to the method in Scherer, Gahinet, and Chilali [113], let $\Gamma_1 = \begin{bmatrix} X & I \\ X & 0 \end{bmatrix}$. *Denote* $\bar{\Gamma} = \text{diag}\{\Gamma_1, I, I, I, I\}$, *perform a congruence transformation with* $\bar{\Gamma}$ *on (6.13) and with gain matrices (6.8), (6.13) is equivalent to (6.7). Thus, the proof is complete.*

Remark 6.2 *Theorem 6.1 shows that the non-fragile controller design problem with* $\Delta A_k = 0$, $\Delta B_k = 0$, *and* ΔC_k *with the norm-bounded uncertainty can be converted into a convex one depending on a single parameter* $\varepsilon_c > 0$.

Then, with the designed C_k we will give the non-fragile H_∞ dynamic output feedback controller design method in the following.

To facilitate the presentation, we denote

$$M_0(\Delta A_k, \Delta B_k, \Delta C_k) = \begin{bmatrix} \Xi_1 & \Xi_2 & 0 & \Xi_4 & S^T A & S^T B_1 \\ * & \Xi_3 & 0 & \Xi_5 & \Xi_6 & \Xi_7 \\ * & * & -I & \Xi_8 & C_1 & 0 \\ * & * & * & -\bar{P}_{11} & -\bar{P}_{12} & 0 \\ * & * & * & * & -\bar{P}_{22} & 0 \\ * & * & * & * & * & -\gamma^2 I \end{bmatrix} \tag{6.15}$$

where

$$
\begin{aligned}
\Xi_1 &= \bar{P}_{11} - S - S^T, \Xi_2 = \bar{P}_{12} - S - S^T, \\
\Xi_3 &= \bar{P}_{22} - S - S^T + N + N^T, \\
\Xi_4 &= S^T A + S^T B_2 (C_k + \Delta C_k) \\
\Xi_5 &= (S - N)^T A + F_B C_2 + N^T \Delta B_k C_2 + F_A \\
&\quad + N^T \Delta A_k + (S - N)^T B_2 (C_k + \Delta C_k), \\
\Xi_6 &= (S - N)^T A + F_B C_2 + N^T \Delta B_k C_2, \\
\Xi_7 &= (S - N)^T B_1 + F_B D_{21} + N^T \Delta B_k D_{21}, \\
\Xi_8 &= C_1 + D_{12} (C_k + \Delta C_k).
\end{aligned}
\tag{6.16}
$$

Then the following theorem presents a sufficient condition for the solvability of the non-fragile H_∞ control problem with additive uncertainty.

Theorem 6.2 *Consider system (6.1). Let scalars $\gamma > 0, \theta_a > 0$ and gain matrix C_k be given. If there exist matrices $F_A, F_B, S, N, \bar{P}_{12}$, and $\bar{P}_{11} > 0, \bar{P}_{22} > 0$, such that the following LMIs hold:*

$$
\begin{aligned}
M_0(\Delta A_k, \Delta B_k, \Delta C_k) &< 0, \quad \theta_{aij}, \theta_{bik}, \theta_{clj} \in \{-\theta_a, \theta_a\}, \\
i, j &= 1, \cdots, n; \ k = 1, \cdots, p; \ l = 1, \cdots, q,
\end{aligned}
\tag{6.17}
$$

then controller (6.2) with additive uncertainty described by (6.3), C_k, and

$$
A_k = (N^T)^{-1} F_A, \quad B_k = (N^T)^{-1} F_B,
\tag{6.18}
$$

solves the non-fragile H_∞ control problem for system (6.1).

Proof 6.2 *By Lemma 2.5, it is sufficient to show that there exist a matrix G with structure (2.14) and a symmetric positive matrix $P = \begin{bmatrix} P_{11} & P_{12} \\ P_{12}^T & P_{22} \end{bmatrix} > 0$ such that*

$$
M_1 = \begin{bmatrix}
P - G - G^T & 0 & G^T A_e & G^T B_e \\
* & -I & C_e & 0 \\
* & * & -P & 0 \\
* & * & * & -\gamma^2 I
\end{bmatrix} < 0
\tag{6.19}
$$

holds for all $\theta_{aij}, \theta_{bik}$, and θ_{clj} satisfying (6.3).
    Denote

$$
S = Y + N, \quad \Gamma_1 = \begin{bmatrix} I & I \\ I & 0 \end{bmatrix}, \bar{\Gamma}_1 = diag\{\Gamma_1, I, \Gamma_1, I\},
$$

$$
\bar{P}_{11} = P_{11} + P_{12} + P_{12}^T + P_{22}, \bar{P}_{12} = P_{11} + P_{12}^T, \bar{P}_{22} = P_{11}.
$$

Then (6.19) is equivalent to

$$M_2 = \bar{\Gamma}_1 M_1 \bar{\Gamma}_1^T$$

$$= \begin{bmatrix} \Xi_1 & \Xi_2 & 0 & \Xi_4 & S^T A & S^T B_1 \\ * & \Xi_3 & 0 & \Pi_1 & \Pi_2 & \Pi_3 \\ * & * & -I & \Xi_8 & C_1 & 0 \\ * & * & * & -\bar{P}_{11} & -\bar{P}_{12} & 0 \\ * & * & * & * & -\bar{P}_{22} & 0 \\ * & * & * & * & * & -\gamma^2 I \end{bmatrix} < 0 \qquad (6.20)$$

which holds for all θ_{aij}, θ_{bik}, and θ_{clj} satisfying (6.3), where $\Xi_1, \Xi_2, \Xi_3, \Xi_4$ are defined by (6.16), and

$$\begin{aligned} \Pi_1 &= (S - N)^T A + N^T B_k C_2 + N^T \Delta B_k C_2 + N^T A_k \\ &\quad + N^T \Delta A_k + (S - N)^T B_2 (C_k + \Delta C_k), \\ \Pi_2 &= (S - N)^T A + N^T B_k C_2 + N^T \Delta B_k C_2, \\ \Pi_3 &= (S - N)^T B_1 + N^T B_k D_{21} + N^T \Delta B_k D_{21}. \end{aligned}$$

By (6.18), (6.19), and the Schur complement, it concludes that (6.20) is equivalent to (6.17).

Thus, the proof is complete.

Remark 6.3 *Theorem 6.2 presents a sufficient condition in terms of solutions to a set of LMIs for the solvability of the non-fragile H_∞ control problem. By the proofs of Theorem 6.2 and Lemma 2.5, Theorem 6.2 also shows that the non-fragile H_∞ control problem becomes a convex one when the gain matrix C_k is known and the state-space realizations of the designed controllers with gain variations admit the slack variable matrix G with the structure of (2.14).*

However, for the non-fragile H_∞ controller design method, it should be noted that the number of LMIs involved in (6.17) is $2^{n^2 + np + nq}$, which results in the difficulty of implementing the LMI constraints in computation. For example, when $n = 6$ and $p = q = 1$, the number of LMIs involved in (6.17) is 2^{48}, which already exceeds the capacity of the current LMI solver in MATLAB. To overcome the difficulty arising from implementing the design condition given in Theorem 6.2, the following method is developed.

Denote

$$\begin{aligned} F_{a1} &= [f_{a11}\ f_{a12}\ \cdots\ f_{a1l_a}], \\ F_{a2} &= [f_{a21}^T\ f_{a22}^T\ \cdots\ f_{a2l_a}^T]^T \end{aligned} \qquad (6.21)$$

where $l_a = n^2 + np + nq$, and

$$f_{a1k} = \begin{bmatrix} \mathbf{0}_{1\times n} & (N^T e_i)^T & \mathbf{0}_{1\times q} & \mathbf{0}_{1\times n} & \mathbf{0}_{1\times n} & \mathbf{0}_{1\times r} \end{bmatrix}^T,$$

$$f_{a2k} = \begin{bmatrix} \mathbf{0}_{1\times n} & \mathbf{0}_{1\times n} & \mathbf{0}_{1\times q} & e_j^T & \mathbf{0}_{1\times n} & \mathbf{0}_{1\times r} \end{bmatrix},$$

for $k = (i-1)n + j$, $i, j = 1, \cdots, n$.

$$f_{a1k} = \begin{bmatrix} \mathbf{0}_{1\times n} & (N^T e_i)^T & \mathbf{0}_{1\times q} & \mathbf{0}_{1\times n} & \mathbf{0}_{1\times n} & \mathbf{0}_{1\times r} \end{bmatrix}^T,$$

$$f_{a2k} = \begin{bmatrix} \mathbf{0}_{1\times n} & \mathbf{0}_{1\times n} & \mathbf{0}_{1\times q} & h_j^T C_2 & h_j^T C_2 & h_j^T D_{21} \end{bmatrix},$$

for $k = n^2 + (i-1)p + j$, $i = 1, \cdots, n, j = 1, \cdots, p$.

$$f_{a1k} = \begin{bmatrix} (S^T B_2 g_i)^T & [(S-N)^T B_2 g_i]^T & (D_{12} g_i)^T & \mathbf{0}_{1\times n} & \mathbf{0}_{1\times n} & \mathbf{0}_{1\times r} \end{bmatrix}^T,$$

$$f_{a2k} = \begin{bmatrix} \mathbf{0}_{1\times n} & \mathbf{0}_{1\times n} & \mathbf{0}_{1\times q} & e_j^T & \mathbf{0}_{1\times n} & \mathbf{0}_{1\times r} \end{bmatrix},$$

for $k = n^2 + np + (i-1)n + j$, $i = 1, \cdots, q, j = 1, \cdots, n$.

Let $k_0, k_1, \cdots, k_{s_a}$ be integers satisfying

$$k_0 = 0 < k_1 < \cdots < k_{s_a} = l_a$$

and matrix Σ have the following structure:

$$\Sigma = \begin{bmatrix} diag[\sigma_{11}^1 \cdots \sigma_{11}^{s_a}] & diag[\sigma_{12}^1 \cdots \sigma_{12}^{s_a}] \\ diag[\sigma_{12}^1 \cdots \sigma_{12}^{s_a}]^T & diag[\sigma_{22}^1 \cdots \sigma_{22}^{s_a}] \end{bmatrix}, \tag{6.22}$$

where $\sigma_{11}^i, \sigma_{12}^i$, and $\sigma_{22}^i \in R^{(k_i - k_{i-1}) \times (k_i - k_{i-1})}, i = 1, \cdots, s_a$. Then, we have the following theorem.

Theorem 6.3 *Consider system (6.1). Let scalars $\gamma > 0, \theta_a > 0$, and gain matrix C_k be given. If there exist matrices $F_A, F_B, S, N, \bar{P}_{12}, \bar{P}_{11} > 0, \bar{P}_{22} > 0$, and symmetric matrix Σ with the structure described by (6.22) such that the following LMIs hold:*

$$\begin{bmatrix} Q & F_{a1} \\ F_{a1}^T & 0 \end{bmatrix} + \begin{bmatrix} F_{a2} & 0 \\ 0 & I \end{bmatrix}^T \Sigma \begin{bmatrix} F_{a2} & 0 \\ 0 & I \end{bmatrix} < 0, \tag{6.23}$$

$$\begin{bmatrix} I \\ diag[\theta_{k_{i-1}+j} \cdots \theta_{k_i}] \end{bmatrix}^T \begin{bmatrix} \sigma_{11}^i & \sigma_{12}^i \\ (\sigma_{12}^i)^T & \sigma_{22}^i \end{bmatrix}$$
$$\times \begin{bmatrix} I \\ diag[\theta_{k_{i-1}+j} \cdots \theta_{k_i}] \end{bmatrix} \geq 0, \text{ for all } \theta_{k_{i-1}+j} \in \{-\theta_a, \theta_a\}, \tag{6.24}$$
$$j = 1, \cdots, k_i - k_{i-1}, i = 1, \cdots, s_a,$$

where

$$Q = \begin{bmatrix} \Xi_1 & \Xi_2 & 0 & \Psi_1 & S^T A & S^T B_1 \\ * & \Xi_3 & 0 & \Psi_2 & \Psi_3 & \Psi_4 \\ * & * & -I & C_1 + D_{12} C_k & C_1 & 0 \\ * & * & * & -\bar{P}_{11} & -\bar{P}_{12} & 0 \\ * & * & * & * & -\bar{P}_{22} & 0 \\ * & * & * & * & * & -\gamma^2 I \end{bmatrix} \tag{6.25}$$

with Ξ_1, Ξ_2, Ξ_3 defined by (6.16) and

$$\Psi_1 = S^T A + S^T B_2 C_k$$
$$\Psi_2 = (S - N)^T A + F_B C_2 + F_A + (S - N)^T B_2 C_k,$$
$$\Psi_3 = (S - N)^T A + F_B C_2,$$
$$\Psi_4 = (S - N)^T B_1 + F_B D_{21},$$

then controller (6.2) with additive uncertainty described by (6.3) and the controller gain parameters given by (6.18) solve the non-fragile H_∞ control problem for system (6.1).

Proof 6.3 *By (6.15), we have*

$$M_0 = Q + \Delta Q + \Delta Q^T < 0, \tag{6.26}$$

where

$$\Delta Q = \begin{bmatrix} 0 & 0 & 0 & \Delta Q_1 & 0 & 0 \\ 0 & 0 & 0 & \Delta Q_2 & \Delta Q_3 & \Delta Q_4 \\ 0 & 0 & 0 & \Delta Q_5 & 0 & 0 \\ 0 & 0 & 0 & 0 & 0 & 0 \\ 0 & 0 & 0 & 0 & 0 & 0 \\ 0 & 0 & 0 & 0 & 0 & 0 \end{bmatrix}$$

with

$$\Delta Q_1 = S^T B_2 \sum_{i=1}^{q} \sum_{j=1}^{n} \theta_{cij} g_i e_j^T,$$

$$\Delta Q_2 = \sum_{i,j=1}^{n} \theta_{aij} N^T e_i e_j^T + \sum_{i=1}^{n} \sum_{j=1}^{p} \theta_{bij} N^T e_i h_j^T C_2$$
$$+ (S - N)^T B_2 \sum_{i=1}^{q} \sum_{j=1}^{n} \theta_{cij} g_i e_j^T,$$

$$\Delta Q_3 = \sum_{i=1}^{n} \sum_{j=1}^{p} \theta_{bij} N^T e_i h_j^T C_2,$$

$$\Delta Q_4 = \sum_{i=1}^{n} \sum_{j=1}^{p} \theta_{bij} N^T e_i h_j^T D_{21},$$

$$\Delta Q_5 = D_{12} \sum_{i=1}^{q} \sum_{j=1}^{n} \theta_{cij} g_j e_j^T.$$

By (6.21) and (6.26), it follows that (6.17) is equivalent to

$$
\begin{aligned}
M_0 &= Q + \sum_{i=1}^{l_a} \theta_i f_{a1i} f_{a2i} + \left(\sum_{i=1}^{l_a} \theta_i f_{a1i} f_{a2i}\right)^T \\
&= Q + F_{a1} \tilde{\Delta}_a F_{a2} + (F_{a1} \tilde{\Delta}_a F_{a2})^T < 0
\end{aligned}
\tag{6.27}
$$

which holds for all $|\theta_i| \le \theta_a$, *where* $\tilde{\Delta}_a = diag[\theta_1, \cdots, \theta_{l_a}]$. *By Lemma 2.5, it follows that (6.27) holds if and only if there exists a symmetric matrix* $\Sigma \in R^{l_a \times l_a}$ *such that (6.23) and*

$$
\begin{bmatrix} I \\ \tilde{\Delta}_a \end{bmatrix}^T \Sigma \begin{bmatrix} I \\ \tilde{\Delta}_a \end{bmatrix} \ge 0
\tag{6.28}
$$

hold for all $\theta_i \in \{-\theta_a, \theta_a\}$, $i = 1, \cdots, l_a$. *Notice that the set of* Σ *satisfying (6.22) is a subset of the set of* Σ *satisfying (6.28), hence the conclusion follows.*

Remark 6.4 *From the proof of Theorem 6.3, it follows that when* $s_a = 1$, *the set of* Σ *satisfying (6.22) is equal to the set of* Σ *satisfying (6.28), and the design conditions given in Theorem 6.3 and Theorem 6.2 are equivalent.* Σ *satisfying (6.23) and (6.28) (or (6.24) with* $s_a = 1$) *is said to be a vertex separator [67]. Notice that the number of LMIs involved in (6.28) or (6.24) with* $s_a = 1$ *still is* $2^{n^2 + np + nq}$, *so that the difficulty of implementing the LMI constraints remains. To overcome the difficulty, Theorem 6.3 presents a sufficient condition for the non-fragile* H_∞ *controller design in terms of separator* Σ *with the structure described by (6.22), where the number of LMIs involved in (6.24) is* $\sum_{i=1}^{s_a} 2^{k_i - k_{i-1}}$, *which can be controlled not to grow exponentially by reducing the value of max* $k_i - k_{i-1}$: $i = 1, \cdots, s_a$. *Compared with the* Σ *being of full block in (6.23) and (6.28),* Σ *with the structure described by (6.22) satisfying (6.23) and (6.24) is said to be a structured vertex separator. However, it should be noted that the design condition given in Theorem 6.3 may be more conservative than that given in Theorem 6.2 because of the structure constraint on* Σ. *But the smaller the value of* s_a *is, the less conservativeness is introduced.*

To illustrate the comparison between the proposed method and the existing design method, the result of a non-fragile H_∞ controller design with norm-bounded gain variations is introduced in the following.

Similar to Yang, Wang, and Lin [149] and Mahmoud [103] for non-fragile problems with norm-bounded uncertainty, the norm-bounded type of gain variations $\Delta A_k, \Delta B_k$, and ΔC_k can be over-bounded [107] by the following norm-bounded uncertainty:

$$
\begin{aligned}
\Delta A_k &= M_a F_1(t) E_a, \Delta B_k = M_b F_2(t) E_b, \\
\Delta C_k &= M_c F_3(t) E_c,
\end{aligned}
\tag{6.29}
$$

where

$$
M_a = [M_{a1} \cdots M_{an^2}], E_a = [E_{a1}^T \cdots E_{an^2}^T]^T,
$$

$$M_b = [M_{b1} \cdots M_{bnp}], E_b = [E_{b1}^T \cdots E_{bnp}^T]^T,$$
$$M_c = [M_{c1} \cdots M_{cnq}], E_c = [E_{c1}^T \cdots E_{cnq}^T]^T,$$

with

$$M_{ak} = e_i, E_{ak} = e_j^T$$
for $k = (i-1)n + j, \ i, j = 1, \cdots, n,$
$$M_{bk} = e_i, E_{bk} = h_j^T$$
for $k = n^2 + (i-1)p + j, \ i = 1, \cdots, n, \ j = 1, \cdots, p,$
$$M_{ck} = g_i, E_{ck} = e_j^T$$
for $k = n^2 + np + (i-1)n + j, \ i = 1, \cdots, q, \ j = 1, \cdots, n$

and $F_i^T(t)F_i(t) \le \theta_a^2 I$ for $i = 1, 2, 3$, represent the uncertain parameters, here, θ_a is the same as before.

Noting that the problem of non-fragile dynamic output feedback H_∞ controller design with norm-bounded gain variations is also a non-convex problem, and similar to Theorem 6.3, when the controller gain C_k is known, it can be converted to a convex one.

To facilitate the presentation, denote

$$\bar{F}_A = \bar{N}A_k, \bar{F}_B = \bar{N}B_k,$$

$$M_{a1} = \begin{bmatrix} 0 & \bar{S}B_2M_c & 0 \\ \bar{N}M_a & (\bar{S}-\bar{N})B_2M_c & \bar{N}M_b \\ 0 & D_{12}M_c & 0 \\ 0 & 0 & 0 \\ 0 & 0 & 0 \\ 0 & 0 & 0 \end{bmatrix},$$

$$M_{a2} = \begin{bmatrix} 0 & 0 & 0 & E_a & 0 & 0 \\ 0 & 0 & 0 & E_c & 0 & 0 \\ 0 & 0 & 0 & E_bC_2 & E_bC_2 & E_bD_{21} \end{bmatrix}.$$

Assume that C_k is known, by using the method in Yang, Wang, and Lin [149] and Mahmoud [103], the non-fragile H_∞ controller design with norm-bounded gain variations is reduced to solve the following LMI:

$$\begin{bmatrix} \bar{Q} & M_{a1} & \theta_a\varepsilon M_{a2}^T \\ * & -\varepsilon I & 0 \\ * & * & -\varepsilon I \end{bmatrix} < 0, \tag{6.30}$$

with matrix variables $\bar{S} > 0, \bar{N} < 0$, and scalar $\varepsilon > 0$, where

$$\bar{Q} = \begin{bmatrix} -\bar{S} & -\bar{S} & 0 & \bar{S}(A+B_2C_k) & \bar{S}A & \bar{S}B_1 \\ * & -\bar{S}+\bar{N} & 0 & Q_1 & Q_2 & Q_3 \\ * & * & -I & C_1+D_{12}C_k & C_1 & 0 \\ * & * & * & -\bar{S} & -\bar{S} & 0 \\ * & * & * & * & -\bar{S}+\bar{N} & 0 \\ * & * & * & * & * & -\gamma^2 I \end{bmatrix},$$

with

$$Q_1 = (\bar{S} - \bar{N})(A + B_2 C_k) + \bar{F}_A + \bar{F}_B C_2,$$
$$Q_2 = (\bar{S} - \bar{N})A + \bar{F}_B C_2,$$
$$Q_3 = (\bar{S} - \bar{N})B_1 + \bar{F}_B D_{21}.$$

The following lemma will show the relationship between condition (6.30) and the condition for designing non-fragile H_∞ controllers given in Theorem 6.3.

Lemma 6.1 *Consider system (6.1). If condition (6.30) is feasible, then the condition for designing non-fragile H_∞ controllers given in Theorem 6.3 is feasible.*

Proof 6.4 *To proceed, let $\bar{S} = \bar{P}_{11} = \bar{P}_{12} = S, \bar{N} = N, \bar{S} - \bar{N} = \bar{P}_{22} > 0$. It is easy to see that $\bar{Q} = Q, M_{a1} = F_{a1}$, and $M_{a2} = F_{a2}$, that is, condition (6.30) becomes*

$$\begin{bmatrix} Q & F_{a1} & \theta_a \varepsilon F_{a2}^T \\ * & -\varepsilon I & 0 \\ * & * & -\varepsilon I \end{bmatrix} < 0. \tag{6.31}$$

In Theorem 6.3, when $s_a = l_a$, according to (6.31) and $F_i^T(t)F_i(t) \leq \theta_a^2 I$, $i = 1, 2, 3$, and by the Schur complement, there exists Σ with the structure

$$\Sigma = \begin{bmatrix} \varepsilon \theta_a^2 I & 0_{l_a \times l_a} \\ 0_{l_a \times l_a} & -\varepsilon I \end{bmatrix}, \tag{6.32}$$

such that the following LMIs hold:

$$\begin{bmatrix} Q & F_{a1} \\ F_{a1}^T & 0 \end{bmatrix} + \begin{bmatrix} F_{a2} & 0 \\ 0 & I \end{bmatrix}^T \Sigma \begin{bmatrix} F_{a2} & 0 \\ 0 & I \end{bmatrix}$$
$$= \begin{bmatrix} Q + \varepsilon \theta_a^2 F_{a2}^T F_{a2} & F_{a1} \\ F_{a1}^T & -\varepsilon I \end{bmatrix} < 0, \tag{6.33}$$

$$\begin{bmatrix} I \\ \theta_i \end{bmatrix}^T \begin{bmatrix} \sigma_{11}^i & \sigma_{12}^i \\ (\sigma_{12}^i)^T & \sigma_{22}^i \end{bmatrix} \begin{bmatrix} I \\ \theta_i \end{bmatrix} = \varepsilon \theta_a^2 - \varepsilon \theta_i^2 \geq 0,$$
for all $i = 1, \cdots, l_a$. \tag{6.34}

Thus, the proof is complete.

Remark 6.5 *From the proof of Lemma 6.1, it follows that condition (6.30) is more conservative than the non-fragile H_∞ controller existence condition in Theorem 6.3 with $s_a = l_a$. However, as indicated in Remark 6.4, the case of $s_a = l_a$ is the worst case of the new proposed method. So the existing non-fragile H_∞ controller design method with the norm-bounded gain variations is more conservative than the one given by Theorem 6.3.*

Combining Theorem 6.1 and Theorem 6.3, a two-step procedure is summarized as follows:

Algorithm 6.1

Step 1. *Minimize γ subject to $X > 0, Y > 0$, and LMI (6.7). Denote the optimal solutions as $X = X_{opt}$ and $\hat{C} = \hat{C}_{opt}$. Then by (6.8), $C_{kopt} = \hat{C}_{opt} X_{opt}^{-1}$.*

Step 2. *Let $C_k = C_{kopt}$, minimize γ subject to $F_A, F_B, N, S, \bar{P}_{12}, \bar{P}_{11} > 0, \bar{P}_{22} > 0$, and LMIs (6.23) and (6.24). Denote the optimal solutions as $N = N_{opt}$, $F_A = F_{Aopt}$, and $F_B = F_{Bopt}$. Then according to (6.18), we obtain $A_k = (N^T)^{-1} F_{Aopt}$, $B_k = (N^T)^{-1} F_{Bopt}$. The resulting A_k, B_k, and C_k will form the non-fragile dynamic output feedback H_∞ controller gains.*

In Theorem 6.3, we restrict the slack variable matrix G with the structure (2.14) for obtaining the convex design condition, which may result in more conservative evaluation of the H_∞ performance index bound. So, in this section, for a designed controller, the matrix G without the restriction is exploited for obtaining less conservative evaluation of the H_∞ performance index bound.

When the controller parameter matrices A_k, B_k, and C_k are known, the problem of minimizing γ subject to (6.3) for a given $\theta_a > 0$ and $\| G_{zw}(z) \| < \gamma$ can be converted into the one: minimize γ^2 subject to the following LMIs:

$$
\begin{bmatrix}
P - G - G^T & 0 & G^T A_e & G^T B_e \\
* & -I & C_e & 0 \\
* & * & -P & 0 \\
* & * & * & -\gamma^2 I
\end{bmatrix} < 0, \theta_{aij}, \theta_{bik}, \theta_{clj} \in \{-\theta_a, \theta_a\},
$$
$$
i, j = 1, \cdots, n; \; k = 1, \cdots, p; \; l = 1, \cdots, q,
$$
$$\tag{6.35}$$

where A_e, B_e, and C_e are defined as in (6.4).

Similar to the design condition given in Theorem 6.2, the design condition given by (6.35) is also with the numerical computation problem. To solve the problem, the following lemma provides a solution using the structured vertex separator approach.

Denote

$$
\begin{aligned}
G_{a1} &= [g_{a11} \; g_{a12} \; \cdots \; g_{a1l_a}], \\
G_{a2} &= [g_{a21}^T \; g_{a22}^T \; \cdots \; g_{a2l_a}^T]^T
\end{aligned}
\tag{6.36}
$$

where

$$g_{a1k} = \left[\left(\mathbf{0}_{1\times n} \quad e_i^T \right) G \quad \mathbf{0}_{1\times q} \quad \mathbf{0}_{1\times 2n} \quad \mathbf{0}_{1\times r} \right]^T,$$

$$g_{a2k} = \left[\mathbf{0}_{1\times 2n} \quad \mathbf{0}_{1\times q} \quad \mathbf{0}_{1\times n} \quad e_j^T \quad \mathbf{0}_{1\times n} \right],$$

for $k = (i-1)n + j$, $i, j = 1, \cdots, n$.

$$g_{a1k} = \left[\left(\mathbf{0}_{1\times n} \quad e_i^T \right) G \quad \mathbf{0}_{1\times q} \quad \mathbf{0}_{1\times 2n} \quad \mathbf{0}_{1\times r} \right]^T,$$

$$g_{a2k} = \left[\mathbf{0}_{1\times 2n} \quad \mathbf{0}_{1\times q} \quad h_j^T C_2 \quad \mathbf{0}_{1\times n} \quad h_j^T D_{21} \right],$$

for $k = n^2 + (i-1)p + j$, $i = 1, \cdots, n, j = 1, \cdots, p$.

$$g_{a1k} = \left[\left((B_2 g_i)^T \quad \mathbf{0}_{1\times n} \right) G \quad (D_{12} g_i)^T \quad \mathbf{0}_{1\times 2n} \quad \mathbf{0}_{1\times r} \right]^T,$$

$$g_{a2k} = \left[\mathbf{0}_{1\times 2n} \quad \mathbf{0}_{1\times q} \quad \mathbf{0}_{1\times n} \quad e_j^T \quad \mathbf{0}_{1\times n} \right],$$

for $k = n^2 + np + (i-1)n + j$, $i = 1, \cdots, q, j = 1, \cdots, n$.

Then we have the following lemma.

Lemma 6.2 *Consider system (6.1). Let $\gamma > 0, \theta_a > 0$ be constants, and let controller parameter matrices $A_k, B_k,$ and C_k be given. Then $\parallel G_{zw} \parallel < \gamma$ holds for all θ_{aij}, θ_{bit}, and θ_{clj} satisfying (6.3), if there exist a matrix G, a positive-definite matrix $P > 0$, and a symmetric matrix Σ with the structure described by (6.22) such that (6.24) and the following LMI hold:*

$$\begin{bmatrix} Q_s & G_{a1} \\ G_{a1}^T & 0 \end{bmatrix} + \begin{bmatrix} G_{a2} & 0 \\ 0 & I \end{bmatrix}^T \Sigma \begin{bmatrix} G_{a2} & 0 \\ 0 & I \end{bmatrix} < 0 \qquad (6.37)$$

where

$$Q_s = \begin{bmatrix} P - G - G^T & 0 & G^T A_{e0} & G^T B_{e0} \\ * & -I & C_{e0} & 0 \\ * & * & -P & 0 \\ * & * & * & -\gamma^2 I \end{bmatrix}$$

with $A_{e0}, B_{e0},$ and C_{e0} defined by (2.8).

Proof 6.5 *By using (6.35) and (6.36), it is similar to the proof of Theorem 6.3, and omitted here.*

Remark 6.6 *For evaluating the H_∞ performance bound of the transfer function from w to z, the condition given in Lemma 6.2 usually is less conservative than that given in Theorem 6.3 because no structure constraint on the slack variable matrix G in Lemma 6.2 is imposed.*

6.2.3 Example

In the following, an example is given to illustrate the effectiveness of the proposed methods. At the same time, a comparison is made between the proposed non-fragile H_∞ controller design method and the existing non-fragile H_∞ controller design method.

Consider a discrete-time linear system of the form (6.1) with

$$A = \begin{bmatrix} 0 & -1 & 0 \\ 0.5 & -1 & 1 \\ 0.5 & -1 & 1 \end{bmatrix}, B_1 = \begin{bmatrix} -0.5 & 0 \\ -0.5 & 0 \\ -1 & 0 \end{bmatrix}, B_2 = \begin{bmatrix} 1 \\ 1 \\ -1 \end{bmatrix},$$

$$C_1 = \begin{bmatrix} 1 & -1 & -1 \\ 0 & 0 & 0 \end{bmatrix}, C_2 = \begin{bmatrix} -1 & 1 & -3 \end{bmatrix},$$

$$D_{21} = \begin{bmatrix} 0 & 1 \end{bmatrix}, D_{12} = \begin{bmatrix} 0 \\ 1 \end{bmatrix}.$$

By the standard H_∞ controller design method [24], we obtain the optimal H_∞ performance index for the system as $\gamma_{opt} = 2.1622$.

On the other hand, assume that the designed controller is with form (6.5). Let $\theta_a = 0.05$, by Theorem 6.1 with $\varepsilon_c = 155.9999$, and we obtain

$$C_{k_{ini}} = \begin{bmatrix} 0.2573 & -0.2351 & 0.3380 \end{bmatrix}.$$

Here ε_c is obtained by searching such that the H_∞ performance index is optimal. θ_a is chosen large appropriately such that the designed $C_{k_{ini}}$ can guarantee that Step 2 is feasible.

First, we design an H_∞ controller by condition (6.30) with $C_k = C_{k_{ini}}$.

Assume that the designed controller is with norm-bounded additive uncertainties described by (6.29), by applying condition (6.30) with $\theta_a = 0.006$ to design a non-fragile controller. The obtained gains are given by

$$A_{k_{nm}} = \begin{bmatrix} 0.3855 & -1.3850 & 0.7002 \\ 0.2214 & -1.2223 & 0.7067 \\ -0.1037 & -0.5537 & -0.4515 \end{bmatrix},$$

$$B_{k_{nm}} = \begin{bmatrix} 0.0766 & -0.0874 & -0.4793 \end{bmatrix}^T,$$

and the H_∞ performance index of the obtained non-fragile controller is 2.8645.

Then, in the following, we design an H_∞ controller by Theorem 6.3 with $C_k = C_{k_{ini}}$.

Assume that the designed controller is with the additive uncertainties described by (6.3). For this case with $C_k = C_{k_{ini}}$, it is difficult to apply Theorem 6.2 to design a non-fragile H_∞ controller because the number of the LMI constraints involved in (6.17) is 2^{15}. However, Theorem 6.3 is applicable for solving this problem. By applying Theorem 6.3 with $\theta_a = 0.006$ and $k_i = i, i = 1, \cdots, 15$, that is, $s_a = 15$ as well as $k_i = 3i, i = 1, \cdots, 5$, that is, $s_a = 5$ to design a non-fragile H_∞ controller, the obtained gains are as follows.

TABLE 6.1

Performance Index by Design with $\theta_a = 0.006$

γ	Condition (6.30) 2.8645	Theorem 6.3 ($s_a = 15$) 2.5204	Theorem 6.3 ($s_a = 5$) 2.5099

For $s_a = 15$,

$$A_k = \begin{bmatrix} 0.3270 & -1.3894 & 0.5548 \\ 0.1699 & -1.2336 & 0.5885 \\ -0.1696 & -0.5686 & -0.6114 \end{bmatrix},$$

$$B_k = \begin{bmatrix} -0.0027 & -0.1566 & -0.5843 \end{bmatrix}^T.$$

For $s_a = 5$,

$$A_k = \begin{bmatrix} 0.3261 & -1.3872 & 0.5532 \\ 0.1699 & -1.2336 & 0.5885 \\ -0.1696 & -0.5686 & -0.6114 \end{bmatrix},$$

$$B_k = \begin{bmatrix} -0.0027 & -0.1566 & -0.5843 \end{bmatrix}^T.$$

Correspondingly, the H_∞ performance indexes of the obtained non-fragile controllers are $\gamma = 2.5204$ and $\gamma = 2.5099$, respectively.

In this part, for the above designed controllers, we will illustrate that Lemma 6.2 can give better evaluations of the H_∞ performance index bounds.

First, to facilitate the presentation, denote the controller designed by condition (6.30) as K_{nm} and denote the controllers designed by Theorem 6.3 as K_{in15} ($s_a = 15$) and K_{in5} ($s_a = 5$).

By Lemma 6.2, the H_∞ performance indices of the controller K_{nm} are $\gamma = 2.6507$ ($s_a = 15$) and $\gamma = 2.6501$ ($s_a = 5$) while the H_∞ performance indices of the controllers K_{in15} and K_{in5} are $\gamma = 2.4934$ ($s_a = 15$) and $\gamma = 2.4903$ ($s_a = 5$), respectively.

Here, tables are given to provide a comparison between two methods.

First, Table 6.1 shows the H_∞ performance indices achieved by the designs of the existing method (Condition (6.30)) and the proposed method (Theorem 6.3).

From Table 6.1, it is easy to see that compared with the optimal H_∞ performance index bound $\gamma_{opt} = 2.1622$, the performance index of the controller designed by condition (6.30) is degraded 32.48%. The performance indices of the controllers designed by Theorem 6.3 are degraded 16.57% ($s_a = 15$) and 16.08% ($s_a = 5$), which are much more improved than 32.48%.

Second, for the designed non-fragile controllers, Lemma 6.2 gives better performance indices shown in Table 6.2.

Obviously, compared with $\gamma_{opt} = 2.1622$, by using Lemma 6.2, the H_∞ performance indices of the controller K_{nm} are degraded 22.59% for $s_a = 15$ and degraded 22.56% for $s_a = 5$. In contrast, the performance indices for the controllers K_{in15} and K_{in5} are degraded 15.32% for $s_a = 15$ and 15.17% for

TABLE 6.2
Performance Evaluation by Lemma 6.2
with $\theta_a = 0.006$

	K_{nm}	K_{in15}	K_{in5}
$\gamma(s_a = 15)$	2.6507	2.4934	—
$\gamma(s_a = 5)$	2.6501	—	2.4903

$s_a = 5$, respectively. The result shows the advantage of the new proposed method.

Finally, we will give a simulation to illustrate the effectiveness of the proposed method.

Let $x(0) = \begin{bmatrix} -1 & -0.5 & 0.5 \end{bmatrix}^T, \xi(0) = \begin{bmatrix} -0.5 & -0.5 & 0.5 \end{bmatrix}^T$, and let the disturbance $w(k)$ be

$$w(k) = \begin{cases} \begin{bmatrix} -5 \\ -5 \end{bmatrix}, & 21 \le t \le 40 \text{ (step)}, \\ 0 & \text{otherwise}. \end{cases}$$

Then Figure 6.1 shows the regulated output responses of z_1 controlled by the non-fragile controllers designed by the existing method (condition (6.30)) and

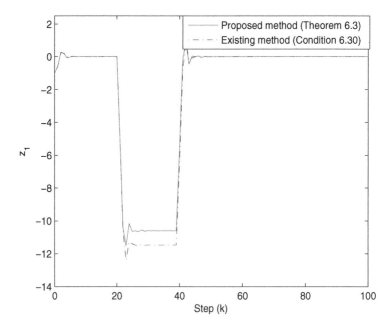

FIGURE 6.1
Comparison of the regulated output responses of $z_1(t)$ with $\theta_a = 0.006$.

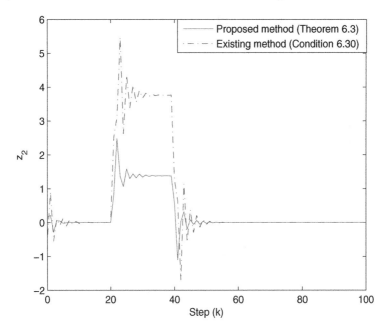

FIGURE 6.2
Comparison of the regulated output responses of $z_2(t)$ with $\theta_a = 0.006$.

the proposed method (Theorem 6.3) with $s_a = 15$, respectively. The responses of z_2 controlled by the two non-fragile controllers are shown in Figure 6.2.

From this example, we can see that the worst case ($s_a = 15$) of Theorem 6.3 also is more effective than the non-fragile H_∞ controller existence condition (6.30).

6.3 Non-Fragile H_∞ Controller Design for Continuous-Time Systems

In this section, the non-fragile dynamic output feedback H_∞ controller design for continuous-time systems with the interval type of gain variations is considered.

6.3.1 Problem Statement

Consider a continuous-time linear time-invariant model described by

$$
\begin{aligned}
\dot{x}(t) &= Ax(t) + B_1\omega(t) + B_2u(t), \\
z(t) &= C_1x(t) + D_{12}u(t), \\
y(t) &= C_2x(t) + D_{21}\omega(t),
\end{aligned}
\tag{6.38}
$$

where $x(t) \in R^n$ is the state, $u(t) \in R^m$ is the control input, $y(t) \in R^p$ is the measured output, $\omega(t) \in R^r$ is the disturbance input, $z(t) \in R^q$ is the regulated output, and $A, B_1, B_2, C_1, C_2, D_{12}$, and D_{21} are known constant matrices of appropriate dimensions.

And for system (6.38), consider a dynamic output feedback controller with gain variations of the following form:

$$
\begin{aligned}
\dot{\xi}(t) &= (A_k + \Delta A_k)\xi(t) + (B_k + \Delta B_k)y(t), \\
u(t) &= (C_k + \Delta C_k)\xi(t),
\end{aligned}
\tag{6.39}
$$

where $\xi(t) \in R^n$ is the controller state, and A_k, B_k, and C_k are controller gain matrices of appropriate dimensions to be designed. $\Delta A_k, \Delta B_k$, and ΔC_k represent the additive gain variations of the interval type (6.3).

6.3.2 Non-Fragile H_∞ Controller Design Methods

Note that the proofs of the main results are similar to those of discrete-time systems, so here we only give the main results for continuous-time systems without proofs.

First, we give the design method of the initial controller gain C_k, which will be used in the non-fragile H_∞ controller design.

Consider the controller (6.39) with $\Delta A_k = 0$ and $\Delta B_k = 0$, which is described by

$$
\begin{aligned}
\dot{\xi}(t) &= A_k\xi(t) + B_ky(t), \\
u(t) &= (C_k + \Delta C_k)\xi(t),
\end{aligned}
\tag{6.40}
$$

where ΔC_k is with the norm-bounded form as in the discrete-time case of the former section.

Combining the controller (6.40) with the system (6.38), we obtain the following closed-loop system:

$$
\begin{aligned}
\dot{x}_e(t) &= A_{edc}x_e(t) + B_{edc}\omega(t), \\
z(t) &= C_ex_e(t).
\end{aligned}
\tag{6.41}
$$

Then the following theorem gives a design method of the initial controller gain C_k.

Theorem 6.4 *Consider system (6.38), where $\gamma > 0$ and $\theta_a > 0$ are constants. If there exist matrices $\hat{A}$, $\hat{B}$, $\hat{C}$, $X > 0$, and $Y > 0$, and a constant $\varepsilon_c > 0$ such that the following LMI holds:*

$$
\begin{bmatrix}
\Sigma_1 & A + \hat{A} & B_1 & XC_1^T + \hat{C}^T D_{12}^T \\
* & \Sigma_2 & YB_1 + \hat{B}D_{21} & C_1^T \\
* & * & -\gamma^2 I & 0 \\
* & * & * & -I \\
* & * & * & * \\
* & * & * & *
\end{bmatrix}
$$

$$
\begin{bmatrix}
B_2 M_c & \varepsilon_c \theta_a X E_c^T \\
YB_2 M_c & 0 \\
0 & 0 \\
D_{12} M_c & 0 \\
-\varepsilon_c I & 0 \\
* & -\varepsilon_c I
\end{bmatrix} < 0, \tag{6.42}
$$

where $\Sigma_1 = AX + B_2\hat{C} + XA^T + \hat{C}^T B_2^T$ and $\Sigma_2 = YA + \hat{B}C_2 + A^TY + C_2^T\hat{B}^T$, then the controller (6.40) with

$$
A_k = (X^{-1} - Y)^{-1}(\hat{A} - YAX - \hat{B}C_2X - YB_2\hat{C})X^{-1},
$$
$$
B_k = (X^{-1} - Y)^{-1}\hat{B}, \quad C_k = \hat{C}X^{-1} \tag{6.43}
$$

solves the non-fragile H_∞ control problem for system (6.38).

Then, with the initial controller gain C_k designed above, we give the non-fragile dynamic output feedback H_∞ controller design methods in the following.

Theorem 6.5 *Consider the plant (6.38), where $\gamma > 0$ and $\theta_a > 0$ are given constants, and the gain matrix C_k is given. If there exist matrices F_A, F_B, $S_a > 0, N_a < 0$, such that the following LMIs hold:*

$$
M_0(\Delta A_k, \Delta B_k, \Delta C_k) < 0, \quad \theta_{aij}, \theta_{bit}, \theta_{clj} \in \{-\theta_a, \theta_a\},
$$
$$
i, j = 1, \cdots, n; \ t = 1, \cdots, p; \ l = 1, \cdots, m, \tag{6.44}
$$

then the controller (6.39) with additive uncertainty described by (6.3), C_k, and

$$
A_k = N_a^{-1}F_A, \quad B_k = N_a^{-1}F_B \tag{6.45}
$$

solves the non-fragile H_∞ control problem for system (6.38), where

$$
M_0(\Delta A_k, \Delta B_k, \Delta C_k) = \begin{bmatrix} M_{0a} + M_{0a}^T & M_{0b} & M_{0c}^T \\ * & -\gamma^2 I & 0 \\ * & * & -I \end{bmatrix}, \tag{6.46}
$$

with

$$M_{0a} = \begin{bmatrix} S_a A + S_a B_2 (C_k + \Delta C_k) \\ \Phi_1 + (S_a - N_a)A + F_B C_2 + N_a \Delta B_k C_2 \\ S_a A \\ (S_a - N_a)A + F_B C_2 + N_a \Delta B_k C_2 \end{bmatrix},$$

$$M_{0b} = \begin{bmatrix} S_a B_1 \\ (S_a - N_a)B_1 + F_B D_{21} + N_a \Delta B_k D_{21} \end{bmatrix},$$

$$M_{0c} = \begin{bmatrix} C_1 + D_{12}(C_k + \Delta C_k) & C_1 \end{bmatrix},$$

and $\Phi_1 = F_A + N_a \Delta A_k + (S_a - N_a)B_2(C_k + \Delta C_k)$.

Similar to the discrete-time case, it should be noted that the number of LMIs involved in (6.44) is $2^{n^2 + np + nq}$, which results in the difficulty of implementing the LMI constraints in computation. To overcome the difficulty arising from implementing the design condition given in Theorem 6.5, the following method is developed.

Denote

$$\begin{aligned} F_{a1} &= [f_{a11} \ f_{a12} \ \cdots \ f_{a1l_a}], \\ F_{a2} &= [f_{a21}^T \ f_{a22}^T \ \cdots \ f_{a2l_a}^T]^T \end{aligned} \tag{6.47}$$

where $l_a = n^2 + np + nm$, and

$$f_{a1k} = \begin{bmatrix} \mathbf{0}_{1 \times n} & (N_a e_i)^T & \mathbf{0}_{1 \times r} & \mathbf{0}_{1 \times q} \end{bmatrix}^T,$$
$$f_{a2k} = \begin{bmatrix} e_j^T & \mathbf{0}_{1 \times n} & \mathbf{0}_{1 \times r} & \mathbf{0}_{1 \times q} \end{bmatrix},$$
for $k = (i-1)n + j$, $i, j = 1, \cdots, n$.

$$f_{a1k} = \begin{bmatrix} \mathbf{0}_{1 \times n} & (N_a e_i)^T & \mathbf{0}_{1 \times r} & \mathbf{0}_{1 \times q} \end{bmatrix}^T,$$
$$f_{a2k} = \begin{bmatrix} h_j^T C_2 & h_j^T C_2 & h_j^T D_{21} & \mathbf{0}_{1 \times q} \end{bmatrix},$$
for $k = n^2 + (i-1)p + j$, $i = 1, \cdots, n, j = 1, \cdots, p$.

$$f_{a1k} = \begin{bmatrix} (S_a B_2 g_i)^T & [(S_a - N_a)B_2 g_i]^T & \mathbf{0}_{1 \times r} & (D_{12} g_i)^T \end{bmatrix}^T,$$
$$f_{a2k} = \begin{bmatrix} e_j^T & \mathbf{0}_{1 \times n} & \mathbf{0}_{1 \times r} & \mathbf{0}_{1 \times q} \end{bmatrix},$$
for $k = n^2 + np + (i-1)n + j$, $i = 1, \cdots, m, j = 1, \cdots, n$.

Let $k_0, k_1, \cdots, k_{s_a}$ be integers satisfying

$$k_0 = 0 < k_1 < \cdots < k_{s_a} = l_a \tag{6.48}$$

and matrix Σ have the following structure,

$$\Sigma = \begin{bmatrix} diag[\sigma_{11}^1 \cdots \sigma_{11}^{s_a}] & diag[\sigma_{12}^1 \cdots \sigma_{12}^{s_a}] \\ diag[\sigma_{12}^1 \cdots \sigma_{12}^{s_a}]^T & diag[\sigma_{22}^1 \cdots \sigma_{22}^{s_a}] \end{bmatrix}, \tag{6.49}$$

where σ_{11}^i, σ_{12}^i, and $\sigma_{22}^i \in R^{(k_i - k_{i-1}) \times (k_i - k_{i-1})}$, $i = 1, \cdots, s_a$.

Theorem 6.6 *Consider the plant (6.38), where $\gamma > 0$, $\theta_a > 0$, and $s_a > 0$ are given constants, and the gain matrix C_k is given. If there exist matrices $F_A, F_B, S_a > 0, N_a < 0$, and symmetric matrix Σ with the structure described by (6.49) such that the following LMIs hold:*

$$\begin{bmatrix} Q & F_{a1} \\ F_{a1}^T & 0 \end{bmatrix} + \begin{bmatrix} F_{a2} & 0 \\ 0 & I \end{bmatrix}^T \Sigma \begin{bmatrix} F_{a2} & 0 \\ 0 & I \end{bmatrix} < 0, \tag{6.50}$$

$$\begin{bmatrix} I \\ diag[\theta_{k_{i-1}+j} \cdots \theta_{k_i}] \end{bmatrix}^T \begin{bmatrix} \sigma_{11}^i & \sigma_{12}^i \\ (\sigma_{12}^i)^T & \sigma_{22}^i \end{bmatrix}$$
$$\times \begin{bmatrix} I \\ diag[\theta_{k_{i-1}+j} \cdots \theta_{k_i}] \end{bmatrix} \geq 0, \quad for\ all\ \theta_{k_{i-1}+j} \in \{-\theta_a, \theta_a\},$$
$$j = 1, \cdots, k_i - k_{i-1}, i = 1, \cdots, s_a, \tag{6.51}$$

where

$$Q = \begin{bmatrix} M_{as} + M_{as}^T & M_{bs} & M_{cs}^T \\ * & -\gamma^2 I & 0 \\ * & * & -I \end{bmatrix}, \tag{6.52}$$

$$M_{as} = \begin{bmatrix} S_a A + S_a B_2 C_k & S_a A \\ (S_a - N_a)A + F_B C_2 + \Phi_3 & (S_a - N_a)A + F_B C_2 \end{bmatrix},$$

$$M_{bs} = \begin{bmatrix} S_a B_1 \\ (S_a - N_a)B_1 + F_B D_{21} \end{bmatrix}, M_{cs} = \begin{bmatrix} C_1 + D_{12}C_k & C_1, \end{bmatrix},$$

with $\Phi_3 = F_A + (S_a - N_a)B_2 C_k$.

Then (6.39) with additive uncertainty described by (6.3), C_k, and the controller gain parameters given by (6.45) solve the non-fragile H_∞ control problem for system (6.38).

Remark 6.7 *In Theorem 6.6, in some cases, the magnitude of the designed A_k (B_k) may be too large to be applied in practice. For solving the problem, in Theorem 6.6, by adding the following constraints*

$$N_a < -\alpha I, \quad F_A F_A^T < \beta I, \tag{6.53}$$

then the magnitude of A_k can be reduced. In fact, by $A_k = N_a^{-1} F_A$ and (6.53), it follows that

$$\| A_F \| < 1/\alpha \| F_A \| \leq \sqrt{\beta}/\alpha.$$

The same method can be applied for the gain B_k.

In the following, we will introduce the result of non-fragile H_∞ controller design with norm-bounded gain variations with the form (6.29), and the relationship with our result is discussed.

Noting that the problem of non-fragile dynamic output feedback H_∞ controller design with norm-bounded gain variations is also a non-convex problem, and similar to Theorem 6.6, when the controller gain C_k is known it is converted to a convex one. First, assume C_k is known. Then, by using the method in Yang, Wang, and Lin [149] and Mahmoud [103], the non-fragile H_∞ controller design with norm-bounded gain variations is reduced to solve the following LMI:

$$\begin{bmatrix} Q & M_{a1} & \theta_a\varepsilon M_{a2}^T \\ * & -\varepsilon I & 0 \\ * & * & -\varepsilon I \end{bmatrix} < 0, \tag{6.54}$$

with $S_a > 0$ and $N_a < 0$, where Q is defined by (6.52) with known C_k, $\varepsilon > 0$ is a given scalar, and

$$M_{a1} = \begin{bmatrix} S_aB_2M_c & 0 & 0 & 0 \\ 0 & N_aM_a & N_aM_b & S_aB_2M_c \\ 0 & 0 & 0 & 0 \\ 0 & 0 & 0 & 0 \end{bmatrix}$$
$$\begin{matrix} 0 & 0 \\ -N_aB_2M_c & 0 \\ 0 & 0 \\ 0 & D_{12}M_c \end{matrix},$$

$$M_{a2} = \begin{bmatrix} E_c & 0 & 0 & 0 \\ E_a & 0 & 0 & 0 \\ E_bC_2 & E_bC_2 & E_bD_{21} & 0 \\ E_c & 0 & 0 & 0 \\ E_c & 0 & 0 & 0 \\ E_c & 0 & 0 & 0 \end{bmatrix}.$$

The following lemma will show the relationship between condition (6.54) and the condition for designing non-fragile H_∞ controllers given in Theorem 6.6.

Lemma 6.3 *Consider the system (6.38). If condition (6.54) is feasible, then the condition for designing non-fragile H_∞ controllers given in Theorem 6.6 is feasible.*

Combining the results in Theorem 6.4 and Theorem 6.6, we have the following algorithm.

Algorithm 6.2
    Step 1. *Minimize γ subject to $X > 0$, $Y > 0$, and LMI (6.42). Denote the optimal solutions as $X = X_{opt}$ and $\hat{C} = \hat{C}_{opt}$. Then by (6.43), $C_{kopt} = \hat{C}_{opt}X_{opt}^{-1}$.*

Step 2. *Let $C_k = C_{kopt}$, minimize γ subject to $F_A, F_B, N_a > 0$, $S_a > 0$, and LMIs (6.50) and (6.51). Denote the optimal solutions as $N_a = N_{aopt}$, $F_A = F_{Aopt}$, and $F_B = F_{Bopt}$. Then according to (6.45), we obtain $A_k = N_a^{-1}F_{Aopt}$, and $B_k = N_a^{-1}F_{Bopt}$. The resulting A_k, B_k, and C_k will form the non-fragile dynamic output feedback H_∞ controller gains.*

In Theorem 6.6, we restrict the Lyapunov function matrix P with the structure (2.14) for obtaining the convex design condition, which may result in more conservative evaluation of the H_∞ performance index bound. So, in this section, for a designed controller, the Lyapunov matrix P without the restriction is exploited for obtaining less conservative evaluation of the H_∞ performance index bound.

When the controller parameter matrices A_k, B_k, and C_k are known, the problem of minimizing γ subject to (6.3) for a given $\theta_a > 0$ and $\| T_{z_e\omega} \| < \gamma$ can be converted into the one: minimize γ^2 subject to the following LMIs:

$$
\begin{bmatrix} PA_e + A_e^T P & PB_e & C_e^T \\ * & -\gamma^2 I & 0 \\ * & * & -I \end{bmatrix} < 0, \theta_{aij}, \ \theta_{bit}, \ \theta_{clj} \in \{-\theta_a, \theta_a\},
$$

$$
i, j = 1, \cdots, n; \ t = 1, \cdots, p; \ l = 1, \cdots, m.
$$

(6.55)

Similar to the design condition given in Theorem 6.5, the above optimization problem is also with the numerical computational problem, to solve the problem, the following lemma provides a solution using the structured vertex separator approach.

Denote

$$
G_{a1} = [g_{a11} \ g_{a12} \ \cdots \ g_{a1l_a}],
$$
$$
G_{a2} = [g_{a21}^T \ g_{a22}^T \ \cdots \ g_{a2l_a}^T]^T,
$$

(6.56)

where

$$
g_{a1k} = \left[\begin{bmatrix} \mathbf{0}_{1\times n} & e_i^T \end{bmatrix} P \quad \mathbf{0}_{1\times r} \quad \mathbf{0}_{1\times q} \right]^T,
$$
$$
g_{a2k} = \begin{bmatrix} \mathbf{0}_{1\times n} & e_j^T & \mathbf{0}_{1\times r} & \mathbf{0}_{1\times q} \end{bmatrix},
$$
for $k = (i-1)n + j, \ i, j = 1, \cdots, n$.
$$
g_{a1k} = \left[\begin{bmatrix} \mathbf{0}_{1\times n} & e_i^T \end{bmatrix} P \quad \mathbf{0}_{1\times r} \quad \mathbf{0}_{1\times q} \right]^T,
$$
$$
g_{a2k} = \begin{bmatrix} h_j^T C_2 & \mathbf{0}_{1\times n} & h_j^T D_{21} & \mathbf{0}_{1\times q} \end{bmatrix},
$$
for $k = n^2 + (i-1)p + j, \ i = 1, \cdots, n, j = 1, \cdots, p$.
$$
g_{a1k} = \left[\begin{bmatrix} (B_2 g_i)^T & \mathbf{0}_{1\times n} \end{bmatrix} P \quad \mathbf{0}_{1\times n} \quad \mathbf{0}_{1\times r} \quad (D_{12}g_i)^T \right]^T,
$$
$$
g_{a2k} = \begin{bmatrix} \mathbf{0}_{1\times n} & e_j^T & \mathbf{0}_{1\times r} & \mathbf{0}_{1\times q} \end{bmatrix}
$$
for $k = n^2 + np + (i-1)n + j, \ i = 1, \cdots, m, j = 1, \cdots, n$.

Then we have the following lemma.

Lemma 6.4 *Consider the system (6.38). Let $\gamma > 0$ and $\theta_a > 0$ be constants, and let controller parameter matrices $A_k, B_k,$ and C_k be given. Then $\| T_{zw} \| < \gamma$ holds for all $\theta_{aij},$ $\theta_{bit},$ and θ_{clj} satisfying (6.3), if there exist matrices $P > 0$ and symmetric matrix Σ with the structure described by (6.49) such that (6.51) and the following LMI hold:*

$$\begin{bmatrix} Q_s & G_{a1} \\ G_{a1}^T & 0 \end{bmatrix} + \begin{bmatrix} G_{a2} & 0 \\ 0 & I \end{bmatrix}^T \Sigma \begin{bmatrix} G_{a2} & 0 \\ 0 & I \end{bmatrix} < 0 \qquad (6.57)$$

where

$$Q_s = \begin{bmatrix} P\begin{bmatrix} A & B_2 C_k \\ B_k C_2 & A_k \end{bmatrix} + \begin{bmatrix} A & B_2 C_k \\ B_k C_2 & A_k \end{bmatrix}^T P & & \\ & * & \\ & * & \\ P\begin{bmatrix} B_1 \\ B_k D_{21} \end{bmatrix} & \begin{bmatrix} C_1^T \\ (D_{12}C_k)^T \end{bmatrix} \\ -\gamma^2 I & 0 \\ * & -I \end{bmatrix}. \qquad (6.58)$$

Remark 6.8 *For evaluating the H_∞ performance bound of the transfer function from w to z, the condition given in Lemma 6.4 usually is less conservative than that given in Theorem 6.6 because no structure constraint on the Lyapunov function matrix in Lemma 6.4 is imposed.*

6.3.3　Example

To illustrate the effectiveness of the proposed non-fragile H_∞ controller, an example is given to provide a comparison between the proposed non-fragile H_∞ controller design method and the existing non-fragile H_∞ controller design method with the norm-bounded gain variations. Consider a linear system of the form (6.38) with

$$A = \begin{bmatrix} 0 & 1 & 0 \\ 0 & -1 & -1 \\ 0 & 1 & -1 \end{bmatrix}, B_1 = \begin{bmatrix} -1 & 0 \\ 1 & 0 \\ 1 & 0 \end{bmatrix}, B_2 = \begin{bmatrix} 1 \\ 2 \\ -2 \end{bmatrix},$$

$$C_1 = \begin{bmatrix} 2 & -1 & -2 \\ 0 & 0 & 0 \end{bmatrix}, C_2 = \begin{bmatrix} -2 & 2 & -1 \end{bmatrix},$$

$$D_{21} = \begin{bmatrix} 0 & 0.5 \end{bmatrix}, D_{12} = \begin{bmatrix} 0 \\ 1 \end{bmatrix}.$$

By the standard H_∞ controller design method [113], we obtain the optimal H_∞ performance index for the system as $\gamma_{opt} = 2.7069$.

On the other hand, assume that the designed controller is in the form of (6.40). Let $\theta_a = 0.05$, then by Theorem 6.3 with $\varepsilon_c = 0.0023$, we obtain

$$C_{k_{ini}} = \begin{bmatrix} -3.0525 & 0.0998 & 1.3036 \end{bmatrix}.$$

Here, ε_c is obtained by searching such that the H_∞ performance index is optimal. The value for θ_a is chosen appropriately large such that the designed $C_{k_{ini}}$ can guarantee that Step 2 is feasible.

In the following, we design an H_∞ controller by condition (6.54) with $C_k = C_{k_{ini}}$.

Assume that the designed controller is with norm-bounded additive uncertainties described by (6.29), by applying condition (6.54) with $\theta_a = 0.02$ and $\varepsilon_c = 0.0040$ to design a non-fragile controller. The obtained gains are given by

$$A_{k_{nm}} = 10^8 \times \begin{bmatrix} -2.0220 & 2.0588 & -0.9508 \\ -0.9435 & 0.9607 & -0.4436 \\ 1.0268 & -1.0455 & 0.4828 \end{bmatrix},$$

$$B_{k_{nm}} = 10^8 \times \begin{bmatrix} -1.0030 & -0.4680 & 0.5093 \end{bmatrix}^T,$$

and the H_∞ performance index of the obtained non-fragile controller is 3.3719. Note that the ε_c used here is obtained by the searching method to guarantee the H_∞ performance index is optimal.

Notice that the gains of the above controller are too large to be applied in practice. As indicated in Remark 6.7, applying condition (6.54) with (6.53), and with $\alpha = 0.4$ and $\beta = 1000000$, the controller gains are reduced as:

$$A_{k_{nm}} = \begin{bmatrix} -93.7030 & 93.1259 & -41.7578 \\ -45.8612 & 39.1903 & -16.4768 \\ 48.3981 & -42.3883 & 16.5820 \end{bmatrix},$$

$$B_{k_{nm}} = \begin{bmatrix} -45.0240 & -19.3069 & 21.1626 \end{bmatrix}^T,$$

and the H_∞ performance index is 3.4514.

Then, we design an H_∞ controller by Theorem 6.6 with $C_k = C_{k_{ini}}$.

Assume that the designed controller is with the additive interval type of uncertainties described by (6.3). For this case with $C_k = C_{k_{ini}}$, it is difficult to apply Theorem 6.5 to design a non-fragile H_∞ controller because the number of the LMI constraints involved in (6.44) is 2^{15}. However, Theorem 6.6 is applicable for solving this problem. By applying Theorem 6.6 with $\theta_a = 0.02$ and $k_i = i, i = 1, \cdots, 15$, that is, $s_a = 15$ as well as $k_i = 3i, i = 1, \cdots, 5$, that is, $s_a = 5$ to design a non-fragile controller, the obtained gains are as follows. For $s_a = 15$,

$$A_k = 10^7 \times \begin{bmatrix} -9.5028 & 9.4661 & -4.4679 \\ -3.3994 & 3.3863 & -1.5983 \\ 6.2383 & -6.2142 & 2.9330 \end{bmatrix},$$

$$B_k = 10^7 \times \begin{bmatrix} -4.6909 & -1.6781 & 3.0794 \end{bmatrix}^T.$$

For $s_a = 5$,

$$A_k = 10^8 \times \begin{bmatrix} -4.7744 & 4.7522 & -2.2444 \\ -1.7067 & 1.6988 & -0.8023 \\ 3.1388 & -3.1243 & 1.4755 \end{bmatrix},$$

$$B_k = 10^8 \times \begin{bmatrix} -2.3556 & -0.8421 & 1.5487 \end{bmatrix}^T.$$

Correspondingly, the H_∞ performance indexes of the obtained non-fragile controllers are $\gamma = 3.0598$ and $\gamma = 3.0578$, respectively.

Similarly, the gains of the above two controllers are too large to be applied in practice. Applying Theorem 6.6 with (6.53), and with $\alpha = 0.4$ and $\beta = 1000000$, the controller gains are reduced as follows.

For $s_a = 15$,

$$A_k = \begin{bmatrix} -83.1053 & 81.2701 & -37.0045 \\ -31.4088 & 24.1061 & -9.5972 \\ 56.8175 & -50.1947 & 20.7228 \end{bmatrix},$$

$$B_k = \begin{bmatrix} -39.8799 & -12.1129 & 25.4119 \end{bmatrix}^T.$$

For $s_a = 5$,

$$A_k = \begin{bmatrix} -82.9818 & 81.1005 & -36.9408 \\ -31.3580 & 24.0411 & -9.5724 \\ 56.7847 & -50.1297 & 20.7037 \end{bmatrix},$$

$$B_k = \begin{bmatrix} -39.8034 & -12.0838 & 25.3848 \end{bmatrix}^T.$$

The H_∞ performance indices of the obtained non-fragile controllers are $\gamma = 3.0913$ and $\gamma = 3.0898$, respectively. Compared with the above design without (6.53), the design results in a slight performance degradation, but the magnitude of A_k and B_k has been significantly reduced.

In the following part, for the above designed controllers, Lemma 6.4 can give better evaluations of the H_∞ performance index bounds.

First, to facilitate the presentation, denote the controller designed by condition (6.54) with (6.53) as $K_{nm\alpha\beta}$, while denoting the controllers designed by Theorem 6.6 with (6.53) as $K_{in15\alpha\beta}$ ($s_a = 15$) and $K_{in5\alpha\beta}$ ($s_a = 5$).

By Lemma 6.4, the H_∞ performance indices of the controller $K_{nm\alpha\beta}$ are $\gamma = 3.2945$ ($s_a = 15$) and $\gamma = 3.2940$ ($s_a = 5$), while the H_∞ performance indices of the controllers $K_{in15\alpha\beta}$ and $K_{in5\alpha\beta}$ are $\gamma = 3.0872$ ($s_a = 15$) and $\gamma = 3.0849$ ($s_a = 5$), respectively.

Then, tables are given to provide a comparison between the two methods.

First, Table 6.3 shows the H_∞ performance indices achieved by the designs directly.

From Table 6.3, it is easy to see that compared with the optimal H_∞ performance index bound $\gamma_{opt} = 2.7069$, the performance index of the controller designed by condition (6.54) is degraded 24.57%. The performance indices of

TABLE 6.3

Performance Index by Design with $\theta_a = 0.02$

	Condition (6.54)	Theorem 6.6 ($s_a = 15$)	Theorem 6.6 ($s_a = 5$)
γ	3.3719	3.0598	3.0578

TABLE 6.4

Performance Index by Design with $\theta_a = 0.02$, $\alpha = 0.4$, $\beta = 1000000$

	Condition (6.54)	Theorem 6.6 ($s_a = 15$)	Theorem 6.6 ($s_a = 5$)
γ	3.4514	3.0913	3.0898

the controllers designed by Theorem 6.6 are degraded 13.04% ($s_a = 15$) and 12.96% ($s_a = 5$), which are much more improved than 24.57%.

Then, Table 6.4 shows the H_∞ performance indices achieved by the design conditions with (6.53).

We can see from Table 6.4 that compared with $\gamma_{opt} = 2.7069$, the performance index of the controller designed by (6.54) with (6.53) is degraded 27.51%. The performance indices of the controllers designed by Theorem 6.6 with (6.53) are degraded 14.20% ($s_a = 15$) and 14.15% ($s_a = 5$).

Finally, for the designed non-fragile controllers, Lemma 6.4 gives better performance indices, shown in Table 6.5.

Obviously, compared with $\gamma_{opt} = 2.7069$, by using Lemma 6.4, the H_∞ performance indices of the controller $K_{nm\alpha\beta}$ are degraded 21.71% for $s_a = 15$ and 21.69% for $s_a = 5$. Correspondingly, the performance indices for the controllers $K_{in15\alpha\beta}$ and $K_{in5\alpha\beta}$ are degraded 14.05% for $s_a = 15$ and 13.96% for $s_a = 5$, respectively. The result shows the advantage of the new proposed method.

In this part, we will give a simulation to illustrate the effectiveness of the presented design method. Let $x(0) = \begin{bmatrix} -0.1 & 0.6 & -0.2 \end{bmatrix}$, $\xi(0) = \begin{bmatrix} -0.1 & 0.5 & 0.5 \end{bmatrix}$. And let the disturbance $\omega(t)$ be

$$\omega(t) = \begin{cases} 0.5, & 15 \le t \le 25 \text{ (second)} \\ 0 & \text{otherwise.} \end{cases}$$

Then Figure 6.3 shows the regulated output responses of z_1 controlled by the non-fragile controllers designed by condition (6.54) with (6.53) and Theorem

TABLE 6.5

Performance Evaluation by Lemma 6.4
with $\theta_a = 0.02$

	$K_{nm\alpha\beta}$	$K_{in15\alpha\beta}$	$K_{in5\alpha\beta}$
γ ($s_a = 15$)	3.2945	3.0872	—
γ ($s_a = 5$)	3.2940	—	3.0849

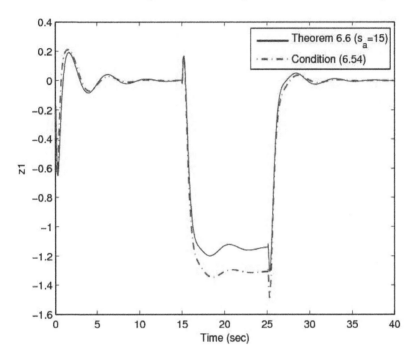

FIGURE 6.3
Comparison of the regulated output responses of $z_1(t)$ with $\theta_a = 0.02$.

6.6 with (6.53), and with $\alpha = 0.4$, $\beta = 1000000$. The responses of z_2 controlled by the two non-fragile controllers are the same, and omitted here.

From this example, we can see that the worst case ($s_a = 15$) of Theorem 6.6 also is more effective than the non-fragile H_∞ controller existence condition (6.54).

6.4   Non-Fragile H_∞ Controller Designs with Sparse Structures

6.4.1   Problem Statement

In this section, we define a class of sparse structures for controllers to be designed.

For system (6.38), consider the dynamic output feedback controller described as:

$$\dot{\xi}(t) = A_k\xi(t) + B_ky(t),$$
$$u(t) = C_k\xi(t), \tag{6.59}$$

where $\xi(t) \in R^n$ is the controller state, and A_k, B_k, and C_k are gain matrices of appropriate dimensions.

Then, a method for finding a class of feasible sparse structures from a given fully parameterized H_∞ controller is presented as follows.

By Lemma 2.7, without loss of generality, we may assume that A_k, B_k, and C_k satisfy the following assumption:

Assumption 6.1 *There exists a symmetric matrix $P > 0$ with the structure*
$P = \begin{bmatrix} Y & N \\ N & -N \end{bmatrix}$ *such that*

$$A_{e0}^T P + P A_{e0} + \frac{1}{\gamma^2} P B_{e0} B_{e0}^T P + C_{e0}^T C_{e0} < 0. \tag{6.60}$$

Moreover, for the designed fully parameterized H_∞ controller gains A_k, B_k, and C_k, the characteristic polynomial of A_k is described as

$$det(sI - A_k) = s^n + \alpha_{n-1}s^{n-1} + \ldots + \alpha_1 + \alpha_0, \tag{6.61}$$

where $\alpha_0, \alpha_1, \ldots, \alpha_{n-1}$ are scalars. Assume there exists a row vector c such that $Q = [(cA_k^{n-1})^T \cdots (cA_k)^T \ c^T]^T$ is nonsingular. Construct the following transformation matrix:

$$T = \begin{bmatrix} 1 & \alpha_{n-1} & \cdots & & \alpha_1 \\ & \ddots & \ddots & & \vdots \\ & & \ddots & & \alpha_{n-1} \\ \mathbf{0} & & & & 1 \end{bmatrix} Q. \tag{6.62}$$

Then we have that $\bar{A}_k, \bar{B}_k$, and $\bar{C}_k$ are with the following sparse structured form

$$\begin{aligned} \bar{A}_k &= TA_kT^{-1} = A_{kc} + f_A\phi, \\ \bar{B}_k &= TB_k, \quad \bar{C}_k = C_kT, \end{aligned} \tag{6.63}$$

where

$$A_{kc} = \begin{bmatrix} 0 & \cdots & & & 0 \\ 1 & & & & 0 \\ & \ddots & & & \vdots \\ \mathbf{0} & & & 1 & 0 \end{bmatrix}, \tag{6.64}$$

$$f_A = \begin{bmatrix} \alpha_0 & \alpha_1 & \cdots & \alpha_{n-1} \end{bmatrix}^T, \phi = \begin{bmatrix} 0 & \cdots & -1 \end{bmatrix}.$$

Hereafter, a controller described by

$$\dot{\bar{\xi}}(t) = \bar{A}_k \bar{\xi}(t) + \bar{B}_k y(t),$$
$$u(t) = \bar{C}_k \bar{\xi}(t), \qquad (6.65)$$

with structure (6.63) is said to be a sparse structured controller.

Remark 6.9 *When the gain matrices $\bar{A}_k$, $\bar{B}_k$, and $\bar{C}_k$ have the structure of (6.63), the nontrivial elements are only present in f_A, $\bar{B}_k$, and $\bar{C}_k$. So the number of additive uncertain parameters in the controller gains with any of the two structures is reduced from $n^2 + np + mn$ for full parameterized controller (6.59) to $n + np + mn$ for the sparse structured controller (6.65). Especially, when the vector c in the matrix Q is chosen from a row of C_k, denoted as C_{ki}, then $\bar{C}_{ki} = e_n^T$, that is, the elements of the ith row of $\bar{C}_k$ become trivial.*

Consider a sparse structured controller with gain variations described as:

$$\dot{\bar{\xi}}(t) = (\bar{A}_k + \Delta \bar{A}_k)\bar{\xi}(t) + (\bar{B}_k + \Delta \bar{B}_k)y(t),$$
$$u(t) = (\bar{C}_k + \Delta \bar{C}_k)\bar{\xi}(t), \qquad (6.66)$$

where $\bar{\xi}(t) \in R^n$ is the controller state, and $\bar{A}_k$, $\bar{B}_k$, and $\bar{C}_k$ are with the structure described by (6.63). The additive gain variations $\Delta \bar{A}_k$, $\Delta \bar{B}_k$, and $\Delta \bar{C}_k$ are with the following form:

$$\Delta \bar{A}_k = \Delta \bar{A}_{k_a} = [\theta_{a\bar{a}i}]_{n \times 1} v, |\theta_{a\bar{a}i}| \le \theta_a, i = 1, \cdots, n,$$
$$\Delta \bar{B}_k = \Delta \bar{B}_{k_a} = E_{Ba} diag[\theta_{a\bar{b}1}, \cdots, \theta_{a\bar{b}r_B}]E_{Bb}, |\theta_{a\bar{b}i}| \le \theta_a, i = 1, \cdots, r_B,$$
$$\Delta \bar{C}_k = \Delta \bar{C}_{k_a} = E_{Ca} diag[\theta_{a\bar{c}1}, \cdots, \theta_{a\bar{c}r_C}]E_{Cb}, |\theta_{a\bar{c}i}| \le \theta_a, i = 1, \cdots, r_C$$
$$(6.67)$$

where E_{Ba}, E_{Bb}, E_{Ca}, and E_{Cb} are constant matrices.

Remark 6.10 *For the additive case, the description of the gain variations in $\bar{B}_k$ and $\bar{C}_k$ given by (6.67) can cover the cases where $\bar{B}_k$ and $\bar{C}_k$ are with or without trivial elements.*

Applying controller (6.66) to system (6.38), the closed-loop system is illustrated by Figure 6.4 and given by

$$\dot{\bar{x}}_e(t) = \bar{A}_e \bar{x}_e(t) + \bar{B}_e w(t),$$
$$z(t) = \bar{C}_e \bar{x}_e(t), \qquad (6.68)$$

where $\bar{x}_e(t) = [x(t)^T, \bar{\xi}(t)^T]^T$, and

$$\bar{A}_e = \begin{bmatrix} A & B_2(\bar{C}_k + \Delta \bar{C}_k) \\ (\bar{B}_k + \Delta \bar{B}_k)C_2 & \bar{A}_k + \Delta \bar{A}_k \end{bmatrix},$$
$$\bar{B}_e = \begin{bmatrix} B_1 \\ (\bar{B}_k + \Delta \bar{B}_k)D_{21} \end{bmatrix}, \bar{C}_e = [C_1 \quad D_{12}(\bar{C}_k + \Delta \bar{C}_k)]. \qquad (6.69)$$

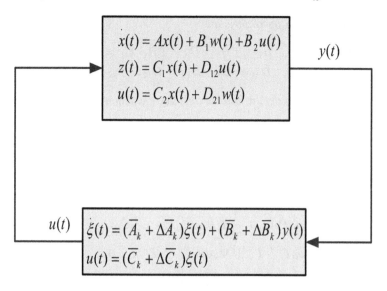

FIGURE 6.4
Control block diagram for Equation (6.68).

The transfer function matrix of the closed-loop system (6.68) from w to z is given by

$$\bar{T}_{zw} = \bar{C}_e(sI - \bar{A}_e)^{-1}\bar{B}_e.$$

Then the problem under consideration in this chapter is as follows.

Non-fragile H_∞ control problem with sparse structure: Given positive constants γ and θ_a, design a controller described by (6.66) with additive gain variations of form (6.67) such that the resulting system (6.68) is asymptotically stable and $\|\bar{T}_{zw}\| < \gamma$.

The following preliminaries and lemmas are required in this chapter.

Lemma 6.5 *Let the controller gain matrices A_k, B_k, and C_k satisfy Assumption 6.1, and T is defined by (6.62), then there exists a symmetric matrix $\bar{P} > 0$ with $\bar{P} = \begin{bmatrix} \bar{Y} & \bar{N} \\ \bar{N} & -\bar{N} \end{bmatrix}$ such that*

$$\bar{A}_{e0}^T\bar{P} + \bar{P}\bar{A}_{e0} + \frac{1}{\gamma^2}\bar{P}\bar{B}_{e0}\bar{B}_{e0}^T\bar{P} + \bar{C}_{e0}^T\bar{C}_{e0} < 0, \tag{6.70}$$

where

$$\bar{A}_{e0} = \begin{bmatrix} \bar{A} & \bar{B}_2\bar{C}_k \\ \bar{B}_k\bar{C}_2 & \bar{A}_k \end{bmatrix},$$

$$\bar{B}_{e0} = \begin{bmatrix} \bar{B}_1 \\ \bar{B}_k\bar{D}_{21} \end{bmatrix}, \bar{C}_{e0} = \begin{bmatrix} \bar{C}_1 & \bar{D}_{12}\bar{C}_k \end{bmatrix}, \tag{6.71}$$

with $\bar{A}_k$, $\bar{B}_k$, and $\bar{C}_k$ given by (6.63), and

$$\bar{A} = TAT^{-1}, \quad \bar{B}_1 = TB_1, \quad \bar{B}_2 = TB_2 \quad \bar{C}_1 = C_1T^{-1},$$
$$\bar{C}_2 = C_2T^{-1}, \quad \bar{D}_{21} = D_{21} \quad \bar{D}_{12} = D_{12}. \tag{6.72}$$

Proof 6.6 *By Assumption 6.1, it follows that there exists a symmetric matrix* $P > 0$ *with* $P = \begin{bmatrix} Y & N \\ N & -N \end{bmatrix}$ *such that (6.60) holds. Let*

$$\bar{P} = diag[T^{-1} \ T^{-1}]^T P \, diag[T^{-1} \ T^{-1}],$$

then from (6.60), it follows that

$$\bar{A}_{e0}^T \bar{P} + \bar{P} \bar{A}_{e0} + \frac{1}{\gamma^2} \bar{P} \bar{B}_{e0} \bar{B}_{e0}^T \bar{P} + \bar{C}_{e0}^T \bar{C}_{e0}$$
$$= diag[T^{-1} \ T^{-1}]^T \Psi_1 \, diag[T^{-1} \ T^{-1}] < 0,$$

where $\Psi_1 = A_{e0}^T P + P A_{e0} + \frac{1}{\gamma^2} P B_{e0} B_{e0}^T P + C_{e0}^T C_{e0}$.
Thus, the proof is complete.

Lemma 6.6 *Let* γ *and* θ_a *be given positive constants. If there exists a symmetric matrix* $\bar{P} > 0$ *with* $\bar{P} = \begin{bmatrix} \bar{Y} & \bar{N} \\ \bar{N} & -\bar{N} \end{bmatrix}$ *such that*

$$\bar{A}_{ea}^T \bar{P} + \bar{P} \bar{A}_{ea} + \frac{1}{\gamma^2} \bar{P} \bar{B}_{ea} \bar{B}_{ea}^T \bar{P} + \bar{C}_{ea}^T \bar{C}_{ea} < 0 \tag{6.73}$$

holds for all $\Delta \bar{A}_k$, $\Delta \bar{B}_k$, *and* $\Delta \bar{C}_k$ *satisfying form (6.67), where*

$$\bar{A}_{ea} = \begin{bmatrix} \bar{A} & \bar{B}_2(\bar{C}_k + \Delta \bar{C}_k) \\ (\bar{B}_k + \Delta \bar{B}_k)\bar{C}_2 & \bar{A}_k + \Delta \bar{A}_k \end{bmatrix},$$

$$\bar{B}_{ea} = \begin{bmatrix} \bar{B}_1 \\ (\bar{B}_k + \Delta \bar{B}_k)\bar{D}_{21} \end{bmatrix}, \quad \bar{C}_{ea} = \begin{bmatrix} \bar{C}_1 & \bar{D}_{12}(\bar{C}_k + \Delta \bar{C}_k) \end{bmatrix}, \tag{6.74}$$

$\bar{A}$, $\bar{B}_1$, $\bar{B}_2$, $\bar{C}_1$, $\bar{C}_2$, $\bar{D}_{12}$, *and* $\bar{D}_{21}$ *are defined by (6.72). Then the controller described by (6.66) with additive gain variations (6.67) solves the non-fragile* H_∞ *control problem with sparse structure for system (6.38).*

Proof 6.7 *Let* $\bar{P}_s = diag[T \ I]^T \bar{P} diag[T \ I]$. *According to (6.71)–(6.74), it follows that*

$$\bar{A}_e^T \bar{P}_s + \bar{P}_s \bar{A}_e + \frac{1}{\gamma^2} \bar{P}_s \bar{B}_e \bar{B}_e^T \bar{P}_s + \bar{C}_e^T \bar{C}_e$$
$$= diag[T \ I]^T \Psi_2 \, diag[T \ I] < 0,$$

where $\Psi_2 = \bar{A}_{ea}^T \bar{P} + \bar{P} \bar{A}_{ea} + \frac{1}{\gamma^2} \bar{P} \bar{B}_{ea} \bar{B}_{ea}^T \bar{P} + \bar{C}_{ea}^T \bar{C}_{ea}$ *and* $\bar{A}_e$, $\bar{B}_e$, $\bar{C}_e$ *are defined by (6.69). Then, by Lemma 2.6, the conclusion follows.*

6.4.2 Sparse Structured Controller Design

In this section, a three-step procedure for designing non-fragile H_∞ controllers with the sparse structure with respect to additive gain uncertainties is presented.

First, we will give a method for designing a non-fragile H_∞ controller with a sparse structure (6.63) and with additive gain variations (6.67) under the assumption that the controller gain $\bar{C}_k$ is known.

To facilitate the presentation of Theorem 6.7, we denote

$$M_{0s}(\Delta\bar{A}_k, \Delta\bar{B}_k, \Delta\bar{C}_k)) = \begin{bmatrix} M_{0sa} + M_{0sa}^T & M_{0sb} & M_{0sc}^T \\ * & -\gamma^2 I & 0 \\ * & * & -I \end{bmatrix}, \qquad (6.75)$$

where

$$M_{0as} = \begin{bmatrix} \bar{S}_a(\bar{A} + \bar{B}_{2e}e_n^T) + \bar{S}_a\bar{B}_{2r}(\bar{C}_{kr} + \Delta\bar{C}_{kr}) \\ (\bar{S}_a - \bar{N}_a)(\bar{A} + \bar{B}_{2e}e_n^T) + (\bar{F}_B + \bar{N}_a\Delta\bar{B}_k)\bar{C}_2 + \Phi_1 \\ \bar{S}_a\bar{A} \\ (\bar{S}_a - \bar{N}_a)\bar{A} + (\bar{F}_B + \bar{N}_a\Delta\bar{B}_k)\bar{C}_2 \end{bmatrix},$$

$$M_{0bs} = \begin{bmatrix} \bar{S}_a\bar{B}_1 \\ (\bar{S}_a - \bar{N}_a)\bar{B}_1 + (\bar{F}_B + \bar{N}_a\Delta\bar{B}_k)\bar{D}_{21} \end{bmatrix},$$

$$M_{0cs} = \begin{bmatrix} \bar{C}_1 + \bar{D}_{12e}e_n^T + \bar{D}_{12r}\bar{C}_{kr} + \bar{D}_{12r}\Delta\bar{C}_{kr} & \bar{C}_1 \end{bmatrix}.$$

Here, $\Phi_1 = \bar{N}_a A_{kc} + (\bar{F}_A + \bar{N}_a\Delta f_A)\phi + (\bar{S}_a - \bar{N}_a)\bar{B}_{2r}(\bar{C}_{kr} + \Delta\bar{C}_{kr})$ and $\bar{A}$, $\bar{B}_1$, $\bar{B}_2$, $\bar{C}_1$, $\bar{C}_2$, $\bar{D}_{12}$, and $\bar{D}_{21}$ are defined by (6.72) with T defined by (6.62).

Then, by using Lemma 6.6, the following theorem presents a sufficient condition for the solvability of the non-fragile H_∞ control problem with sparse structure (6.63).

Theorem 6.7 *Consider system (6.38). Let $\gamma > 0$ and $\theta_a > 0$ be given constants. If there exist matrices $\bar{F}_A$, $\bar{F}_B$, $\bar{S}_a > 0$, and $\bar{N}_a < 0$ such that the following condition holds:*

$$M_{0s}(\Delta\bar{A}_{k_a}, \Delta\bar{B}_{k_a}, \Delta\bar{C}_{k_a}) < 0,$$
$$\text{for all } \theta_{a\bar{a}i}, \; \theta_{a\bar{b}j}, \; \theta_{a\bar{c}r} \in \{-\theta_a, \theta_a\}, \qquad (6.76)$$
$$i = 1, \cdots, n; \; j = 1, \cdots, r_B, r = 1, \cdots, r_C$$

then controller (6.66) with the gain variations described by (6.67), and

$$\bar{A}_k = A_{kc} + \bar{N}_a^{-1}\bar{F}_A\phi, \; \bar{B}_k = \bar{N}_a^{-1}\bar{F}_B, \; \bar{C}_k \qquad (6.77)$$

solves the non-fragile H_∞ control problem with sparse structure for system (6.38).

Proof 6.8 *By Lemma 6.6, it is sufficient to show that there exists a symmetric matrix* $\bar{P} > 0$ *with structure (2.14), namely* $\bar{P} = \begin{bmatrix} \bar{Y} & \bar{N}_a \\ \bar{N}_a & -\bar{N}_a \end{bmatrix}$ *such that*

$$M_1 = \bar{P}\bar{A}_{ea} + \bar{A}_{ea}^T\bar{P} + \frac{1}{\gamma^2}\bar{P}\bar{B}_{ea}\bar{B}_{ea}^T\bar{P} + \bar{C}_{ea}^T\bar{C}_{ea} < 0 \qquad (6.78)$$

holds for all gain variations satisfying (6.67), where $\bar{A}_{ea}$, $\bar{B}_{ea}$, *and* $\bar{C}_{ea}$ *are defined by (6.74). Denote* $\bar{S}_a = \bar{Y} + \bar{N}_a$, $\Gamma_1 = \begin{bmatrix} I & I \\ I & 0 \end{bmatrix}$, *then* $\bar{P} > 0$ *with the structure described by (2.14) is equivalent to* $\bar{S}_a > 0$ *and* $\bar{N}_a < 0$. *Furthermore,* Γ_1 *is nonsingular for* $I > 0$, *and (6.78) is equivalent to*

$$
\begin{aligned}
M_2 &= \Gamma_1 M_1 \Gamma_1^T \\
&= M_{2a} + M_{2a}^T + \frac{1}{\gamma^2}M_{2b}M_{2b}^T + M_{2c}^T M_{2c} < 0,
\end{aligned}
\qquad (6.79)
$$

where

$$M_{2a} = \begin{bmatrix} \bar{S}_a\bar{A} + \bar{S}_a\bar{B}_{2e}e_n^T + \bar{S}_a\bar{B}_{2r}(\bar{C}_{kr} + \Delta\bar{C}_{kr}) \\ \bar{N}_a(\bar{B}_k + \Delta\bar{B}_k)\bar{C}_2 + (\bar{S}_a - \bar{N}_a)\bar{A} + \Phi_2 \\ \bar{S}_a\bar{A} \\ (\bar{S}_a - \bar{N}_a)\bar{A} + \bar{N}_a(\bar{B}_k + \Delta\bar{B}_k)\bar{C}_2 \end{bmatrix},$$

$$M_{2b} = \begin{bmatrix} \bar{S}_a\bar{B}_1 \\ (\bar{S}_a - \bar{N}_a)\bar{B}_1 + \bar{N}_a(\bar{B}_k + \Delta\bar{B}_k)\bar{D}_{21} \end{bmatrix},$$

$$M_{2c} = \begin{bmatrix} \bar{C}_1 + \bar{D}_{12e}e_n^T + \bar{D}_{12r}(\bar{C}_{kr} + \Delta\bar{C}_{kr}) & \bar{C}_1 \end{bmatrix}$$

with $\Phi_2 = \bar{N}_a(A_{kc} + f_A\phi + \Delta f_A\phi) + (\bar{S}_a - \bar{N}_a)\bar{B}_{2e}e_n^T + (\bar{S}_a - \bar{N}_a)\bar{B}_{2r}(\bar{C}_{kr} + \Delta\bar{C}_{kr})$.

Then, by (6.75), (6.77), and the Schur complement, it follows that (6.79) is equivalent to (6.76) with respect to additive gain variations (6.67). Thus, the proof is complete.

Remark 6.11 *In Theorem 6.7, convex sufficient conditions for the solvability of the non-fragile* H_∞ *control problem with sparse structure (6.63) are given in terms of solutions to a set of LMIs. In fact, the result provides an LMI-based method for designing non-fragile* H_∞ *controllers with the sparse structure from a given fully parameterized* H_∞ *controller. When the designed controller contains no gain variations, from Lemma 2.7, it follows that the design condition given in Theorem 6.7 reduces to a necessary and sufficient condition for the standard* H_∞ *control problem, which means that the structure constraint (2.14) on Lyapunov function matrices does not result in any conservativeness for the standard* H_∞ *controller design.*

Here, we will show that the problem of the special case of single-input–single-output (SISO) is a convex one.

Consider the designed controller with gain matrices described by (6.63) and with additive gain variations (6.67).

To facilitate the presentation, denote

$$
M_{0s}(\Delta \bar{A}_k, \Delta \bar{B}_k, \Delta \bar{C}_k)) = \begin{bmatrix} M_{0sa} + M_{0sa}^T & M_{0sb} & M_{0sc}^T \\ * & -\gamma^2 I & 0 \\ * & * & -I \end{bmatrix}, \quad (6.80)
$$

where

$$
M_{0as} = \begin{bmatrix} \bar{S}_a \bar{A} + \bar{S}_a \bar{B}_2 (\bar{C}_k + \Delta \bar{C}_k) \\ (\bar{S}_a - \bar{N}_a) \bar{A} + (\bar{F}_B + \bar{N}_a \Delta \bar{B}_k) \bar{C}_2 + \Phi_1 \end{bmatrix}
$$

$$
\begin{bmatrix} \bar{S}_a \bar{A} \\ (\bar{S}_a - \bar{N}_a) \bar{A} + (\bar{F}_B + \bar{N}_a \Delta \bar{B}_k) \bar{C}_2 \end{bmatrix},
$$

$$
M_{0bs} = \begin{bmatrix} \bar{S}_a \bar{B}_1 \\ (\bar{S}_a - \bar{N}_a) \bar{B}_1 + (\bar{F}_B + \bar{N}_a \Delta \bar{B}_k) \bar{D}_{21} \end{bmatrix},
$$

$$
M_{0cs} = \begin{bmatrix} \bar{C}_1 + \bar{D}_{12} (\bar{C}_k + \Delta \bar{C}_k) & \bar{C}_1. \end{bmatrix},
$$

Here, $\Phi_1 = \bar{N}_a A_{kc} + (\bar{F}_A + \bar{N}_a \Delta f_A)\phi + (\bar{S}_a - \bar{N}_a)\bar{B}_2(\bar{C}_k + \Delta \bar{C}_k)$ and $\bar{A}, \bar{B}_1, \bar{B}_2, \bar{C}_1, \bar{C}_2, \bar{D}_{12}$, and $\bar{D}_{21}$ are defined by (6.72) with T defined by (6.62).

Then, according to Theorem 6.7, the following corollary presents a sufficient condition for the solvability of the non-fragile H_∞ control problem with the sparse structure (6.63).

Corollary 6.1 *Consider system (6.38). Let $\gamma > 0$ and $\theta_a > 0$ be given constants. If there exist matrices $\bar{F}_A$, $\bar{F}_B$, $\bar{S}_a > 0$, and $\bar{N}_a < 0$ such that the following conditions hold:*

$$
\begin{aligned}
& M_{0s}(\Delta \bar{A}_{k_a}, \Delta \bar{B}_{k_a}, \Delta \bar{C}_{k_a}) < 0, \\
& \text{for all } \theta_{a\bar{a}i}, \ \theta_{a\bar{b}j} \in \{-\theta_a, \theta_a\}, \\
& i = 1, \cdots, n; \ j = 1, \cdots, r_B,
\end{aligned} \quad (6.81)
$$

then the controller (6.66) with the gain variations described by form (6.67), and

$$
\begin{aligned}
\bar{A}_k &= A_{kc} + \bar{N}_a^{-1} \bar{F}_A \phi, \\
\bar{B}_k &= \bar{N}_a^{-1} \bar{F}_B, \bar{C}_k = \begin{bmatrix} 0 & \cdots & 0 & 1 \end{bmatrix}
\end{aligned} \quad (6.82)
$$

solves the non-fragile H_∞ control problem with sparse structure for system (6.38).

Remark 6.12 *For SISO systems, the problem of non-fragile dynamic output feedback H_∞ controller design with the sparse structure can be converted to a convex one. In fact, if we construct the transformation matrix T by using A_k and C_k, we can obtain the sparse structured controller with $\bar{C}_k = \begin{bmatrix} 1 & 0 & \cdots & 0 \end{bmatrix}$, namely, $\bar{C}_k$ is known. So, by Theorem 6.7, the design conditions are given directly.*

Due to the fact that the design method given by Theorem 6.7 is based on Assumption 6.1, we need to give methods for designing H_∞ controllers satisfying Assumption 6.1. First, the following standard H_∞ controller design method [113] is introduced.

Lemma 6.7 *[113] Consider system (6.38). Let $\gamma > 0$ be a given scalar. If there exist matrices $\hat{A}, \hat{B}, \hat{C}, X > 0$, and $Y > 0$ such that the following LMI holds:*

$$\begin{bmatrix} AX + XA^T + B_2\hat{C} + (B_2\hat{C})^T & \hat{A}^T + A & \\ * & A^TY + YA + \hat{B}C_2 + (\hat{B}C_2)^T & \\ * & * & \\ * & * & \\ B_1 & (C_1X + D_{12}\hat{C})^T & \\ YB_1 + \hat{B}D_{21} & C_1^T & \\ -\gamma^2 I & 0 & \\ * & -I & \end{bmatrix} < 0,$$

(6.83)

then controller (6.59) with

$$\begin{aligned} A_k &= N^{-1}(\hat{A} - YAX - \hat{B}C_2X - YB_2\hat{C})M^{-1}, \\ B_k &= N^{-1}\hat{B}, \quad C_k = \hat{C}M^{-1} \end{aligned}$$

(6.84)

(2.19) holds, where $MN^T = I - XY$.

Then, by using Lemma 6.6, we present a design method to design H_∞ controllers satisfying Assumption 6.1.

Lemma 6.8 *In Lemma 6.6, let $M = X$, then controller (6.59) with*

$$\begin{aligned} A_k &= (X^{-1} - Y)^{-1}(\hat{A} - YAX - \hat{B}C_2X - YB_2\hat{C})X^{-1}, \\ B_k &= (X^{-1} - Y)^{-1}\hat{B}, \quad C_k = \hat{C}X^{-1} \end{aligned}$$

(6.85)

(6.60) holds with $P = \begin{bmatrix} Y & X^{-1} - Y \\ X^{-1} - Y & -(X^{-1} - Y) \end{bmatrix}$.

Proof 6.9 *Let $M = X$, then, by using the arguments developed in Scherer, Gahinet, and Chilali [113], the conclusion follows.*

Remark 6.13 *Lemma 6.8 presents a method of designing H_∞ controllers for satisfying Assumption 6.1, which is the initial step for the following algorithm.*

Based on Lemma 6.7 and Lemma 6.8, the following algorithm is presented to solve the non-fragile H_∞ control problem with sparse structures described by (6.63).

Algorithm 6.3 *Let $\gamma > 0$ be a given scalar.*

Step 1. *Minimize γ subject to LMIs $X > 0, Y > 0$, and (6.83). Denote the optimal solutions as $X = X_{opt}, Y = Y_{opt}, \hat{A} = \hat{A}_{opt}, \hat{B} = \hat{B}_{opt}$, and $\hat{C} = \hat{C}_{opt}$. Substitute the matrices $(X_{opt}, Y_{opt}, \hat{A}_{opt}, \hat{B}_{opt}, \hat{C}_{opt})$ to (6.85), compute*

$$A_k = (X_{opt}^{-1} - Y_{opt})^{-1}(\hat{A} - Y_{opt}AX_{opt} - \hat{B}_{opt}C_2X_{opt} - Y_{opt}B_2\hat{C}_{opt})X_{opt}^{-1},$$

$$B_k = (X_{opt}^{-1} - Y_{opt})^{-1}\hat{B}_{opt}, \quad C_k = \hat{C}_{opt}X_{opt}^{-1},$$

then go to Step 2.

Step 2. *Combining A_k with the a row vector c (it can be a row vector C_{ki} of C_k) such that $Q = [(cA_k^{n-1})^T \cdots (cA_k)^T c^T]^T$ is nonsingular, by using (6.62), we construct a transformation matrix T. Then, compute $\bar{A}, \bar{B}_1, \bar{B}_2, \bar{C}_1, \bar{C}_2, \bar{D}_{12}$, and $\bar{D}_{21}$ according to (6.72), and go to Step 3.*

Step 3. *Let $\bar{C}_k = C_kT^{-1}$. Minimize γ subject to $\bar{F}_A, \bar{F}_B, \bar{N}_a < 0, \bar{S}_a > 0$, and LMI (6.76) for additive gain variations. Denote the optimal solutions as $\bar{N}_a = \bar{N}_{aopt}, F_A = F_{Aopt}$, and $F_B = F_{Bopt}$. Then, according to (6.77),*

$$\bar{A}_k = A_{kc} + \bar{N}_{aopt}^{-1}\bar{F}_{Aopt}\phi, \bar{B}_k = \bar{N}_{aopt}^{-1}\bar{F}_{Bopt}, \bar{C}_k = \bar{C}_k.$$

The resulting $\bar{A}_k, \bar{B}_k$, and $\bar{C}_k$ form the sparse structured non-fragile H_∞ controller gains.

Remark 6.14 *Algorithm 6.3 gives a method of designing the non-fragile H_∞ controller with sparse structure described by (6.63). In Step 1, an H_∞ controller satisfying Assumption 6.1 is designed, followed by determining the sparse structure, and then a non-fragile H_∞ controller with the sparse structure is obtained. In order to clarify the algorithm, Figure 6.5 is given to illustrate the flowchart of Algorithm 6.3.*

For obtaining the convex design conditions, we restrict the Lyapunov function matrix P with structure (2.14) in Theorem 6.7, which may result in a more conservative evaluation of the H_∞ performance bounds. So, in this section, for a designed controller with the sparse structure, the Lyapunov matrix P without the restriction is exploited for obtaining a less conservative evaluation of the H_∞ performance bounds.

When the controller parameter matrices $\bar{A}_k, \bar{B}_k$, and $\bar{C}_k$ are known, the following lemma presents sufficient conditions for measuring the non-fragile H_∞ performance of system (6.68).

Denote

$$M_{0ss}(\Delta\bar{A}_k, \Delta\bar{B}_k, \Delta\bar{C}_k) = \begin{bmatrix} P\bar{A}_e + \bar{A}_e^T P & P\bar{B}_e & \bar{C}_e^T \\ * & -\gamma^2 I & 0 \\ * & * & -I \end{bmatrix}, \qquad (6.86)$$

where $\bar{A}_e, \bar{B}_e$, and $\bar{C}_e$ are defined by (6.69).

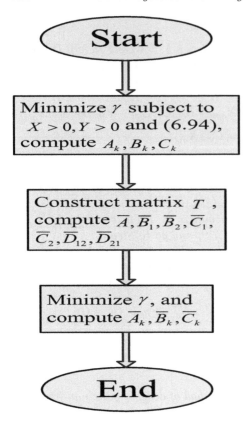

FIGURE 6.5
Flowchart of Algorithm 6.3.

Lemma 6.9 *Consider system (6.38). Let $\gamma > 0$, $\theta_a > 0$ be constants, and let controller parameter matrices $\bar{A}_k$, $\bar{B}_k$, and $\bar{C}_k$ be given. Then $\| \bar{T}_{zw} \| < \gamma$ holds for all gain variations described by (6.67), if there exists a symmetric matrix $P > 0$ such that the following condition holds:*

$$M_{0ss}(\Delta \bar{A}_{k_a}, \Delta \bar{B}_{k_a}, \Delta \bar{C}_{k_a}) < 0,$$
$$\text{for all }\ \theta_{a\bar{a}i},\ \theta_{a\bar{b}j},\ \in \{-\theta_a, \theta_a\}$$
$$i = 1, \cdots, n;\ j = 1, \cdots, r_B. \tag{6.87}$$

6.4.3   Example

In this section, we will illustrate the effectiveness of our design methods of the non-fragile H_∞ controller with the sparse structure.

Example 6.1 *Consider a linear system of form (6.38) with*

$$A = \begin{bmatrix} 0 & 1 & 0 \\ 0 & -1 & -1 \\ 0 & 1 & -1 \end{bmatrix}, B_1 = \begin{bmatrix} -1 & 0 \\ 0.5 & 0 \\ 1.5 & 0 \end{bmatrix}, B_2 = \begin{bmatrix} 1 \\ 2 \\ -2.5 \end{bmatrix},$$

$$C_1 = \begin{bmatrix} 1 & -1 & -3 \\ 0 & 0 & 0 \end{bmatrix}, C_2 = \begin{bmatrix} -3 & 2 & -1 \end{bmatrix},$$

$$D_{21} = \begin{bmatrix} 0 & 0.5 \end{bmatrix}, D_{12} = \begin{bmatrix} 0 & 1 \end{bmatrix}^T.$$

First, by Step 1 of Algorithm 6.3, we obtain the H_∞ controller K_{so} with gains as

$$A_k = \begin{bmatrix} -20.0503 & 13.3586 & -2.7130 \\ -4.9211 & 1.3024 & 4.8168 \\ 26.2416 & -14.4317 & -1.8096 \end{bmatrix},$$

$$B_k = \begin{bmatrix} -6.0622 & -0.4246 & 7.2303 \end{bmatrix}^T,$$

$$C_k = \begin{bmatrix} -1.7241 & 0.6165 & 2.9543 \end{bmatrix},$$

and the optimal H_∞ performance is $\gamma_{opt} = 2.9246$.
It is easy to see that the matrix $Q = [(C_k A_k^{n-1})^T \cdots (C_k A_k)^T C_k^T]^T$ is nonsingular, so in Step 2 of Algorithm 6.3, let the row vector $c = C_k$. The transformation matrix is obtained as

$$T_o = \begin{bmatrix} 65.4937 & 137.8543 & 67.8067 \\ 73.6172 & -52.1905 & 63.0340 \\ -1.7241 & 0.6165 & 2.9543 \end{bmatrix}.$$

Then, by Step 3, when the designed sparse structured controller is assumed to be with additive uncertainties described by (6.67), by solving (6.76) with $\theta_a = 0.65$, the non-fragile sparse structured H_∞ controller K_{noa} is obtained with gains as:

$$\bar{A}_k = \begin{bmatrix} 0 & 0 & 56.8811 \\ 1 & 0 & -261.3554 \\ 0 & 1 & -40.7364 \end{bmatrix}, \bar{C}_k = \begin{bmatrix} 0 & 0 & 1 \end{bmatrix},$$

$$\bar{B}_k = \begin{bmatrix} 42.6526 & 37.2199 & 33.9007 \end{bmatrix}^T,$$

and the H_∞ performance index is 3.4547.
On the other hand, for the designed controller K_{so} in Step 1 with the optimal H_∞ performance $\gamma_{opt} = 2.9246$, according to (6.63) and transforming K_{so} with the transformation matrix T_o, we obtain the sparse structured H_∞ controller K_{sol} directly with the gains as

$$\bar{A}_k = \begin{bmatrix} 0 & 0 & 123.0840 \\ 1 & 0 & -214.2598 \\ 0 & 1 & -20.5575 \end{bmatrix}, \bar{C}_k = \begin{bmatrix} 0 & 0 & 1 \end{bmatrix},$$

TABLE 6.6

Performance Evaluation of
Lemma 6.9 with $\theta_a = 0.65$

Controller	K_{noa}	K_{sol}
γ	3.2643	4.2225

TABLE 6.7

H_∞ Performance Level for Variant Values of θ_a

θ_a	0.1	0.2	0.3	0.4	0.55	0.60	0.65
γ	2.9721	3.0516	3.1377	3.2268	3.3631	3.4089	3.4547

$$\bar{B}_k = \begin{bmatrix} 34.6939 & 31.6321 & 31.5505 \end{bmatrix}^T.$$

In this part, for the above designed controllers, Lemma 6.9 gives better evaluations of the H_∞ performance index bounds, which are compared and shown in the above tables.

For the sparse structured controllers K_{noa} (obtained by our design Algorithm 6.3) and K_{sol} (obtained by transformation with T_o directly), with $\theta_a = 0.65$, Lemma 6.9 gives better evaluations shown in Table 6.6.

Obviously, compared with $\gamma_{opt} = 2.9246$, the H_∞ performance index of the controller K_{noa} is degraded 11.62%, while the performance index of the controller K_{sol} is degraded 44.38%.

Then, a simulation is given to illustrate the effectiveness of the design method. Let $x(0) = [0.8 \ -0.6 \ 0.4]^T, \xi(0) = [0.3 \ -0.5 \ 0.8]^T$. And let the disturbance $w(t)$ be

$$w(t) = \begin{cases} \begin{bmatrix} 1 \\ 1 \end{bmatrix}, & 40 \leq t \leq 40.5 \ (second), \\ 0, & otherwise. \end{cases}$$

Then Figure 6.6 shows the regulated output responses of z_1 controlled by the sparse structured controller K_{sol} (obtained by transformation with T_o directly) and the non-fragile sparse structured controller K_{noa} (obtained by our design Algorithm 6.3) with $\theta_a = 0.65$, respectively. The responses of z_2 controlled by the two controllers with $\theta_a = 0.65$ are shown in Figure 6.7. From the two figures, we can see the superiority of our proposed methods.

In the following, for variant values of θ_a, we obtain variant H_∞ performance levels by using our proposed design method as shown in Table 6.7. Figure 6.8 further presents the relationship between the H_∞ performance level and the value of θ_a clearly. It can be seen that, the larger the value of θ_a is, the worse the H_∞ performance level is.

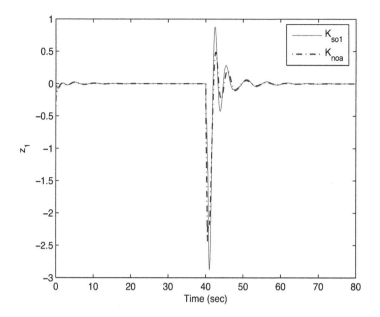

FIGURE 6.6
Responses of $z_1(t)$ with $\theta_a = 0.65$.

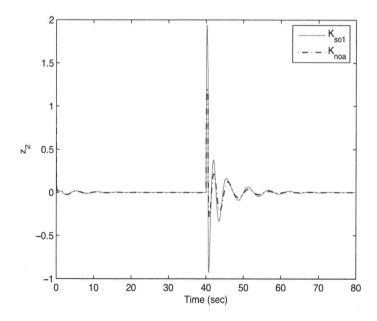

FIGURE 6.7
Responses of $z_2(t)$ with $\theta_a = 0.65$.

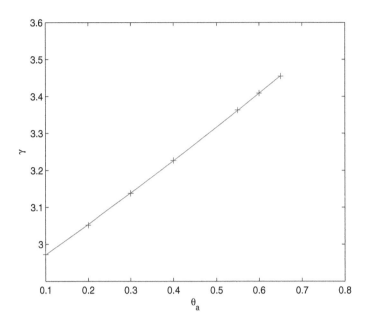

FIGURE 6.8

Response of H_∞ performance (γ) with respect to θ_a.

6.5 Conclusion

The full parameterized and sparse structured non-fragile H_∞ controller design problems have been investigated in this chapter. The controller to be designed is assumed to be with the additive gain variations of interval type, which are due to the FWL effects when the controller is implemented. For the full parameterized controller design problem, we consider the discrete-time and continuous-time systems, respectively. And a two-step procedure is adopted to solve this non-convex problem. In addition, the notion of a structured vertex separator is proposed to approach the numerical computational problem resulting from the interval type of gain uncertainties, and exploited to develop sufficient conditions for the non-fragile H_∞ controller design in terms of solutions to a set of LMIs. For the sparse structured controller design problem, a class of sparse structures is specified. Then, a three-step procedure for non-fragile H_∞ controller design under the restriction of the sparse structure is provided. The contribution of this method is that it not only reduces the number of nontrivial parameters but also designs the sparse structured controllers with non-fragility. The resulting designs of the two cases guarantee that the closed-loop system is asymptotically stable and the H_∞ performance from the disturbance to the regulated output is less than a prescribed level. Numerical examples are given to illustrate the effectiveness of the proposed design methods.

7

Non-Fragile H_∞ Filtering with Interval-Bounded Coefficient Variations

7.1 Introduction

The previous chapter studied the *non-fragile* controller design problem with the *additive* interval-bounded coefficient variations consideration. Similar to the control problem, the analysis and design problems of H_∞ filtering have been investigated extensively over the last two decades and many approaches have been developed [see, for example, 29, 38, 41, 138, and references therein]. An implicit assumption in the above mentioned works is that the filter will be implemented exactly. However, similar to the controller implementation, in the course of filter implementation based on different design algorithms, it also turns out that the filters can be sensitive with respect to errors in the filter coefficients [27, 147]. Also, due to the requirement of computational efficiency in real-time applications, it is highly desirable for a filter to have a sparse structure, which contains many trivial parameters.

In this chapter, the *full parameterized* and *sparse structured* non-fragile filtering problems with the additive interval type of uncertainties are considered, respectively. For the full parameter filter design, first, a *linear matrix inequality* (LMI)-based sufficient condition is given for the solvability of the non-fragile H_∞ filtering problem, which requires checking all of the vertices of the set of uncertain parameters that grows exponentially with the number of uncertain parameters. It will be difficult to apply the result to systems with high orders. To overcome the difficulty, the structured vertex separator proposed in the previous chapter is exploited to develop sufficient conditions for the *non-fragile H_∞ filter* design in terms of solutions to a set of LMIs. For the sparse structured filter design, first, a class of sparse structures is specified from a given filter, which renders the augmented system to be *asymptotically stable* and meets an H_∞ performance requirement, but it is fully parameterized. Then, an LMI-based procedure for non-fragile H_∞ filter design under the restriction of the *sparse structure* is provided. It is worth mentioning that the method proposed here not only reduces the number of nontrivial parameters but also designs the *sparse structured filters* with non-fragility. The resulting design guarantees that the augmented system is *asymptotically stable* and the

H_∞ performance of the system from the exogenous signals to the estimation errors is less than a prescribed level.

7.2 Non-Fragile H_∞ Filtering for Discrete-Time Systems

7.2.1 Problem Statement

Consider a linear time invariant (LTI) discrete-time model described by

$$\begin{aligned}
x(k+1) &= Ax(k) + B_1\omega(k),\\
z(k) &= C_1 x(k),\\
y(k) &= C_2 x(k) + D_{21}\omega(k),
\end{aligned} \tag{7.1}$$

where $x(k) \in R^n$ is the state, $y(k) \in R^p$ is the measured output, $\omega(k) \in R^r$ is the disturbance input, and $z(k) \in R^q$ is the regulated output, respectively. A, B_1, C_1, C_2, and D_{21} are known constant matrices of appropriate dimensions.

To cope with the filtering problem, we consider a discrete-time filter with *gain variations* of the following form:

$$\begin{aligned}
\xi(k+1) &= (A_F + \Delta A_F)\xi(k) + (B_F + \Delta B_F)y(k),\\
z_F(k) &= (C_F + \Delta C_F)\xi(k),
\end{aligned} \tag{7.2}$$

where $\xi(k) \in R^n$ is the filter state, $z_F(k)$ is the estimation of $z(k)$, and the constant matrices A_F, B_F, and C_F are filter matrices to be designed. $\Delta A_F, \Delta B_F$, and ΔC_F represent the interval type of additive gain variations with the following form:

$$\begin{aligned}
\Delta A_F &= [\theta_{aij}]_{n\times n}, |\theta_{aij}| \le \theta_a, i, j = 1, \cdots, n,\\
\Delta B_F &= [\theta_{bij}]_{n\times p}, |\theta_{bij}| \le \theta_a, i = 1, \cdots, n, j = 1, \cdots, p,\\
\Delta C_F &= [\theta_{cij}]_{q\times n}, |\theta_{cij}| \le \theta_a, i = 1, \cdots, q, j = 1, \cdots, n.
\end{aligned} \tag{7.3}$$

Let $e_k \in R^n, h_k \in R^p$, and $g_k \in R^q$ denote the column vectors in which the kth element equals one and the others equal zero. Then the gain variations of the form (7.3) can be described as :

$$\Delta A_F = \sum_{i=1}^{n}\sum_{j=1}^{n} \theta_{aij} e_i e_j^T, \quad \Delta B_F = \sum_{i=1}^{n}\sum_{j=1}^{p} \theta_{bij} e_i h_j^T,$$

$$\Delta C_F = \sum_{i=1}^{q}\sum_{j=1}^{n} \theta_{cij} g_i e_j^T.$$

Combining filter (7.2) with system (7.1), we obtain the filter error system as:

$$\begin{aligned}
x_e(k+1) &= A_e x_e(k) + B_e\omega(k),\\
z_e(k) &= C_e x_e(k),
\end{aligned} \tag{7.4}$$

where $x_e(k) = [x(k)^T, \xi(k)^T]^T$, $z_e(k) = z(k) - z_F(k)$ is the estimation error, and

$$A_e = \begin{bmatrix} A & 0 \\ (B_F + \Delta B_F)C_2 & A_F + \Delta A_F \end{bmatrix},$$

$$B_e = \begin{bmatrix} B_1 \\ (B_F + \Delta B_F)D_{21} \end{bmatrix}, C_e = \begin{bmatrix} C_1 & -C_F - \Delta C_F \end{bmatrix}.$$

The transfer function matrix of the augmented system (7.4) from w to z_e is given by

$$G_{z_e w}(z) = C_e(zI - A_e)^{-1}B_e.$$

Then the problem under consideration in this chapter is as follows.

Non-fragile H_∞ filtering problem with filter gain variations: Given a positive constant γ, find a filter described by (7.2) with the gain variations of the form (7.3) such that the resulting system (7.4) is asymptotically stable and $\|G_{z_e w}(z)\| < \gamma$.

In this section, an LMI-based method for designing H_∞ filters with respect to additive uncertainties is presented, and further, a comparison between the new proposed method and the existing method is given.

7.2.2   Non-Fragile H_∞ Filter Design Methods

To facilitate the presentation, we denote

$$M_0(\Delta A_F, \Delta B_F, \Delta C_F)$$
$$= \begin{bmatrix} \Xi_1 & \Xi_2 & 0 & S^T A & S^T A & S^T B_1 \\ * & \Xi_3 & 0 & \Xi_5 & \Xi_6 & \Xi_7 \\ * & * & -I & \Xi_4 & C_1 & 0 \\ * & * & * & -\bar{P}_{11} & -\bar{P}_{12} & 0 \\ * & * & * & * & -\bar{P}_{22} & 0 \\ * & * & * & * & * & -\gamma^2 I \end{bmatrix} \qquad (7.5)$$

where

$$
\begin{aligned}
\Xi_1 &= \bar{P}_{11} - S - S^T, \Xi_2 = \bar{P}_{12} - S - S^T, \\
\Xi_3 &= \bar{P}_{22} - S - S^T + N + N^T, \Xi_4 = C_1 - C_F - \Delta C_F, \\
\Xi_5 &= (S - N)^T A + F_B C_2 + N^T (\Delta B_F C_2 + \Delta A_F) + F_A, \qquad (7.6) \\
\Xi_6 &= (S - N)^T A + F_B C_2 + N^T \Delta B_F C_2, \\
\Xi_7 &= (S - N)^T B_1 + F_B D_{21} + N^T \Delta B_F D_{21}.
\end{aligned}
$$

Then the following theorem presents a sufficient condition for the solvability of the non-fragile H_∞ filtering problem with additive uncertainty.

Theorem 7.1 *Consider system (7.1). Let $\gamma > 0$ and $\theta_a > 0$ be given constants. If there exist matrices $F_A, F_B, C_F, S, N, \bar{P}_{12}$, and $\bar{P}_{11} > 0, \bar{P}_{22} > 0$, such that the following LMIs hold:*

$$M_0(\Delta A_F, \Delta B_F, \Delta C_F) < 0, \quad \theta_{aij}, \theta_{bik}, \theta_{clj} \in \{-\theta_a, \theta_a\},$$
$$i, j = 1, \cdots, n; \ k = 1, \cdots, p; \ l = 1, \cdots, q, \tag{7.7}$$

then the filter (7.2) with additive uncertainty described by (7.3) and

$$A_F = (N^T)^{-1} F_A, \quad B_F = (N^T)^{-1} F_B, \quad C_F = C_F \tag{7.8}$$

solves the non-fragile H_∞ filtering problem for the system (7.1).

Proof 7.1 *By Lemma 2.5, it is sufficient to show that there exist a matrix G with structure (2.14) and a symmetric positive matrix $P = \begin{bmatrix} P_{11} & P_{12} \\ P_{12}^T & P_{22} \end{bmatrix} > 0$ such that*

$$M_1 = \begin{bmatrix} P - G - G^T & 0 & G^T A_e & G^T B_e \\ * & -I & C_e & 0 \\ * & * & -P & 0 \\ * & * & * & -\gamma^2 I \end{bmatrix} < 0 \tag{7.9}$$

holds for all $\theta_{aij}, \theta_{bik}$, and θ_{clj} satisfying (7.3).
　　Denote

$$S = Y + N, \quad \Gamma_1 = \begin{bmatrix} I & I \\ I & 0 \end{bmatrix},$$
$$\bar{\Gamma}_1 = diag\{\Gamma_1, I, \Gamma_1, I\},$$
$$\bar{P}_{11} = P_{11} + P_{12} + P_{12}^T + P_{22},$$
$$\bar{P}_{12} = P_{11} + P_{12}^T, \quad \bar{P}_{22} = P_{11}.$$

Then (7.9) is equivalent to

$$M_2 = \bar{\Gamma}_1 M_1 \bar{\Gamma}_1^T$$
$$= \begin{bmatrix} \Xi_1 & \Xi_2 & 0 & S^T A & S^T A & S^T B_1 \\ * & \Xi_3 & 0 & \Pi_1 & \Pi_2 & \Pi_3 \\ * & * & -I & \Xi_4 & C_1 & 0 \\ * & * & * & -\bar{P}_{11} & -\bar{P}_{12} & 0 \\ * & * & * & * & -\bar{P}_{22} & 0 \\ * & * & * & * & * & -\gamma^2 I \end{bmatrix} < 0 \tag{7.10}$$

holds for all $\theta_{aij}, \theta_{bik}$, and θ_{clj} satisfying (7.3), where $\Xi_1, \Xi_2, \Xi_3, \Xi_4$ are defined by (7.6), and

$$\Pi_1 = (S - N)^T A + N^T (B_F C_2 + \Delta B_F C_2 + A_F + \Delta A_F),$$
$$\Pi_2 = (S - N)^T A + N^T B_F C_2 + N^T \Delta B_F C_2,$$
$$\Pi_3 = (S - N)^T B_1 + N^T B_F D_{21} + N^T \Delta B_F D_{21}.$$

Obviously, M_2 is convex for each θ_i, $\theta_i \in \{\theta_{aij}, \theta_{bik}, \theta_{clj}$ satisfying (7.3)\}, so (7.10) is equivalent to

$$
M_3 = \begin{bmatrix}
\Xi_1 & \Xi_2 & 0 & S^T A & S^T A & S^T B_1 \\
* & \Xi_3 & 0 & \Pi_1 & \Pi_2 & \Pi_3 \\
* & * & -I & \Xi_4 & C_1 & 0 \\
* & * & * & -\bar{P}_{11} & -\bar{P}_{12} & 0 \\
* & * & * & * & -\bar{P}_{22} & 0 \\
* & * & * & * & * & -\gamma^2 I
\end{bmatrix} < 0 \tag{7.11}
$$

for all θ_{aij}, θ_{bik}, $\theta_{clj} \in \{-\theta_a, \theta_a\}$, $i, j = 1, \cdots, n$; $k = 1, \cdots, p$; $l = 1, \cdots, q$. By (7.8), (7.9), and the Schur complement, it concludes that (7.11) is equivalent to (7.7).

Thus, the proof is complete.

Remark 7.1 *Theorem 7.1 presents a sufficient condition in terms of solutions to a set of LMIs for the solvability of the non-fragile H_∞ filtering problem. By the proofs of Theorem 7.1 and Lemma 2.5, Theorem 7.1 also shows that the non-fragile H_∞ filtering problem becomes a convex one when the state-space realizations of the designed filters with gain variations admit the slack variable matrix G with the structure of (2.14). For the case that the designed filter contains no gain variations, from Lemma 2.5, it follows that the design condition given in Theorem 7.1 reduces a necessary and sufficient condition for the standard H_∞ filtering problem, which means that the structure constraint (2.14) on the slack matrix G does not result in any conservativeness for the standard H_∞ filter design.*

For the non-fragile filter design method given by Theorem 7.1, it should be noted that the number of LMIs involved in (7.7) is $2^{n^2+np+nq}$, which results in the difficulty of implementing the LMI constraints in computation. For example, when $n = 5$ and $p = q = 1$, the number of LMIs involved in (7.7) is 2^{35}, which already exceeds the capacity of the current LMI solver in MATLAB. To overcome the difficulty arising from implementing the design condition given in Theorem 7.1, the following method is developed.

Denote

$$
\begin{aligned}
F_{a1} &= [f_{a11}\ f_{a12}\ \cdots\ f_{a1l_a}], \\
F_{a2} &= [f_{a21}^T\ f_{a22}^T\ \cdots\ f_{a2l_a}^T]^T
\end{aligned} \tag{7.12}
$$

where $l_a = n^2 + np + nq$, and

$$
f_{a1k} = \begin{bmatrix} \mathbf{0}_{1\times n} & (N^T e_i)^T & \mathbf{0}_{1\times q} & \mathbf{0}_{1\times n} & \mathbf{0}_{1\times n} & \mathbf{0}_{1\times r} \end{bmatrix}^T,
$$
$$
f_{a2k} = \begin{bmatrix} \mathbf{0}_{1\times n} & \mathbf{0}_{1\times n} & \mathbf{0}_{1\times q} & e_j^T & \mathbf{0}_{1\times n} & \mathbf{0}_{1\times r} \end{bmatrix},
$$
for $k = (i-1)n + j$, $i, j = 1, \cdots, n$.
$$
f_{a1k} = \begin{bmatrix} \mathbf{0}_{1\times n} & (N^T e_i)^T & \mathbf{0}_{1\times q} & \mathbf{0}_{1\times n} & \mathbf{0}_{1\times n} & \mathbf{0}_{1\times r} \end{bmatrix}^T,
$$

$$f_{a2k} = \begin{bmatrix} \mathbf{0}_{1\times n} & \mathbf{0}_{1\times n} & \mathbf{0}_{1\times q} & h_j^T C_2 & h_j^T C_2 & h_j^T D_{21} \end{bmatrix},$$
for $k = n^2 + (i-1)p + j$, $i = 1,\cdots,n, j = 1,\cdots,p$.

$$f_{a1k} = \begin{bmatrix} \mathbf{0}_{1\times n} & \mathbf{0}_{1\times n} & -g_i^T & \mathbf{0}_{1\times n} & \mathbf{0}_{1\times n} & \mathbf{0}_{1\times r} \end{bmatrix}^T,$$

$$f_{a2k} = \begin{bmatrix} \mathbf{0}_{1\times n} & \mathbf{0}_{1\times n} & \mathbf{0}_{1\times q} & e_j^T & \mathbf{0}_{1\times n} & \mathbf{0}_{1\times r} \end{bmatrix},$$
for $k = n^2 + np + (i-1)n + j$, $i = 1,\cdots,q, j = 1,\cdots,n$.

Let $k_0, k_1, \cdots, k_{s_a}$ be integers satisfying

$$k_0 = 0 < k_1 < \cdots < k_{s_a} = l_a$$

and matrix Σ have the following structure

$$\Sigma = \begin{bmatrix} diag[\sigma_{11}^1 \cdots \sigma_{11}^{s_a}] & diag[\sigma_{12}^1 \cdots \sigma_{12}^{s_a}] \\ diag[\sigma_{12}^1 \cdots \sigma_{12}^{s_a}]^T & diag[\sigma_{22}^1 \cdots \sigma_{22}^{s_a}] \end{bmatrix}, \tag{7.13}$$

where $\sigma_{11}^i, \sigma_{12}^i$, and $\sigma_{22}^i \in R^{(k_i - k_{i-1}) \times (k_i - k_{i-1})}, i = 1, \cdots, s_a$. Then, we have the following theorem.

Theorem 7.2 *Consider system (7.1). Let $\gamma > 0$ and $\theta_a > 0$ be given constants. If there exist matrices $F_A, F_B, C_F, S, N, \bar{P}_{12}, \bar{P}_{11} > 0, \bar{P}_{22} > 0$, and symmetric matrix Σ with the structure described by (7.13) such that the following LMIs hold:*

$$\begin{bmatrix} Q & F_{a1} \\ F_{a1}^T & 0 \end{bmatrix} + \begin{bmatrix} F_{a2} & 0 \\ 0 & I \end{bmatrix}^T \Sigma \begin{bmatrix} F_{a2} & 0 \\ 0 & I \end{bmatrix} < 0, \tag{7.14}$$

$$\begin{bmatrix} I \\ diag[\theta_{k_{i-1}+j} \cdots \theta_{k_i}] \end{bmatrix}^T \begin{bmatrix} \sigma_{11}^i & \sigma_{12}^i \\ (\sigma_{12}^i)^T & \sigma_{22}^i \end{bmatrix}$$
$$\times \begin{bmatrix} I \\ diag[\theta_{k_{i-1}+j} \cdots \theta_{k_i}] \end{bmatrix} \geq 0, \text{ for all } \theta_{k_{i-1}+j} \in \{-\theta_a, \theta_a\},$$
$$j = 1, \cdots, k_i - k_{i-1}, i = 1, \cdots, s_a, \tag{7.15}$$

where

$$Q = \begin{bmatrix} \Xi_1 & \Xi_2 & 0 & S^T A & S^T A & S^T B_1 \\ * & \Xi_3 & 0 & (S-N)^T A + F_B C_2 + F_A & (S-N)^T A + F_B C_2 & (S-N)^T B_1 + F_B D_{21} \\ * & * & -I & C_1 - C_F & C_1 & 0 \\ * & * & * & -\bar{P}_{11} & -\bar{P}_{12} & 0 \\ * & * & * & * & -\bar{P}_{22} & 0 \\ * & * & * & * & * & -\gamma^2 I \end{bmatrix} \tag{7.16}$$

with Ξ_1, Ξ_2, Ξ_3 defined by (7.6), then the filter (7.2) with additive uncertainty described by (7.3) and the filter gain parameters given by (7.8) solves the non-fragile H_∞ filtering problem for the system (7.1).

Proof 7.2 *Obviously, (7.7) can be written as*

$$M_0 = Q + \Delta Q + \Delta Q^T < 0, \tag{7.17}$$

where

$$\Delta Q = \begin{bmatrix} 0 & 0 & 0 & 0 & 0 & 0 \\ 0 & 0 & 0 & \Delta Q_1 & \Delta Q_2 & \Delta Q_3 \\ 0 & 0 & 0 & \Delta Q_4 & 0 & 0 \\ 0 & 0 & 0 & 0 & 0 & 0 \\ 0 & 0 & 0 & 0 & 0 & 0 \\ 0 & 0 & 0 & 0 & 0 & 0 \end{bmatrix}$$

with

$$\Delta Q_1 = \sum_{i,j=1}^{n} \theta_{aij} N^T e_i e_j^T + \sum_{i=1}^{n} \sum_{k=1}^{p} \theta_{bik} N^T e_i h_k^T C_2,$$

$$\Delta Q_2 = \sum_{i=1}^{n} \sum_{k=1}^{p} \theta_{bik} N^T e_i h_k^T C_2,$$

$$\Delta Q_3 = \sum_{i=1}^{n} \sum_{k=1}^{p} \theta_{bik} N^T e_i h_k^T D_{21}, \quad \Delta Q_4 = -\sum_{l=1}^{q} \sum_{j=1}^{n} \theta_{clj} g_l e_j^T$$

for all $\theta_{aij}, \theta_{bik}, \theta_{clj} \in \{-\theta_a, \theta_a\}$, $i,j = 1, \cdots, n$; $k = 1, \cdots, p$; $l = 1, \cdots, q$.

By using (7.12), it follows that (7.17) is equivalent to

$$\begin{aligned} M_0 &= Q + \sum_{i=1}^{l_a} \theta_i f_{a1i} f_{a2i} + \left(\sum_{i=1}^{l_a} \theta_i f_{a1i} f_{a2i} \right)^T \\ &= Q + F_{a1} \tilde{\Delta}_a F_{a2} + (F_{a1} \tilde{\Delta}_a F_{a2})^T < 0 \end{aligned} \tag{7.18}$$

holds, where $\tilde{\Delta}_a = diag[\theta_1, \cdots, \theta_{l_a}]$ for all $\theta_i \in \{-\theta_a, \theta_a\}$. By Lemma 2.8, it follows that (7.18) holds if and only if there exists a symmetric matrix $\Sigma \in R^{l_a \times l_a}$ such that (7.14) and

$$\begin{bmatrix} I \\ \tilde{\Delta}_a \end{bmatrix}^T \Sigma \begin{bmatrix} I \\ \tilde{\Delta}_a \end{bmatrix} \geq 0 \tag{7.19}$$

hold for all $\theta_i \in \{-\theta_a, \theta_a\}$, $i = 1, \cdots, l_a$. Notice that the set of Σ satisfying (7.13) is a subset of the set of Σ satisfying (7.19); hence the conclusion follows.

Remark 7.2 *From the proof of Theorem 7.2, it follows that when $s_a = 1$, the set of Σ satisfying (7.13) is equal to the set of Σ satisfying (7.19), and the design conditions given in Theorem 7.2 and Theorem 7.1 are equivalent.*

Σ *satisfying (7.14) and (7.19) (or (7.15) with $s_a = 1$) is said to be a vertex separator [67]. Notice that the number of LMIs involved in (7.19) or (7.15) with $s_a = 1$ still is $2^{n^2+np+nq}$, so that the difficulty of implementing the LMI constraints remains. To overcome the difficulty, Theorem 7.2 presents a sufficient condition for the non-fragile H_∞ filter design in terms of separator Σ with the structure described by (7.13), where the number of LMIs involved in (7.15) is $\sum_{i=1}^{s_a} 2^{k_i-k_{i-1}}$, which can be controlled not to grow exponentially by reducing the value of $\max k_i - k_{i-1} : i = 1, \cdots, s_a$. Compared with the Σ being of full block in (7.14) and (7.19), Σ with the structure described by (7.13) satisfying (7.14) and (7.15) is said to be a structured vertex separator. However, it should be noted that the design condition given in Theorem 7.2 may be more conservative than that given in Theorem 7.1 because of the structure constraint on Σ. But the design condition proposed in Theorem 7.2 solves the numerical computation problem, which cannot be solved by the design condition given in Theorem 7.1. On the other hand, in Theorem 7.2, the smaller the value of s_a is, the less conservativeness is introduced.*

Remark 7.3 *Obviously, the conditions for designing non-fragile H_∞ filters given in Theorem 7.1 and Theorem 7.2 can be easily extended to deal with the robust non-fragile H_∞ filtering problem for systems with polytopic uncertainties too, because the system matrices are affinely involved in the proposed design conditions.*

In the following, we will introduce the result of non-fragile H_∞ filter design with norm-bounded gain variations. And at the same time, the relationship with our result is discussed.

Similar to Yang and Wang [148] and Mahmoud [103] for non-fragile filter design with norm-bounded uncertainty, the norm-bounded type of gain variations $\Delta A_F, \Delta B_F$, and ΔC_F can be overbounded [107] by the following norm-bounded uncertainty:

$$\Delta A_F = M_a F_1(t) E_a, \Delta B_F = M_b F_2(t) E_b,$$
$$\Delta C_F = M_c F_3(t) E_c, \tag{7.20}$$

where

$$M_a = [M_{a1} \cdots M_{an^2}], E_a = [E_{a1}^T \cdots E_{an^2}^T]^T,$$
$$M_b = [M_{b1} \cdots M_{bnp}], E_b = [E_{b1}^T \cdots E_{bnp}^T]^T,$$
$$M_c = [M_{c1} \cdots M_{cnq}], E_c = [E_{c1}^T \cdots E_{cnq}^T]^T,$$

with

$$M_{ak} = e_i, E_{ak} = e_j^T$$
$$\text{for } k = (i-1)n + j, \ i, j = 1, \cdots, n,$$
$$M_{bk} = e_i, E_{bk} = h_j^T$$
$$\text{for } k = n^2 + (i-1)p + j, \ i = 1, \cdots, n, \ j = 1, \cdots, p,$$
$$M_{ck} = g_i, E_{ck} = e_j^T$$
$$\text{for } k = n^2 + np + (i-1)n + j, \ i = 1, \cdots, q, \ j = 1, \cdots, n,$$

and $F_i^T(t)F_i(t) \leq \theta_a^2 I$ for $i = 1, 2, 3$, represent the uncertain parameters. Here, θ_a is the same as before.

To facilitate the presentation, we denote

$$\bar{F}_A = \bar{N}A_F, \bar{F}_B = \bar{N}B_F,$$

$$M_{a1} = \begin{bmatrix} 0 & 0 & 0 \\ \bar{N}M_a & \bar{N}M_b & 0 \\ 0 & 0 & -M_c \\ 0 & 0 & 0 \\ 0 & 0 & 0 \\ 0 & 0 & 0 \end{bmatrix},$$

$$M_{a2} = \begin{bmatrix} 0 & 0 & 0 & E_a & 0 & 0 \\ 0 & 0 & 0 & E_bC_2 & E_bC_2 & E_bD_{21} \\ 0 & 0 & 0 & E_c & 0 & 0 \end{bmatrix}.$$

By using the method in Yang and Wang [147] and Mahmoud [101, 102], the non-fragile H_∞ filter design with norm-bounded gain variations is reduced to solve the following LMI:

$$\begin{bmatrix} \bar{Q} & M_{a1} & \theta_a \varepsilon M_{a2}^T \\ * & -\varepsilon I & 0 \\ * & * & -\varepsilon I \end{bmatrix} < 0, \tag{7.21}$$

with matrix variables $\bar{S} > 0, \bar{N} < 0$, and scalar $\varepsilon > 0$, where

$$\bar{Q} = \begin{bmatrix} -\bar{S} & -\bar{S} & 0 & \bar{S}A & \bar{S}A & \bar{S}B_1 \\ * & -\bar{S}+\bar{N} & 0 & Q_1 & Q_2 & Q_3 \\ * & * & -I & C_1 - C_F & C_1 & 0 \\ * & * & * & -\bar{S} & -\bar{S} & 0 \\ * & * & * & * & -\bar{S}+\bar{N} & 0 \\ * & * & * & * & * & -\gamma^2 I \end{bmatrix},$$

with

$$Q_1 = (\bar{S} - \bar{N})A + \bar{F}_A + \bar{F}_BC_2,$$
$$Q_2 = (\bar{S} - \bar{N})A + \bar{F}_BC_2,$$
$$Q_3 = (\bar{S} - \bar{N})B_1 + \bar{F}_BD_{21}.$$

The following lemma will show the relationship between condition (7.21) and the condition for designing non-fragile H_∞ filters given in Theorem 7.2.

Lemma 7.1 *Consider the system (7.1). If condition (7.21) is feasible, then the condition for designing non-fragile H_∞ filters given in Theorem 7.2 is feasible.*

Proof 7.3 *Let $S = \bar{P}_{11} = \bar{P}_{12} = \bar{S}, N = \bar{N}, \bar{P}_{22} = \bar{S} - \bar{N} > 0$, then it is easy to see that $Q = \bar{Q}, F_{a1} = M_{a1}$, and $F_{a2} = M_{a2}$, that is, condition (7.21) becomes*

$$\begin{bmatrix} Q & F_{a1} & \theta_a \varepsilon F_{a2}^T \\ * & -\varepsilon I & 0 \\ * & * & -\varepsilon I \end{bmatrix} < 0. \tag{7.22}$$

In Theorem 7.2, when $s_a = l_a$, according to (7.22) and $F_i^T(t)F_i(t) \le \theta_a^2 I$, $i = 1, 2, 3$, and by the Schur complement, there exists a matrix Σ with the structure

$$\Sigma = \begin{bmatrix} \varepsilon \theta_a^2 I & \mathbf{0}_{l_a \times l_a} \\ \mathbf{0}_{l_a \times l_a} & -\varepsilon I \end{bmatrix}, \tag{7.23}$$

such that the following LMIs hold:

$$\begin{bmatrix} Q & F_{a1} \\ F_{a1}^T & 0 \end{bmatrix} + \begin{bmatrix} F_{a2} & 0 \\ 0 & I \end{bmatrix}^T \Sigma \begin{bmatrix} F_{a2} & 0 \\ 0 & I \end{bmatrix}$$

$$= \begin{bmatrix} Q + \varepsilon \theta_a^2 F_{a2}^T F_{a2} & F_{a1} \\ F_{a1}^T & -\varepsilon I \end{bmatrix} < 0, \tag{7.24}$$

$$\begin{bmatrix} I \\ \theta_i \end{bmatrix}^T \begin{bmatrix} \sigma_{11}^i & \sigma_{12}^i \\ (\sigma_{12}^i)^T & \sigma_{22}^i \end{bmatrix} \begin{bmatrix} I \\ \theta_i \end{bmatrix} = \varepsilon \theta_a^2 - \varepsilon \theta_i^2 \ge 0,$$

for all $i = 1, \cdots, l_a$. $\tag{7.25}$

Thus, the proof is complete.

Remark 7.4 *From the proof of Lemma 7.1, it follows that condition (7.21) is more conservative than the non-fragile H_∞ filter existence condition given in Theorem 7.2 with $s_a = l_a$. However, as indicated in Remark 7.2, the case of $s_a = l_a$ is the worst case of the method. So, the existing non-fragile H_∞ filter design method with the norm-bounded gain variations is more conservative than the non-fragile H_∞ filter design method proposed in this chapter.*

In Theorem 7.2, we restrict the slack variable matrix G with the structure (2.14) for obtaining the convex design condition, which may result in more conservative evaluation of the H_∞ performance index bound. So, in this section, for a designed filter, the matrix G without the restriction is exploited for obtaining less conservative evaluation of the H_∞ performance index bound.

When the filter parameter matrices A_F, B_F, and C_F are known, the problem of minimizing γ subject to (7.3) for a given $\theta_a > 0$ and $\| G_{z_e\omega}(z) \| < \gamma$ can be converted into the one: minimize γ^2 subject to the following LMIs:

$$\begin{bmatrix} P - G - G^T & 0 & G^T A_e & G^T B_e \\ * & -I & C_e & 0 \\ * & * & -P & 0 \\ * & * & * & -\gamma^2 I \end{bmatrix} < 0,$$

$$\text{for all } \theta_{aij}, \theta_{bik}, \theta_{clj} \in \{-\theta_a, \theta_a\},$$
$$i, j = 1, \cdots, n; \ k = 1, \cdots, p; \ l = 1, \cdots, q. \tag{7.26}$$

Similar to the design condition given in Theorem 7.1, the above method is also with the numerical computation problem. To solve the problem, the following lemma provides a solution using the structured vertex separator approach.

Denote
$$\begin{aligned} G_{a1} &= [g_{a11} \ g_{a12} \ \cdots \ g_{a1l_a}], \\ G_{a2} &= [g_{a21}^T \ g_{a22}^T \ \cdots \ g_{a2l_a}^T]^T \end{aligned} \tag{7.27}$$

where

$$g_{a1k} = \left[\begin{pmatrix} \mathbf{0}_{1\times n} & e_i^T \end{pmatrix} G \quad \mathbf{0}_{1\times q} \quad \mathbf{0}_{1\times 2n} \quad \mathbf{0}_{1\times r} \right]^T,$$
$$g_{a2k} = \left[\mathbf{0}_{1\times 2n} \quad \mathbf{0}_{1\times q} \quad \mathbf{0}_{1\times n} \quad e_j^T \quad \mathbf{0}_{1\times n} \right],$$
$$\text{for } k = (i-1)n + j, \ i, j = 1, \cdots, n.$$

$$g_{a1k} = \left[\begin{pmatrix} \mathbf{0}_{1\times n} & e_i^T \end{pmatrix} G \quad \mathbf{0}_{1\times q} \quad \mathbf{0}_{1\times 2n} \quad \mathbf{0}_{1\times r} \right]^T,$$
$$g_{a2k} = \left[\mathbf{0}_{1\times 2n} \quad \mathbf{0}_{1\times q} \quad h_j^T C_2 \quad \mathbf{0}_{1\times n} \quad h_j^T D_{21} \right],$$
$$\text{for } k = n^2 + (i-1)p + j, \ i = 1, \cdots, n, j = 1, \cdots, p.$$

$$g_{a1k} = \left[\mathbf{0}_{1\times 2n} \quad -g_i^T \quad \mathbf{0}_{1\times 2n} \quad \mathbf{0}_{1\times r} \right]^T,$$
$$g_{a2k} = \left[\mathbf{0}_{1\times 2n} \quad \mathbf{0}_{1\times q} \quad \mathbf{0}_{1\times n} \quad e_j^T \quad \mathbf{0}_{1\times n} \right],$$
$$\text{for } k = n^2 + np + (i-1)n + j, \ i = 1, \cdots, q, j = 1, \cdots, n.$$

Then we have the following lemma.

Lemma 7.2 *Consider the system (7.1). Let $\gamma > 0, \theta_a > 0$ be constants and filter parameter matrices A_F, B_F, and C_F be given. Then $\| G_{z_e\omega}(z) \| < \gamma$ holds for all $\theta_{aij}, \theta_{bik}$ and θ_{clj} satisfying (7.3), if there exist a matrix G, a positive-definite matrix $P > 0$, and a symmetric matrix Σ with the structure described by (7.13) such that (7.15) and the following LMI hold:*

$$\begin{bmatrix} Q_s & G_{a1} \\ G_{a1}^T & 0 \end{bmatrix} + \begin{bmatrix} G_{a2} & 0 \\ 0 & I \end{bmatrix}^T \Sigma \begin{bmatrix} G_{a2} & 0 \\ 0 & I \end{bmatrix} < 0 \tag{7.28}$$

where

$$Q_s = \begin{bmatrix} P - G - G^T & 0 & G^T A_{e0} & G^T B_{e0} \\ * & -I & C_{e0} & 0 \\ * & * & -P & 0 \\ * & * & * & -\gamma^2 I \end{bmatrix}$$

with

$$A_{e0} = \begin{bmatrix} A & 0 \\ B_F C_2 & A_F \end{bmatrix}, \quad B_{e0} = \begin{bmatrix} B_1 \\ B_F D_{21} \end{bmatrix}, \qquad (7.29)$$
$$C_{e0} = \begin{bmatrix} C_1 & -C_F \end{bmatrix}.$$

Proof 7.4 *By using (7.26) and (7.27), it is similar to the proof of Theorem 7.2, and omitted here.*

Remark 7.5 *For evaluating the H_∞ performance bound of the transfer function from w to z_e, the condition given in Lemma 7.2 usually is less conservative than that given in Theorem 7.2 because no structure constraint on the slack variable matrix G in Lemma 7.2 is imposed.*

7.2.3   Example

To illustrate the effectiveness of the designed non-fragile H_∞ filter, an example is given to provide a comparison between the proposed non-fragile H_∞ filter design method and the existing non-fragile H_∞ filter design method with the norm-bounded gain variations. We consider a linear system of the form (7.1) with

$$A = \begin{bmatrix} 0 & 1 & -0.5 \\ -1 & -0.5 & 1 \\ -1 & 0 & 1 \end{bmatrix}, B_1 = \begin{bmatrix} -1 & 0 \\ 0.5 & 0 \\ -1 & 0 \end{bmatrix},$$

$$C_1 = \begin{bmatrix} 1 & -1 & 1 \end{bmatrix}, C_2 = \begin{bmatrix} -1 & 0.5 & 2 \end{bmatrix}, D_{21} = \begin{bmatrix} 0 & 0.9 \end{bmatrix}.$$

For the case that the designed filter contains no gain variations, by the standard H_∞ filtering method [106] for discrete-time systems, the optimal H_∞ performance index of the standard *closed-loop system* is achieved as $\gamma_{opt} = 3.7282$.

First, we design a non-fragile H_∞ filter by using condition (7.21).

Assume that the designed filter is with norm-bounded additive uncertainties described by (7.20), by applying condition (7.21) with $\theta_a = 0.05$ to design a non-fragile H_∞ filter with gains as

$$A_{F_{norm}} = \begin{bmatrix} 0.8889 & 0.9677 & -0.3090 \\ -1.2415 & -0.8559 & 1.4525 \\ 0.0422 & 0.2284 & 0.3613 \end{bmatrix},$$

$$B_{F_{norm}} = \begin{bmatrix} -0.1173 & -0.1478 & 0.2256 \end{bmatrix}^T,$$

$$C_{F_{norm}} = \begin{bmatrix} 0.0799 & 0.0984 & 0.4563 \end{bmatrix},$$

and the H_∞ performance index of the obtained non-fragile H_∞ filter is 4.7319.

Then, we design a non-fragile H_∞ filter by using Theorem 7.2.

Assume that the designed filter is with the additive uncertainties described by (7.3). For this case, it is difficult to apply Theorem 7.1 to design a non-fragile H_∞ filter because the number of the LMI constraints involved in (7.7) is 2^{15}. However, Theorem 7.2 is applicable for this case. By applying Theorem 7.2 with $\theta_a = 0.05$ and $k_i = i, i = 1, \cdots, 15$, that is, $s_a = 15$ as well as $k_i = 3i, i = 1, \cdots, 5$, that is, $s_a = 5$ to design the non-fragile filters with gains as follows.

For $s_a = 15$,

$$A_{Fin15} = \begin{bmatrix} 0.0742 & 0.2788 & -0.1953 \\ -0.3214 & 0.0583 & 0.5724 \\ -0.0112 & 0.2634 & -0.0365 \end{bmatrix},$$

$$B_{Fin15} = \begin{bmatrix} 0.0625 & -0.2398 & 0.4061 \end{bmatrix}^T,$$

$$C_{Fin15} = \begin{bmatrix} 0.0102 & 0.0538 & 0.3678 \end{bmatrix}.$$

For $s_a = 5$,

$$A_{Fin5} = \begin{bmatrix} 0.0769 & 0.2707 & -0.2052 \\ -0.3141 & 0.0521 & 0.5609 \\ -0.0136 & 0.2619 & -0.0379 \end{bmatrix},$$

$$B_{Fin5} = \begin{bmatrix} 0.0753 & -0.2316 & 0.4088 \end{bmatrix}^T,$$

$$C_{Fin5} = \begin{bmatrix} 0.0099 & 0.0541 & 0.3671 \end{bmatrix},$$

and the H_∞ performance indexes of the obtained filters are 4.1791 ($s_a = 15$) and 4.1790 ($s_a = 5$), respectively.

In this part, for the above designed filters, Lemma 7.2 gives better evaluations of the H_∞ performance index bounds.

First, to facilitate the presentation, denote the filter designed by the existing method (7.21) as F_{norm} while denoting the filters designed by the proposed method using Theorem 7.2 as F_{in15} ($s_a = 15$) and F_{in5} ($s_a = 5$).

By applying Lemma 7.2, the H_∞ performance indexes of the non-fragile filter F_{norm} are $\gamma = 4.3539$ ($s_a = 15$) and $\gamma = 4.3439$ ($s_a = 5$) while the H_∞ performance indexes of the non-fragile filters F_{in15} and F_{in5} are $\gamma = 4.1069$ ($s_a = 15$) and $\gamma = 4.1060$ ($s_a = 5$), respectively.

In the following, tables are given to provide a comparison between the non-fragile H_∞ filters designed by the proposed method (Theorem 7.2) and the non-fragile H_∞ filter designed by the existing method (condition (7.21)).

First, the H_∞ performance indexes achieved by the designs are shown in Table 7.1.

From Table 7.1, it is easy to see that compared with the optimal H_∞ performance index bound $\gamma_{opt} = 3.7282$, the performance index of the filter designed by the existing method (condition (7.21)) is degraded 26.92%. The performance indexes of the filters designed by the proposed method (Theorem

TABLE 7.1

Performance Index by Design with $\theta_a = 0.05$

	Existing	Proposed ($s_a = 15$)	Proposed ($s_a = 5$)
γ	4.7319	4.1791	4.1790

TABLE 7.2

Performance Index Evaluation by
Lemma 7.2 with $\theta_a = 0.05$

	F_{norm}	F_{in15}	F_{in5}
$\gamma(s_a = 15)$	4.3539	4.1069	—
$\gamma(s_a = 5)$	4.3439	—	4.1060

7.2) are degraded the same as 12.09% (for $s_a = 15$ or $s_a = 5$), which are much more improved than 26.92%.

For the designed filters, Lemma 7.2 gives better performance indexes as shown in Table 7.2.

Obviously, compared with the optimal H_∞ performance index $\gamma_{opt} = 4.7282$, by using Lemma 7.2, the H_∞ performance indexes of the non-fragile H_∞ filter F_{norm} are degraded 16.78% for $s_a = 15$ and degraded 16.51% for $s_a = 5$. Correspondingly, the performance indexes for the non-fragile H_∞ filters F_{in15} and F_{in5} are degraded 10.16% for $s_a = 15$ and 10.13% for $s_a = 5$, respectively. These numerical results show the superiority of our non-fragile filtering method.

In the following, a simulation is given to illustrate the effectiveness of our design method. Let the system initial state be $x_0 = \begin{bmatrix} 0.5 & -0.5 & 0.5 \end{bmatrix}^T$ and the filter initial state be $\xi_0 = \begin{bmatrix} 1.5 & 1.5 & 0.5 \end{bmatrix}^T$. Let the disturbance $w(k)$ be

$$w(k) = \begin{cases} \begin{bmatrix} -0.5 \\ 5 \end{bmatrix}, & 41 \leq k \leq 45 \text{ (step)}, \\ 0 & \text{otherwise.} \end{cases}$$

Then, Figure 7.1 shows the estimation error responses of the non-fragile filters designed by the proposed method (Theorem 7.2) for $s_a = 15$ and by the existing method, condition (7.21), respectively.

From this example, we can see that the worst case of our result in Theorem 7.2 ($s_a = 15$) is more effective than the existing non-fragile H_∞ filter existence condition (7.21). This phenomenon shows the effectiveness of our design method.

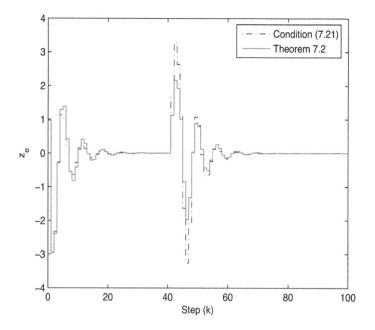

FIGURE 7.1

Comparison of the error responses of $z_e(t)$ with $\theta_a = 0.05$.

7.3 Non-Fragile H_∞ Filter Design for Linear Continuous-Time Systems

In this section, the non-fragile H_∞ filter design for linear continuous-time systems is considered. And an LMI-based method for designing H_∞ filters with respect to additive uncertainties is presented, and further, a comparison between the new proposed method and the existing method is given. Due to the fact that the proofs of the main results of this part are similar to the proofs of the discrete-time case, the proofs are omitted here.

7.3.1 Problem Statement

Consider a linear time-invariant model described by

$$
\begin{aligned}
\dot{x}(t) &= Ax(t) + B_1\omega(t), \\
z(t) &= C_1 x(t), \\
y(t) &= C_2 x(t) + D_{21}\omega(t),
\end{aligned}
\tag{7.30}
$$

where $x(t) \in R^n$ is the state, $y(t) \in R^p$ is the measured output, and $z(t) \in R^q$ is the regulated output, respectively. $A, B_1, C_1, C_2,$ and D_{21} are known constant matrices of appropriate dimensions.

To formulate the filtering problem, we consider a filter with gain variations of the following form:

$$\dot{\xi}(t) = (A_F + \Delta A_F)\xi(t) + (B_F + \Delta B_F)y(t),$$
$$z_F(t) = (C_F + \Delta C_F)\xi(t), \tag{7.31}$$

where $\xi(t) \in R^n$ is the filter state, $z_F(t)$ is the estimation of $z(t)$, and the constant matrices $A_F, B_F,$ and C_F are filter matrices to be designed, while $\Delta A_F, \Delta B_F,$ and ΔC_F represent the interval type of gain variations with the form (7.3).

Applying filter (7.31) to system (7.30), the following augmented system is obtained:

$$\dot{x}_e(t) = A_e x_e(t) + B_e w(t)$$
$$z_e(t) = C_e x_e(t), \tag{7.32}$$

where $x_e(t) = [x(t)^T, \xi(t)^T]^T$, $z_e(t) = z(t) - z_F(t)$ is the estimation error, and

$$A_e = \begin{bmatrix} A & 0 \\ (B_F + \Delta B_F)C_2 & A_F + \Delta A_F \end{bmatrix},$$

$$B_e = \begin{bmatrix} B_1 \\ (B_F + \Delta B_F)D_{21} \end{bmatrix}, C_e = \begin{bmatrix} C_1 & -C_F - \Delta C_F \end{bmatrix}.$$

The transfer function matrix of the augmented system (7.32) from w to z_e is given by

$$T_{z_e w} = C_e(sI - A_e)^{-1}B_e.$$

Then the problem under consideration in this chapter is as follows.

Non-fragile H_∞ filtering problem with filter gain variations: Given a positive constant γ, find a filter described by (7.31) with the gain variations of the form (7.3) such that the resulting system (7.32) is asymptotically stable and $\|T_{z_e w}\| < \gamma$.

7.3.2   Non-Fragile H_∞ Filter Design Methods

The proofs of the main results of the continuous case are similar to those of the discrete case, so the main results are given without the proofs.

To facilitate the presentation, we denote

$$M_0(\Delta A_F, \Delta B_F, \Delta C_F) = \begin{bmatrix} M_{0a} + M_{0a}^T & M_{0b} & M_{0c}^T \\ * & -\gamma^2 I & 0 \\ * & * & -I \end{bmatrix}, \tag{7.33}$$

where

$$M_{0a} = \left[\begin{array}{c} SA \\ F_A + N\Delta A_F + F_B C_2 + N\Delta B_F C_2 + (S-N)A \\ SA \\ (S-N)A + F_B C_2 + N\Delta B_F C_2 \end{array} \right],$$

$$M_{0b} = \left[\begin{array}{c} SB_1 \\ (S-N)B_1 + F_B D_{21} + N\Delta B_F D_{21} \end{array} \right],$$

$$M_{0c} = \left[C_1 - (C_F + \Delta C_F) \quad C_1 \right].$$

Then the following theorem presents a sufficient condition for the solvability of the non-fragile H_∞ filtering problem with additive uncertainty.

Theorem 7.3 *Consider the system (7.30). Let $\gamma > 0$ and $\theta_a > 0$ be given constants. If there exist matrices $F_A, F_B, C_F, S > 0$, and $N < 0$ such that the following LMIs hold:*

$$M_0(\Delta A_F, \Delta B_F, \Delta C_F) < 0, \quad \theta_{aij}, \ \theta_{bik}, \ \theta_{clj} \in \{-\theta_a, \theta_a\},$$
$$i, j = 1, \cdots, n; \ k = 1, \cdots, p; \ l = 1, \cdots, q, \qquad (7.34)$$

then the filter (7.31) with additive uncertainty described by (7.3) and

$$A_F = N^{-1}F_A, \quad B_F = N^{-1}F_B, \quad C_F = C_F \qquad (7.35)$$

solves the non-fragile H_∞ filtering problem for the system (7.30).

For the non-fragile filter design method given by Theorem 7.3, it should be noted that the number of LMIs involved in (7.34) is $2^{n^2+np+nq}$, which results in the difficulty of implementing the LMI constraints in computation. For example, when $n = 5$ and $p = q = 1$, the number of LMIs involved in (7.34) is 2^{35}, which already exceeds the capacity of the current LMI solver in MATLAB. To overcome the difficulty arising from implementing the design condition given in Theorem 7.3, the following method is developed.

Denote

$$F_{a1} = [f_{a11} \ f_{a12} \ \cdots \ f_{a1l_a}],$$
$$F_{a2} = [f_{a21}^T \ f_{a22}^T \ \cdots \ f_{a2l_a}^T]^T \qquad (7.36)$$

where $l_a = n^2 + np + nq$, and

$$f_{a1k} = \begin{bmatrix} \mathbf{0}_{1\times n} & (Ne_i)^T & \mathbf{0}_{1\times r} & \mathbf{0}_{1\times q} \end{bmatrix}^T,$$

$$f_{a2k} = \begin{bmatrix} e_j^T & \mathbf{0}_{1\times n} & \mathbf{0}_{1\times r} & \mathbf{0}_{1\times q} \end{bmatrix},$$

for $k = (i-1)n + j,\ i,j = 1, \cdots, n.$

$$f_{a1k} = \begin{bmatrix} \mathbf{0}_{1\times n} & (Ne_i)^T & \mathbf{0}_{1\times r} & \mathbf{0}_{1\times q} \end{bmatrix}^T,$$

$$f_{a2k} = \begin{bmatrix} h_j^T C_2 & h_j^T C_2 & h_j^T D_{21} & \mathbf{0}_{1\times q} \end{bmatrix}$$

for $k = n^2 + (i-1)p + j,\ i = 1, \cdots, n, j = 1, \cdots, p.$

$$f_{a1k} = \begin{bmatrix} \mathbf{0}_{1\times n} & \mathbf{0}_{1\times n} & \mathbf{0}_{1\times r} & -g_i^T \end{bmatrix}^T,$$

$$f_{a2k} = \begin{bmatrix} e_j^T & \mathbf{0}_{1\times n} & \mathbf{0}_{1\times r} & \mathbf{0}_{1\times q} \end{bmatrix}$$

for $k = n^2 + np + (i-1)n + j,\ i = 1, \cdots, q, j = 1, \cdots, n.$

Let $k_0, k_1, \cdots, k_{s_a}$ be integers satisfying

$$k_0 = 0 < k_1 < \cdots < k_{s_a} = l_a$$

and matrix Σ have the following structure:

$$\Sigma = \begin{bmatrix} diag[\sigma_{11}^1 \cdots \sigma_{11}^{s_a}] & diag[\sigma_{12}^1 \cdots \sigma_{12}^{s_a}] \\ diag[\sigma_{12}^1 \cdots \sigma_{12}^{s_a}]^T & diag[\sigma_{22}^1 \cdots \sigma_{22}^{s_a}] \end{bmatrix}, \tag{7.37}$$

where $\sigma_{11}^i, \sigma_{12}^i$, and $\sigma_{22}^i \in R^{(k_i - k_{i-1}) \times (k_i - k_{i-1})}, i = 1, \cdots, s_a$. Then, we have the following theorem.

Theorem 7.4 *Consider the system (7.30). Let $\gamma > 0$ and $\theta_a > 0$ be given constants. If there exist matrices $F_A, F_B, C_F,\ S > 0,\ N < 0$, and symmetric matrix Σ with the structure described by (7.37) such that the following LMIs hold:*

$$\begin{bmatrix} Q & F_{a1} \\ F_{a1}^T & 0 \end{bmatrix} + \begin{bmatrix} F_{a2} & 0 \\ 0 & I \end{bmatrix}^T \Sigma \begin{bmatrix} F_{a2} & 0 \\ 0 & I \end{bmatrix} < 0, \tag{7.38}$$

$$\begin{bmatrix} I \\ diag[\theta_{k_{i-1}+j} \cdots \theta_{k_i}] \end{bmatrix}^T \begin{bmatrix} \sigma_{11}^i & \sigma_{12}^i \\ (\sigma_{12}^i)^T & \sigma_{22}^i \end{bmatrix}$$

$$\times \begin{bmatrix} I \\ diag[\theta_{k_{i-1}+j} \cdots \theta_{k_i}] \end{bmatrix} \geq 0,\ for\ all\ \theta_{k_{i-1}+j} \in \{-\theta_a, \theta_a\},$$

$$j = 1, \cdots, k_i - k_{i-1}, i = 1, \cdots, s_a, \tag{7.39}$$

where

$$Q = \begin{bmatrix} M_{as} + M_{as}^T & M_{bs} & M_{cs}^T \\ * & -\gamma^2 I & 0 \\ * & * & -I \end{bmatrix} \tag{7.40}$$

with

$$M_{as} = \begin{bmatrix} SA & SA \\ (S-N)A + F_A + F_B C_2 & (S-N)A + F_B C_2 \end{bmatrix},$$

$$M_{bs} = \begin{bmatrix} SB_1 \\ (S-N)B_1 + F_B D_{21} \end{bmatrix}, M_{cs} = \begin{bmatrix} C_1 - C_F & C_1 \end{bmatrix},$$

then the filter (7.31) with the additive uncertainty described by (7.3) and the filter gain parameters given by (7.35) solve the non-fragile H_∞ filtering problem for the system (7.30).

Remark 7.6 *Notice that the number of LMIs involved in (7.39) with $s_a = 1$ still is $2^{n^2+np+nq}$, so that the difficulty of implementing the LMI constraints remains. To overcome the difficulty, Theorem 7.4 presents a sufficient condition for the non-fragile H_∞ filter design in terms of separator Σ with the structure described by (7.37), where the number of LMIs involved in (7.39) is $\sum_{i=1}^{s_a} 2^{k_i - k_{i-1}}$, which can be controlled not to grow exponentially by reducing the max $k_i - k_{i-1} : i = 1, \cdots, s_a$. However, it should be noted that the design condition given in Theorem 7.4 may be more conservative than that given in Theorem 7.3 because of the structure constraint on Σ. But the smaller the value of s_a is, the less conservatism is introduced.*

Remark 7.7 *In Theorem 7.4, in some cases, the magnitude of the designed A_F (B_F and C_F) may be too large to be applied in practice. For solving the problem, by adding the following constraints*

$$N < -\alpha I, \quad F_A F_A^T < \beta I, \tag{7.41}$$

then the magnitude of A_F can be reduced. In fact, by $A_F = N^{-1} F_A$ and (7.41), it follows that

$$\| A_F \| < \sqrt{\beta}/\alpha.$$

A similar method can be used for the gains B_F and C_F.

In the following, we introduce the result of non-fragile H_∞ filter design with norm-bounded gain variations. And at the same time, the relationship with our result is discussed.

Similar to Yang and Wang [147] and Mahmoud [103] for non-fragile filter design with norm-bounded uncertainty, the norm-bounded type of gain variations $\Delta A_F, \Delta B_F$, and ΔC_F can be overbounded by the norm-bounded uncertainty (7.20).

By using the method in Yang and Wang [147] and Mahmoud [101, 102], the non-fragile H_∞ filter design with norm-bounded gain variations is reduced to solve the following LMI:

$$\begin{bmatrix} Q & \varepsilon M_{a1} & \theta_a M_{a2}^T \\ * & -\varepsilon I & 0 \\ * & * & -\varepsilon I \end{bmatrix} < 0, \tag{7.42}$$

with $S > 0$ and $N < 0$, where Q is defined by (7.40), $\varepsilon > 0$ is a given scalar, and

$$
M_{a1} = \begin{bmatrix} 0 & 0 & 0 \\ NM_a & NM_b & 0 \\ 0 & 0 & 0 \\ 0 & 0 & -M_c \end{bmatrix},
$$

$$
M_{a2} = \begin{bmatrix} E_a & 0 & 0 & 0 \\ E_bC_2 & E_bC_2 & E_bD_{21} & 0 \\ E_c & 0 & 0 & 0 \end{bmatrix}.
$$

The following lemma will show the relationship between condition (7.42) and the condition for designing non-fragile H_∞ filters given in Theorem 7.4.

Lemma 7.3 *Consider system (7.30). If condition (7.42) is feasible, then the condition for designing non-fragile H_∞ filters given in Theorem 7.4 is feasible.*

In Theorem 7.4, we restrict the Lyapunov function matrix P with the structure (2.14) for obtaining the convex design condition, which may result in a more conservative evaluation of the H_∞ performance index bound. So, in this section, for a designed filter, the Lyapunov matrix P without the restriction is exploited for obtaining less conservative evaluation of the H_∞ performance index bound.

When the filter parameter matrices A_F, B_F, and C_F are known, the problem of minimizing γ subject to (7.3) for a given $\theta_a > 0$ and $\parallel T_{z_e\omega} \parallel < \gamma$ can be converted into the one: minimize γ^2 subject to the following LMIs:

$$
\begin{bmatrix} PA_e + A_e^T P & PB_e & C_e^T \\ * & -\gamma^2 I & 0 \\ * & * & -I \end{bmatrix} < 0, \theta_{aij}, \ \theta_{bik}, \ \theta_{clj} \in \{-\theta_a, \theta_a\},
$$

$$
i, j = 1, \cdots, n; \ k = 1, \cdots, p; \ l = 1, \cdots, q, \tag{7.43}
$$

where A_e, B_e, and C_e are defined as in (7.32).

Similar to the design condition given in Theorem 7.3, the above method is also with the numerical computation problem. To solve the problem, the following lemma provides a solution using the structured vertex separator approach.

Denote

$$
G_{ai} = [g_{ai1} \ g_{ai2} \ \cdots \ g_{ail_a}], i = 1, 2 \tag{7.44}
$$

where

$$g_{a1k} = \left[\left(\mathbf{0}_{1\times n} \quad e_i^T\right) P \quad \mathbf{0}_{1\times r} \quad \mathbf{0}_{1\times q}\right]^T,$$

$$g_{a2k} = \left[\mathbf{0}_{1\times n} \quad e_j^T \quad \mathbf{0}_{1\times r} \quad \mathbf{0}_{1\times q}\right],$$

for $k = (i-1)n + j$, $i, j = 1, \cdots, n$.

$$g_{a1k} = \left[\left(\mathbf{0}_{1\times n} \quad e_i^T\right) P \quad \mathbf{0}_{1\times r} \quad \mathbf{0}_{1\times q}\right]^T,$$

$$g_{a2k} = \left[h_j^T C_2 \quad \mathbf{0}_{1\times n} \quad h_j^T D_{21} \quad \mathbf{0}_{1\times q}\right],$$

for $k = n^2 + (i-1)p + j$, $i = 1, \cdots, n, j = 1, \cdots, p$.

$$g_{a1k} = \left[\mathbf{0}_{1\times n} \quad \mathbf{0}_{1\times n} \quad \mathbf{0}_{1\times r} \quad -g_i^T\right]^T,$$

$$g_{a2k} = \left[\mathbf{0}_{1\times n} \quad e_j^T \quad \mathbf{0}_{1\times r} \quad \mathbf{0}_{1\times q}\right],$$

for $k = n^2 + np + (i-1)n + j$, $i = 1, \cdots, q, j = 1, \cdots, n$.

Then we have the following lemma.

Lemma 7.4 *Consider the system (7.30). Let $\gamma > 0$ and $\theta_a > 0$ be constants, and filter parameter matrices A_F, B_F, and C_F be given. Then $\| T_{z_e\omega} \| < \gamma$ holds for all θ_{aij}, θ_{bik}, and θ_{clj} satisfying (7.3), if there exist a matrix $P > 0$ and a symmetric matrix Σ with the structure described by (7.37) such that (7.39) and the following LMI hold:*

$$\begin{bmatrix} Q_s & G_{a1} \\ G_{a1}^T & 0 \end{bmatrix} + \begin{bmatrix} G_{a2} & 0 \\ 0 & I \end{bmatrix}^T \Sigma \begin{bmatrix} G_{a2} & 0 \\ 0 & I \end{bmatrix} < 0 \tag{7.45}$$

where

$$Q_s = \begin{bmatrix} P\begin{pmatrix} A & 0 \\ B_F C_2 & A_F \end{pmatrix} + \begin{pmatrix} A & 0 \\ B_F C_2 & A_F \end{pmatrix}^T P & & \\ * & & \\ & * & \\ & P\begin{pmatrix} B_1 \\ B_F D_{21} \end{pmatrix} & \begin{pmatrix} C_1^T \\ -C_F^T \end{pmatrix} \\ & -\gamma^2 I & 0 \\ & * & -I \end{bmatrix}.$$

Remark 7.8 *For evaluating the H_∞ performance bound of the transfer function from ω to z_e, the condition given in Lemma 7.4 usually is less conservative than that given in Theorem 7.4 because no structure constraint on the Lyapunov function matrix in Lemma 7.4 is imposed.*

7.3.3 Example

To illustrate the effectiveness of the proposed non-fragile H_∞ filter, an example is given to provide a comparison between the proposed non-fragile H_∞ filter design method and the existing non-fragile H_∞ filter design method with the norm-bounded gain variations. The considered system is a linear system of the form (7.30) with

$$A = \begin{bmatrix} 0 & 1 & 0 \\ 0 & 0 & 1 \\ -1 & -2.5 & -1 \end{bmatrix}, B_1 = \begin{bmatrix} -1.5 & 0 \\ 1 & 0 \\ 2 & 0 \end{bmatrix},$$

$$C_1 = \begin{bmatrix} 3 & -2 & 1 \end{bmatrix}, C_2 = \begin{bmatrix} -2 & 1 & 2 \end{bmatrix}, D_{21} = \begin{bmatrix} 0 & 0.9 \end{bmatrix}.$$

For the case that the designed filter contains no gain variations, by Theorem 7.3, the optimal H_∞ performance index of the standard closed-loop system is achieved as $\gamma_{opt} = 4.6536$.

First, we design a non-fragile H_∞ filter by condition (7.42).

Assume that the designed filter is with norm-bounded additive uncertainties described by (7.20), by applying the condition (7.42) with $\theta_a = 0.02$ and $\varepsilon = 0.0086$ to design a non-fragile filter with gains as

$$A_{F_{norm}} = \begin{bmatrix} -7.1074 & 3.2284 & 4.6952 \\ -4.8358 & 3.5381 & 7.9641 \\ 11.5712 & -10.9681 & -17.9224 \end{bmatrix},$$

$$B_{F_{norm}} = \begin{bmatrix} -2.8669 & -3.0257 & 7.8866 \end{bmatrix}^T,$$

$$C_{F_{norm}} = \begin{bmatrix} 3.0002 & -1.9999 & 1.0001 \end{bmatrix},$$

and the H_∞ performance index of the obtained non-fragile filter is 5.8160. Note that the ε used here is the one obtained by searching such that the H_∞ performance index is optimal.

Then, we design a non-fragile H_∞ filter by Theorem 7.4.

Assume that the designed filter is with the additive uncertainties described by (7.3). For this case, it is difficult to apply Theorem 7.3 to design a non-fragile H_∞ filter because the number of the LMI constraints involved in (7.34) is 2^{15}. However, Theorem 7.4 is applicable for this case. By applying Theorem 7.4 with $\theta_a = 0.02$ and $k_i = i, i = 1, \cdots, 15$, that is, $s_a = 15$ as well as $k_i = 3i, i = 1, \cdots, 5$, that is, $s_a = 5$ to design the non-fragile filters and the filter gains are obtained as follows.

For $s_a = 15$,

$$A_{F_{in15}} = 10^3 \times \begin{bmatrix} -1.2682 & 0.5679 & 1.1437 \\ 0.0018 & -0.0001 & 0.0008 \\ -0.0072 & -0.0039 & -0.0035 \end{bmatrix},$$

$$B_{F_{in15}} = \begin{bmatrix} -588.9287 & 0.4478 & 0.0681 \end{bmatrix}^T,$$

$$C_{F_{in15}} = \begin{bmatrix} 3.0000 & -2.0000 & 1.0000 \end{bmatrix}.$$

For $s_a = 5$,

$$A_{F_{in5}} = \begin{bmatrix} -739.0366 & 332.4686 & 669.9035 \\ 1.8166 & -0.0200 & 0.8219 \\ -7.1592 & -4.1882 & -3.9043 \end{bmatrix},$$

$$B_{F_{in5}} = \begin{bmatrix} -344.4121 & 0.4258 & 0.2033 \end{bmatrix}^T,$$

$$C_{F_{in5}} = \begin{bmatrix} 3.0000 & -2.0000 & 1.0000 \end{bmatrix},$$

the H_∞ performance indexes of the obtained non-fragile filters are 5.3126 and 5.3068, respectively.

The gains of the two filters achieved by Theorem 7.4 are too large and they are not desired, as indicated in Remark 7.7. By applying Theorem 7.4 with (7.41) and with $\alpha = 0.5$, $\beta = 10000$, the filters are obtained with the gains as follows.

For $s_a = 15$,

$$A_{F_{in15\alpha\beta}} = \begin{bmatrix} -63.8295 & 28.4892 & 55.4688 \\ 0.7940 & -0.3607 & 0.3089 \\ 3.1374 & -2.7461 & -1.7502 \end{bmatrix},$$

$$B_{F_{in15\alpha\beta}} = \begin{bmatrix} -28.9571 & 0.4458 & 0.0752 \end{bmatrix}^T,$$

$$C_{F_{in15\alpha\beta}} = \begin{bmatrix} 1.3222 & -2.4778 & 0.2493 \end{bmatrix}.$$

For $s_a = 5$,

$$A_{F_{in5\alpha\beta}} = \begin{bmatrix} -71.2150 & 23.4192 & 47.1944 \\ 1.8273 & -0.0392 & 0.7857 \\ -7.2598 & -4.0785 & -3.6940 \end{bmatrix},$$

$$B_{F_{in5\alpha\beta}} = \begin{bmatrix} -27.2270 & 0.4399 & 0.1153 \end{bmatrix}^T,$$

$$C_{F_{in5\alpha\beta}} = \begin{bmatrix} 2.9757 & -2.0069 & 0.9891 \end{bmatrix},$$

and the corresponding H_∞ performance indexes remain as 5.3126 and 5.3068, respectively. Comparing the above design without (7.41), the magnitude of A_F has been significantly reduced.

In this part, for the above designed filters, Lemma 7.4 gives better evaluations of the H_∞ performance index bounds.

First, to facilitate the presentation, denote the filter designed by the existing method (7.42) as $filter_{norm}$ while denoting the filters designed by Theorem 7.4 as $filter_{in15}$ ($s_a = 15$) and $filter_{in5}$ ($s_a = 5$). And correspondingly, the filters designed by Theorem 7.4 with (7.41) are denoted as $filter_{in15\alpha\beta}$ ($s_a = 15$) and $filter_{in5\alpha\beta}$ ($s_a = 5$).

By applying Lemma 7.4, the H_∞ performance indexes of the non-fragile filter $filter_{norm}$ are $\gamma = 5.5800$ ($s_a = 15$) and $\gamma = 5.5747$ ($s_a = 5$) while the H_∞ performance indexes of the non-fragile filters $filter_{in15}$ and $filter_{in5}$ are $\gamma = 5.2996$ ($s_a = 15$) and $\gamma = 5.2832$ ($s_a = 5$), respectively. Correspondingly,

TABLE 7.3

Performance Index by Design with $\theta_a = 0.02$

	Condition (7.42)	Theorem 7.3 ($s_a = 15$)	Theorem 7.3 ($s_a = 5$)
γ	5.8160	5.3126	5.3068

TABLE 7.4

Performance Index by Lemma 7.4 with $\theta_a = 0.02$

	$filter_{in15}$	$filter_{in5}$	$filter_{in15\alpha\beta}$	$filter_{in5\alpha\beta}$	$filter_{norm}$
$\gamma(s_a = 15)$	5.2996	—	5.3036	—	5.5800
$\gamma(s_a = 5)$	—	5.2832	—	5.2993	5.5747

for the filters $filter_{in15\alpha\beta}$ and $filter_{in5\alpha\beta}$, Lemma 7.4 gives H_∞ performance indexes as $\gamma = 5.3036$ ($s_a = 15$) and $\gamma = 5.2993$ ($s_a = 5$), respectively.

Tables are given above to provide a comparison between the non-fragile H_∞ filter with interval gain variations and the non-fragile H_∞ filter with the norm-bounded gain variations.

First, the H_∞ performance indexes achieved by the designs are shown in Table 7.3.

From Table 7.3, it is easy to see that compared with the optimal H_∞ performance index bound $\gamma_{opt} = 4.6536$, the performance index of the filter designed by (7.42) is degraded 24.98%. And the performance indexes of the filters designed by Theorem 7.4 are degraded 14.16% ($s_a = 15$) and 13.95% ($s_a = 5$), which are much more improved than 24.98%.

Then, for the designed filters, Lemma 7.4 gives better performance indexes as shown in Table 7.4.

Obviously, compared with the optimal H_∞ performance index $\gamma_{opt} = 4.6536$, by using Lemma 7.4, the H_∞ performance indexes of the non-fragile filter $filter_{norm}$ are degraded 19.91% for $s_a = 15$ and degraded 19.79% for $s_a = 5$. Correspondingly, the performance indexes for the non-fragile filters $filter_{in15}$ and $filter_{in5}$ are degraded 13.88% for $s_a = 15$ and 13.53% for $s_a = 5$, respectively. And the performance indexes for the filters $filter_{in15\alpha\beta}$ and $filter_{in5\alpha\beta}$ are degraded 13.97% for $s_a = 15$ and 13.88% for $s_a = 5$, respectively. This phenomenon indicates the superiority of our proposed method.

Finally, a simulation is given to illustrate the superiority of the presented method. Let the system initial state be $x_0 = [-0.1; 0.5; 0.5]$ and the filter initial state be $\xi_0 = [-1; 0.5; -0.5]$. And let the disturbance $w(t)$ be

$$w(t) = \begin{cases} -0.3, & 20 \le t \le 30 \text{ (second)} \\ 0 & \text{otherwise.} \end{cases}$$

Then, Figure 7.2 shows the estimation error responses of the non-fragile filters designed by the condition (7.42) and by Theorem 7.4 with constraint (7.41), with $\alpha = 0.5, \beta = 10000$.

From this example, we can see that the worst case of our result using

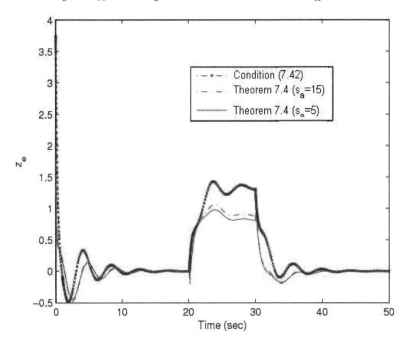

FIGURE 7.2
Comparison of the error responses of $z_e(t)$ with $\theta_a = 0.02$.

Theorem 7.4 ($s_a=15$) is more effective than the non-fragile H_∞ filter existence condition (7.42). And we can also see that the smaller the value of s_a is, the larger non-fragility of the filter is achieved.

7.4   Sparse Structured H_∞ Filter Design

In this section, the non-fragile H_∞ filtering problem with *sparse structure* is studied.

7.4.1   Problem Statement

First, a sparse structure is defined. Consider system (7.1) and a full-order filter given by

$$\xi(k+1) = A_F\xi(k) + B_Fy(k),$$
$$z_F(k) = C_F\xi(k), \tag{7.46}$$

where $\xi(k) \in R^n$ is the filter state, $z_F(k)$ is the estimation of $z(k)$, and the constant matrices $A_F, B_F,$ and C_F are filter matrices to be designed.

First, a method of finding a class of feasible sparse structures for the designed filter to meet a prescribed H_∞ performance requirement is presented as follows.

Denote

$$G_{0z_e w} = C_{e0}(zI - A_{e0})^{-1}B_{e0}, \tag{7.47}$$

where

$$A_{e0} = \begin{bmatrix} A & 0 \\ B_F C_2 & A_F \end{bmatrix},$$

$$B_{e0} = \begin{bmatrix} B_1 \\ B_F D_{21} \end{bmatrix}, C_{e0} = \begin{bmatrix} C_1 & -C_F - \Delta C_F \end{bmatrix}.$$

Let filter gain matrices $A_F, B_F,$ and C_F be given, and such that

$$\|G_{0z_e w}(z) = C_{e0}(zI - A_{e0})^{-1}B_{e0}\| < \gamma. \tag{7.48}$$

By Lemma 2.5, without loss of generality, we may assume that $A_F, B_F,$ and C_F satisfy the following assumption.

Assumption 7.1 *There exist a symmetric matrix $P > 0$ and matrix $G = \begin{bmatrix} Y & N \\ N & -N \end{bmatrix}$ such that*

$$\begin{bmatrix} P - G - G^T & 0 & G^T A_{e0} & G^T B_{e0} \\ * & -I & C_{e0} & 0 \\ * & * & -P & 0 \\ * & * & * & -\gamma^2 I \end{bmatrix} < 0. \tag{7.49}$$

Moreover, for the designed fully parameterized H_∞ filter gains $A_F, B_F,$ and C_F, the characteristic polynomial of A_F is described as

$$det(sI - A_F) = s^n + \alpha_{n-1}s^{n-1} + \ldots + \alpha_1 + \alpha_0, \tag{7.50}$$

where $\alpha_0, \alpha_1, \ldots, \alpha_{n-1}$ are scalars. Assume there exists a row vector c such that $Q = [(cA_F^{n-1})^T \cdots (cA_F)^T c^T]^T$ is nonsingular. Construct the following transformation matrix:

$$T = \begin{bmatrix} 1 & \alpha_{n-1} & \cdots & & \alpha_1 \\ & \ddots & \ddots & & \vdots \\ & & & \ddots & \alpha_{n-1} \\ & \mathbf{0} & & & 1 \end{bmatrix} Q. \tag{7.51}$$

Then, by using the transformation matrix T, we have the following filter gains with the sparse structure

$$\bar{A}_F = TA_F T^{-1} = A_{Fc} + f_A v$$
$$\bar{B}_F = TB_F, \quad \bar{C}_F = C_F T^{-1}, \tag{7.52}$$

where

$$A_{Fc} = \begin{bmatrix} 0 & & \cdots & 0 & 0 \\ 1 & & & & 0 \\ & & \ddots & & \vdots \\ \mathbf{0} & & & 1 & 0 \end{bmatrix},$$

$$f_A = \begin{bmatrix} \alpha_0 & \alpha_1 & \cdots & \alpha_{n-1} \end{bmatrix}^T, \quad v = \begin{bmatrix} 0 & \cdots & -1 \end{bmatrix}.$$

Hereafter, a filter described by

$$\bar{\xi}(k+1) = \bar{A}_F \bar{\xi}(k) + \bar{B}_F \bar{y}(k)$$
$$\bar{z}_F(k) = \bar{C}_F \bar{\xi}(k), \tag{7.53}$$

with the structure described by (7.52) is said to be a sparse structured filter. Then, consider a sparse structured filter with gain variations described by

$$\dot{\bar{\xi}}(k) = (\bar{A}_F + \Delta\bar{A}_F)\bar{\xi}(k) + (\bar{B}_F + \Delta\bar{B}_F)\bar{y}(k)$$
$$\bar{z}_F(k) = (\bar{C}_F + \Delta\bar{C}_F)\bar{\xi}(k), \tag{7.54}$$

where $\bar{A}_F$, $\bar{B}_F$, and $\bar{C}_F$ are with the structure described by (7.52). $\Delta\bar{A}_F$, $\Delta\bar{B}_F$, and $\Delta\bar{C}_F$ represent the additive gain variations with the following form:

$$\Delta\bar{A}_F = \Delta f_A v = [\theta_{a\bar{a}i}]_{n\times 1} v, |\theta_{a\bar{a}i}| \leq \theta_a, i = 1, \cdots, n,$$
$$\Delta\bar{B}_F = E_{Ba} diag[\theta_{a\bar{b}1}, \cdots, \theta_{a\bar{b}r_B}] E_{Bb}, |\theta_{a\bar{b}i}| \leq \theta_a, i = 1, \cdots, r_B,$$
$$\Delta\bar{C}_F = E_{Ca} diag[\theta_{a\bar{c}1}, \cdots, \theta_{a\bar{c}r_C}] E_{Cb}, |\theta_{a\bar{c}i}| \leq \theta_a, i = 1, \cdots, r_C \tag{7.55}$$

where E_{Ba}, E_{Bb}, E_{Ca}, and E_{Cb} are constant matrices.

Remark 7.9 *For the additive case, the description of the gain variations in $\bar{B}_F$ and $\bar{C}_F$ given by (7.55) can cover the cases where $\bar{B}_F$ and $\bar{C}_F$ are with or without trivial elements.*

Applying filter (7.54) to system (7.1), the following augmented system is obtained:

$$\dot{\bar{x}}_e(k) = \bar{A}_e \bar{x}_e(k) + \bar{B}_e w(k)$$
$$\bar{z}_e(k) = \bar{C}_e \bar{x}_e(k) \tag{7.56}$$

where $\bar{x}_e(k) = [x(k)^T, \ \bar{\xi}(k)^T]^T$, $\bar{z}_e(k) = z(k) - \bar{z}_F(k)$ is the estimation error, and

$$\bar{A}_e = \begin{bmatrix} A & 0 \\ (\bar{B}_F + \Delta\bar{B}_F)C_2 & \bar{A}_F + \Delta\bar{A}_F \end{bmatrix}, \quad \bar{B}_e = \begin{bmatrix} B_1 \\ (\bar{B}_F + \Delta\bar{B}_F)D_{21} \end{bmatrix}$$

$$\bar{C}_e = \begin{bmatrix} C_1 & -\bar{C}_F - \Delta\bar{C}_F \end{bmatrix}.$$

The transfer matrix of the augmented system (7.56) from ω to $\bar{z}_e$ is given by

$$\bar{G}_{\bar{z}_e\omega}(z) = \bar{C}_e(zI - \bar{A}_e)^{-1}\bar{B}_e.$$

Then the problem under consideration in this section is as follows.

Non-fragile H_∞ filtering problem with sparse structure: Given positive constants γ and θ_a, find a filter described by (7.54) with the additive gain variations of form (7.55) such that the resulting system (7.56) is asymptotically stable and $\|G_{\bar{z}_e\omega}(z)\| < \gamma$.

The following preliminaries and lemmas will be used in the sequel.

Lemma 7.5 *Let the filter matrices A_F, B_F, and C_F satisfy Assumption 7.1, and let T satisfy (7.51), then there exist a symmetric matrix $\bar{P} > 0$ and matrix $\bar{G} = \begin{bmatrix} \bar{Y} & \bar{N} \\ \bar{N} & -\bar{N} \end{bmatrix}$ such that*

$$\begin{bmatrix} \bar{P} - \bar{G} - \bar{G}^T & 0 & \bar{G}^T\bar{A}_{e0} & \bar{G}^T\bar{B}_{e0} \\ * & -I & \bar{C}_{e0} & 0 \\ * & * & -\bar{P} & 0 \\ * & * & * & -\gamma^2 I \end{bmatrix} < 0 \qquad (7.57)$$

where

$$\bar{A}_{e0} = \begin{bmatrix} \bar{A} & 0 \\ \bar{B}_F\bar{C}_2 & \bar{A}_F \end{bmatrix}, \quad \bar{B}_{e0} = \begin{bmatrix} \bar{B}_1 \\ \bar{B}_F\bar{D}_{21} \end{bmatrix}$$

$$\bar{C}_{e0} = \begin{bmatrix} \bar{C}_1 & -\bar{C}_F \end{bmatrix} \qquad (7.58)$$

with $\bar{A}_F$, $\bar{B}_F$, and $\bar{C}_F$ given by (7.52), and

$$\bar{A} = TAT^{-1}, \quad \bar{B}_1 = TB_1, \quad \bar{C}_1 = C_1T^{-1},$$
$$\bar{C}_2 = C_2T^{-1}, \quad \bar{D}_{21} = D_{21}. \qquad (7.59)$$

Proof 7.5 *By Assumption 7.1, it follows that there exists a symmetric matrix $P > 0$ and matrix $G = \begin{bmatrix} Y & N \\ N & -N \end{bmatrix}$ such that (7.49) holds. Denote $\Gamma_1 = diag[T^{-1} \ T^{-1}]$, $\bar{\Gamma}_1 = diag[\Gamma_1 \ I \ \Gamma_1 \ I]$. Let*

$$\bar{G} = \begin{bmatrix} \bar{Y} & \bar{N} \\ \bar{N} & -\bar{N} \end{bmatrix} = \Gamma_1^T G\Gamma_1, \bar{P} = \Gamma_1^T P\Gamma_1.$$

Then from (7.49), it follows that

$$
\begin{bmatrix} \bar{P} - \bar{G} - \bar{G}^T & 0 & \bar{G}^T \bar{A}_{e0} & \bar{G}^T \bar{B}_{e0} \\ * & -I & \bar{C}_{e0} & 0 \\ * & * & -\bar{P} & 0 \\ * & * & * & -\gamma^2 I \end{bmatrix}
$$

$$
= \bar{\Gamma}_1^T \begin{bmatrix} P - G - G^T & 0 & G^T A_{e0} & G^T B_{e0} \\ * & -I & C_{e0} & 0 \\ * & * & -P & 0 \\ * & * & * & -\gamma^2 I \end{bmatrix} \bar{\Gamma}_1 < 0
$$

Thus, the proof is complete.

Lemma 7.5 shows that the state-space realization (7.58) of system (7.56) without the filter gain variations admits a slack variable matrix with the structure $\bar{G} = \begin{bmatrix} \bar{Y} & \bar{N} \\ \bar{N} & -\bar{N} \end{bmatrix}$ satisfying (7.57). As an extension of the fact, we have the following result for system (7.56) with the filter gain variations.

Lemma 7.6 *Let γ and θ_a be given positive constants. If there exist a symmetric matrix $\bar{P} > 0$ and matrix $\bar{G} = \begin{bmatrix} \bar{Y} & \bar{N} \\ \bar{N} & -\bar{N} \end{bmatrix}$ such that*

$$
\begin{bmatrix} \bar{P} - \bar{G} - \bar{G}^T & 0 & \bar{G}^T \bar{A}_{ea} & \bar{G}^T \bar{B}_{ea} \\ * & -I & \bar{C}_{ea} & 0 \\ * & * & -\bar{P} & 0 \\ * & * & * & -\gamma^2 I \end{bmatrix} < 0 \qquad (7.60)
$$

holds for all $\Delta \bar{A}_F$, $\Delta \bar{B}_F$, and $\Delta \bar{C}_F$ satisfying (7.55), where

$$
\bar{A}_{ea} = \begin{bmatrix} \bar{A} & 0 \\ (\bar{B}_F + \Delta \bar{B}_F) \bar{C}_2 & \bar{A}_F + \Delta \bar{A}_F \end{bmatrix}, \quad \bar{B}_{ea} = \begin{bmatrix} \bar{B}_1 \\ (\bar{B}_F + \Delta \bar{B}_F) \bar{D}_{21} \end{bmatrix}
$$

$$
\bar{C}_{ea} = \begin{bmatrix} \bar{C}_1 & -\bar{C}_F - \Delta \bar{C}_F \end{bmatrix} \qquad (7.61)
$$

with $\bar{A}$, $\bar{B}_1$, $\bar{C}_1$, $\bar{C}_2$, and $\bar{D}_{21}$ defined by (7.59), then the filter described by (7.54) with the gain variations of form (7.55) solves the non-fragile H_∞ filtering problem with sparse structure.

Proof 7.6 *Denote $\Gamma_2 = \mathrm{diag}[T \ I], \bar{\Gamma}_2 = \mathrm{diag}[\Gamma_2 \ I \ \Gamma_2 \ I]$. Let $\bar{P}_s = \Gamma_2^T \bar{P} \Gamma_2, \bar{G}_s = \Gamma_2^T \bar{G} \Gamma_2$. Then, by (7.56) and (7.59)–(7.61), it follows that*

$$
\begin{bmatrix} \bar{P}_s - \bar{G}_s - \bar{G}_s^T & 0 & \bar{G}_s^T \bar{A}_e & \bar{G}_s^T \bar{B}_e \\ * & -I & \bar{C}_e & 0 \\ * & * & -\bar{P}_s & 0 \\ * & * & * & -\gamma^2 I \end{bmatrix}
$$

$$
= \bar{\Gamma}_2^T \begin{bmatrix} \bar{P} - \bar{G} - \bar{G}^T & 0 & \bar{G}^T \bar{A}_{ea} & \bar{G}^T \bar{B}_{ea} \\ * & -I & \bar{C}_{ea} & 0 \\ * & * & -\bar{P} & 0 \\ * & * & * & -\gamma^2 I \end{bmatrix} \bar{\Gamma}_2 < 0.
$$

From Lemma 2.5, the conclusion follows.

7.4.2   Non-Fragile H_∞ Filter Design with Sparse Structures

In this section, a method of designing non-fragile H_∞ filters with sparse structure is presented by using Lemma 7.6.

$$
M_{0s}(\Delta A_F, \Delta B_F, \Delta C_F)
$$

$$
= \begin{bmatrix}
\Xi_1 & \Xi_2 & 0 & \bar{S}^T\bar{A} & \bar{S}^T\bar{A} & \bar{S}^T\bar{B}_1 \\
* & \Xi_3 & 0 & \Xi_5 & \Xi_6 & \Xi_7 \\
* & * & -I & \Xi_4 & \bar{C}_1 & 0 \\
* & * & * & -\bar{P}_{11} & -\bar{P}_{12} & 0 \\
* & * & * & * & -\bar{P}_{22} & 0 \\
* & * & * & * & * & -\gamma^2 I
\end{bmatrix} \tag{7.62}
$$

where

$$
\begin{aligned}
\Xi_1 &= \bar{P}_{11} - \bar{S} - \bar{S}^T, \Xi_2 = \bar{P}_{12} - \bar{S} - \bar{S}^T, \\
\Xi_3 &= \bar{P}_{22} - \bar{S} - \bar{S}^T + \bar{N} + \bar{N}^T, \Xi_4 = \bar{C}_1 - \bar{C}_F - \Delta\bar{C}_F, \\
\Xi_5 &= (\bar{S} - \bar{N})^T\bar{A} + \bar{F}_B\bar{C}_2 + \bar{N}^T(\Delta\bar{B}_F\bar{C}_2 + \Delta\bar{A}_F + A_{Fc}) + \bar{F}_A v, \\
\Xi_6 &= (\bar{S} - \bar{N})^T\bar{A} + \bar{F}_B\bar{C}_2 + \bar{N}^T\Delta\bar{B}_F\bar{C}_2, \\
\Xi_7 &= (\bar{S} - \bar{N})^T\bar{B}_1 + \bar{F}_B\bar{D}_{21} + \bar{N}^T\Delta\bar{B}_F\bar{D}_{21},
\end{aligned} \tag{7.63}
$$

and $\bar{A}$, $\bar{B}_1$, $\bar{C}_1$, $\bar{C}_2$, and $\bar{D}_{21}$ are defined by (7.59).

Then the following theorem presents a sufficient condition for the solvability of the non-fragile H_∞ filtering problem with sparse structure.

Theorem 7.5 *Consider system (7.1). Let $\gamma > 0$ and $\theta_a > 0$ be given constants. If there exist matrices $\bar{F}_A$, $\bar{F}_B$, $\bar{C}_F$, $\bar{S}, \bar{N}$ such that the following conditions hold:*

$$
M_{0s}(\Delta\bar{A}_F, \Delta\bar{B}_F, \Delta\bar{C}_F) < 0, \quad \text{for all } \theta_{a\bar{a}i}, \theta_{a\bar{b}j}, \theta_{a\bar{c}k} \in \{-\theta_a, \theta_a\},
$$
$$
i = 1, \cdots, n; \ j = 1, \cdots, r_B; \ k = 1, \cdots, r_C \tag{7.64}
$$

then filter (7.54) with the additive gain variations described by form (7.55) and

$$
A_F = (\bar{N}^T)^{-1}\bar{F}_A v + A_{Fc}, \quad B_F = (\bar{N}^T)^{-1}\bar{F}_B, \quad C_F = \bar{C}_F \tag{7.65}
$$

solves the non-fragile H_∞ filtering problem with sparse structure for system (7.1).

Proof 7.7 *By Lemma 7.6, it is sufficient to show that there exist a matrix $\bar{G}$*

with structure $\bar{G} = \begin{bmatrix} \bar{Y} & \bar{N} \\ \bar{N} & -\bar{N} \end{bmatrix}$ and a symmetric positive matrix $\bar{P} > 0$ such that

$$
M_1 = \begin{bmatrix} \bar{P} - \bar{G} - \bar{G}^T & 0 & \bar{G}^T \bar{A}_{ea} & \bar{G}^T \bar{B}_{ea} \\ * & -I & \bar{C}_{ea} & 0 \\ * & * & -\bar{P} & 0 \\ * & * & * & -\gamma^2 I \end{bmatrix} < 0 \tag{7.66}
$$

holds for all $\theta_{a\bar{a}i}$, $\theta_{a\bar{b}j}$, and $\theta_{a\bar{c}k}$ satisfying (7.55).

Denote

$$
\bar{S} = \bar{Y} + \bar{N}, \quad \Gamma_1 = \begin{bmatrix} I & I \\ I & 0 \end{bmatrix}, \bar{\Gamma}_1 = diag\{\Gamma_1, I, \Gamma_1, I\}.
$$

Then (7.66) is equivalent to

$$
M_2 = \bar{\Gamma}_1 M_1 \bar{\Gamma}_1^T
$$
$$
= \begin{bmatrix} \Xi_1 & \Xi_2 & 0 & \bar{S}^T \bar{A} & \bar{S}^T \bar{A} & \bar{S}^T \bar{B}_1 \\ * & \Xi_3 & 0 & \Pi_1 & \Pi_2 & \Pi_3 \\ * & * & -I & \Xi_4 & \bar{C}_1 & 0 \\ * & * & * & -\bar{P}_{11} & -\bar{P}_{12} & 0 \\ * & * & * & * & -\bar{P}_{22} & 0 \\ * & * & * & * & * & -\gamma^2 I \end{bmatrix} < 0 \tag{7.67}
$$

holds for all $\theta_{a\bar{a}i}$, $\theta_{a\bar{b}j}$, and $\theta_{a\bar{c}k}$ satisfying (7.55), where $\Xi_1, \Xi_2, \Xi_3, \Xi_4$ are defined by (7.63), and

$$
\Pi_1 = (\bar{S} - \bar{N})^T \bar{A} + \bar{N}^T (\bar{B}_F \bar{C}_2 + \Delta \bar{B}_F \bar{C}_2 + A_{Fc} + f_A v + \Delta \bar{f}_A v),
$$
$$
\Pi_2 = (\bar{S} - \bar{N})^T \bar{A} + \bar{N}^T \bar{B}_F \bar{C}_2 + \bar{N}^T \Delta \bar{B}_F \bar{C}_2,
$$
$$
\Pi_3 = (\bar{S} - \bar{N})^T \bar{B}_1 + \bar{N}^T \bar{B}_F \bar{D}_{21} + \bar{N}^T \Delta \bar{B}_F \bar{D}_{21}.
$$

Obviously, M_2 is convex for each θ_i, $\theta_i \in \{\theta_{a\bar{a}i}, \theta_{a\bar{b}j}, \text{ and } \theta_{a\bar{c}k}\}$ satisfying (7.55), so (7.67) is equivalent to

$$
M_3 = \begin{bmatrix} \Xi_1 & \Xi_2 & 0 & \bar{S}^T \bar{A} & \bar{S}^T \bar{A} & \bar{S}^T \bar{B}_1 \\ * & \Xi_3 & 0 & \Pi_1 & \Pi_2 & \Pi_3 \\ * & * & -I & \Xi_4 & \bar{C}_1 & 0 \\ * & * & * & -\bar{P}_{11} & -\bar{P}_{12} & 0 \\ * & * & * & * & -\bar{P}_{22} & 0 \\ * & * & * & * & * & -\gamma^2 I \end{bmatrix} < 0 \tag{7.68}
$$

for all θ_{aij}, θ_{bik}, $\theta_{clj} \in \{-\theta_a, \theta_a\}$.

By (7.62), (7.65), and the Schur complement, it concludes that (7.68) is equivalent to (7.64) with respect to additive gain variations (7.55).

Remark 7.10 *Theorem 7.5 presents an LMI-based sufficient condition for solving the non-fragile sparse structured H_∞ filtering problem. By the proofs of Theorem 7.5 and Lemma 2.5, Theorem 7.5 also shows that this problem becomes a convex one when the state-space realizations of the designed filters with gain variations admit the slack variable matrix $\bar{G}$ with structure (2.14). For the case that the designed filter contains no gain variations, from Lemma 2.5, it follows that the design condition given in Theorem 7.5 reduces to a necessary and sufficient condition for the standard sparse structured H_∞ filtering problem, which means that the constraint (2.14) on the slack variable $\bar{G}$ does not result in any conservativeness for the standard sparse structured H_∞ filter design.*

For obtaining the convex design conditions, we restrict the slack variable matrix $\bar{G}$ with structure (2.14) in Theorem 7.5, which may result in a more conservative evaluation of the H_∞ performance bounds. So, in this section, for a designed filter with the sparse structure, the slack variable matrix $\bar{G}$ without the restriction is exploited for obtaining a less conservative evaluation of the H_∞ performance bounds.

When the filter parameter matrices $\bar{A}_F, \bar{B}_F$, and $\bar{C}_F$ are known, the following lemma presents a sufficient condition for evaluating the non-fragile H_∞ performance of system (7.56).

Denote

$$M_{0ss}(\Delta\bar{A}_F, \Delta\bar{B}_F, \Delta\bar{C}_F) = \begin{bmatrix} \bar{P} - \bar{G} - \bar{G}^T & 0 & \bar{G}^T\bar{A}_e & \bar{G}^T\bar{B}_e \\ * & -I & \bar{C}_e & 0 \\ * & * & -\bar{P} & 0 \\ * & * & * & -\gamma^2 I \end{bmatrix}, \quad (7.69)$$

where $\bar{A}_e, \bar{B}_e$, and $\bar{C}_e$ are defined by (7.56).

Lemma 7.7 *Consider system (7.1). Let $\gamma > 0, \theta_a > 0$ be constants, and filter parameter matrices $\bar{A}_F, \bar{B}_F$, and $\bar{C}_F$ be given. Then $\| \bar{G}_{\bar{z}_e\omega} \| < \gamma$ holds for all gain variations described by (7.55), if there exist a matrix $\bar{G}$ and a positive-definite matrix $\bar{P} > 0$ such that the following conditions hold:*

$$M_{0ss}(\Delta\bar{A}_F, \Delta\bar{B}_F, \Delta\bar{C}_F) < 0, \quad \text{for all } \theta_{a\bar{a}i}, \theta_{a\bar{b}j}, \theta_{a\bar{c}k} \in \{-\theta_a, \theta_a\},$$
$$i = 1, \cdots, n; \ j = 1, \cdots, r_B; \ k = 1, \cdots, r_C. \quad (7.70)$$

Due to the fact that the design method given by Theorem 7.5 is based on Assumption 7.1, we need to give a method for designing H_∞ filters satisfying Assumption 7.1. First, the following standard H_∞ filter design method is introduced.

Lemma 7.8 *[106] Consider system (7.1). Let $\gamma > 0$ be a given scalar. If* *and only if there exist matrices $Z, M, C_F, X > 0$, and $R > 0$ such that the following LMI holds*

$$
\begin{bmatrix}
-R & -R & 0 & RA & RA & RB_1 \\
* & -X & 0 & XA + ZC_2 + M & XA + ZC_2 & RB_1 + ZD_{21} \\
* & * & -I & C_1 - C_F & L & 0 \\
* & * & * & -R & -R & 0 \\
* & * & * & * & -X & 0 \\
* & * & * & * & * & -\gamma^2 I
\end{bmatrix} < 0,
$$

(7.71)

then filter (7.46) with

$$
A_F = (R - X)^{-1}M, B_F = (R - X)^{-1}Z, \quad C_F = C_F \tag{7.72}
$$

renders (7.48), which holds.

Then, by using Lemma 7.8, we present a design method to design H_∞ filters satisfying Assumption 7.1.

Lemma 7.9 *Let $G = P = \begin{bmatrix} X & R - X \\ R - X & -(R - X) \end{bmatrix}$. Then, filter (7.46) with*

$$
A_F = (R - X)^{-1}M, B_F = (R - X)^{-1}Z, \quad C_F = C_F \tag{7.73}
$$

renders (7.49), which holds.

Proof 7.8 *As in Palhares and Peres [106], partition P^{-1} as $P^{-1} = \begin{bmatrix} Y & N \\ N^T & V \end{bmatrix}$. Let $Y = N$, then, similarly by using the arguments developed in Palhares and Peres [106], the conclusion follows.*

Remark 7.11 *Lemma 7.9 presents a method of designing H_∞ filters for satisfying Assumption 7.1, which is the initial step for the following algorithm.*

Based on Lemma 7.9, the following algorithm is presented to solve the non-fragile H_∞ filtering problem with sparse structure described by (7.52).

Algorithm 7.1 *Let $\gamma > 0$ and $\theta_a > 0$ be given scalars.*

   Step 1. *Minimize γ subject to LMIs $X > 0, R > 0$, and (7.71). Denote the optimal solutions as $X = X_{opt}, R = R_{opt}, Z = Z_{opt}, M = M_{opt}$, and $C_F = C_{Fopt}$. Substitute the matrices $(X_{opt}, R_{opt}, Z_{opt}, M_{opt}, C_{Fopt})$ to (7.73), compute*

$$
A_F = (R_{opt} - X_{opt})^{-1}M_{opt}, B_F = (R_{opt} - X_{opt})^{-1}Z_{opt},
$$

then go to Step 2.

   Step 2. *Combining A_F with a row vector c by using (7.51), we construct* a transformation matrix T. Then, compute $\bar{A}$, $\bar{B}_1$, $\bar{B}_2$, $\bar{C}_1$, $\bar{C}_2$, $\bar{D}_{12}$, and $\bar{D}_{21}$ according to (7.59), and go to Step 3.

Step 3. Minimize γ subject to $\bar{F}_A$, $\bar{F}_B$, $\bar{C}_F$, $\bar{N}$, $\bar{S}$, and (7.64) for additive gain variations. Denote the optimal solutions as $\bar{N} = \bar{N}_{opt}$, $\bar{F}_A = \bar{F}_{Aopt}$, and $\bar{F}_B = \bar{F}_{Bopt}$, $\bar{C}_F = \bar{C}_{Fopt}$. Then, according to (7.65),

$$\bar{A}_F = A_{Fc} + \bar{N}_{opt}^{-1}\bar{F}_{Aopt}v, \bar{B}_F = \bar{N}_{opt}^{-1}\bar{F}_{Bopt}, \bar{C}_F = \bar{C}_F.$$

The resulting $\bar{A}_F$, $\bar{B}_F$, and $\bar{C}_F$ form the sparse structured non-fragile H_∞ filter gains.

Remark 7.12 *Algorithm 7.1 gives a method of designing the non-fragile H_∞ filter with sparse structure described by (7.52). In Step 1, an H_∞ filter satisfying Assumption 7.1 is designed, followed by determining the sparse structure, and then a non-fragile H_∞ filter with the sparse structure is obtained.*

Remark 7.13 *The method of finding a class of feasible sparse structures based on the controllable standard form is similar to the method based on the observable standard form, and omitted here.*

7.4.3   Example

In this section, the effectiveness of the method of designing non-fragile H_∞ filters with sparse structure will be illustrated via a numerical example.

Example 7.1 *Consider a linear system of form (7.1) with*

$$A = \begin{bmatrix} 0 & 1 & -0.5 \\ -1 & -0.5 & 1 \\ -1 & 0 & 1 \end{bmatrix}, B = \begin{bmatrix} -1 & 0 \\ 0.5 & 0 \\ -1 & 0 \end{bmatrix},$$

$$C_1 = \begin{bmatrix} 1 & -1 & 1 \end{bmatrix}, C_2 = \begin{bmatrix} -1 & 0.5 & 2 \end{bmatrix}, D_{21} = \begin{bmatrix} 0 & 0.9 \end{bmatrix}.$$

By Step 1, the standard H_∞ filter F_{stan} is obtained with gains as follows:

$$A_F = \begin{bmatrix} -0.2864 & 1.1432 & 0.0727 \\ -1.4001 & -0.2999 & 1.8003 \\ -1.2622 & 0.1311 & 1.5245 \end{bmatrix},$$

$$B_F = \begin{bmatrix} -0.2863 & -0.4001 & -0.2622 \end{bmatrix}^T, C_F = \begin{bmatrix} 1.0000 & -1.0000 & 1.0000 \end{bmatrix},$$

and the optimal H_∞ performance is $\gamma_{opt} = 3.7282$.

Let $c = C_F = \begin{bmatrix} 1.0000 & -1.0000 & 1.0000 \end{bmatrix}$. Then, by Step 2, we obtain

$$T = \begin{bmatrix} -1.1175 & -2.7938 & 3.3525 \\ -1.0866 & 2.5124 & -1.1413 \\ 1.0000 & -1.0000 & 1.0000 \end{bmatrix}.$$

TABLE 7.5

Performance Index Evaluation by Lemma
7.7 with $\theta_a = 0.01$

γ	F_{nonsp} 3.9986	F_{stansp} 4.8215

On the one hand, according to (7.52), the standard sparse structured filter F_{stansp}, which is obtained from the standard filter F_{stan} via transforming directly by using the transform matrix T, is with gains as follows:

$$\bar{A}_F = \begin{bmatrix} -0.0000 & -0.0000 & 0.0000 \\ 1.0000 & 0.0000 & -0.6484 \\ 0.0000 & 1.0000 & 0.9382 \end{bmatrix},$$

$$\bar{B}_F = \begin{bmatrix} 0.5588 & -0.3949 & -0.1484 \end{bmatrix}^T, \quad \bar{C}_F = \begin{bmatrix} 0 & 0 & 1 \end{bmatrix}.$$

Assume that the standard sparse structured filter F_{stansp} is with additive uncertainty described by (7.55). For this case, by Lemma 7.7 with $\theta_a = 0.01$, the H_∞ performance of filter F_{stansp} is 4.8215.

On the other hand, by applying Theorem 7.5 with $\theta_a = 0.01$, the non-fragile sparse structured H_∞ filter F_{nonsp} is designed with gains as follows:

$$\bar{A}_F = \begin{bmatrix} 0 & 0 & -0.0019 \\ 1.0000 & 0 & -0.5902 \\ 0 & 1.0000 & 0.8824 \end{bmatrix},$$

$$\bar{B}_F = \begin{bmatrix} 0.5871 & -0.4094 & -0.1090 \end{bmatrix}^T, \quad \bar{C}_F = \begin{bmatrix} 0 & 0 & 1 \end{bmatrix},$$

and the corresponding optimal value of γ is 4.0623.

For this non-fragile sparse structured H_∞ filter F_{nonsp}, our result Lemma 7.7 gives better evaluations as shown in Table 7.5.

Obviously, compared with the optimal H_∞ performance $\gamma_{opt} = 3.7282$, by Lemma 7.7, the H_∞ performance index of F_{stansp} is degraded 29.33%. Correspondingly, the performance index of F_{nonsp} is degraded 7.53%, which is improved 21.80% than the H_∞ performance index of F_{stansp}.

Let the system initial state be $x_0 = \begin{bmatrix} 1 & -1 & 1 \end{bmatrix}^T$ and the filter initial state be $\xi_0 = \begin{bmatrix} 3 & 3 & 1 \end{bmatrix}^T$. Let the disturbance $w(k)$ be

$$w(k) = \begin{cases} \begin{bmatrix} -0.5 \\ 5 \end{bmatrix} \cos(k), & 41 \leq k \leq 60 \ (step), \\ 0 & otherwise. \end{cases}$$

Then Figure 7.3 shows the estimation error responses of the augmented system with the non-fragile sparse structured H_∞ filter F_{nonsp} and the standard sparse structured H_∞ filter F_{stansp}, respectively.

From this figure, we can see the superiority of our proposed method.

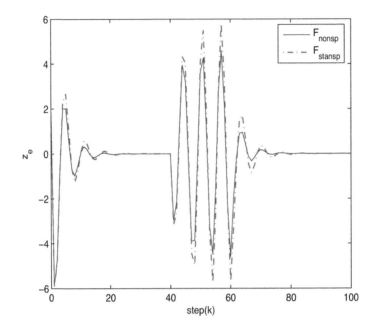

FIGURE 7.3
Responses of $z_e(t)$ with $\theta_a = 0.01$.

7.5 Conclusion

This chapter has studied the problem of non-fragile H_∞ filter design with
the additive interval type of gain uncertainties considered. And two cases of
full parameterized and sparse structured filter design are investigated, respectively. For the full parameter filter design, the structured vertex separator
proposed in the previous chapter is exploited to solve the numerical computational problem and further to develop sufficient conditions for the non-fragile
H_∞ filter design in terms of solutions to a set of linear matrix inequalities
(LMIs). For the sparse structured filter design, first, a class of sparse structures is specified. Then, an LMI-based procedure for non-fragile H_∞ filters
design under the restriction of the sparse structure is provided. The resulting
design renders the augmented system asymptotically stable and the H_∞ performance of the system from the exogenous signals to the estimation errors
are less than a prescribed level. The effectiveness of the proposed methods are
illustrated via some numerical examples and their simulations.

8

Insensitive H_∞ Filtering of Continuous-Time Systems

8.1 Introduction

An important topic in filter design is *filter coefficient sensitivity* [130] because very small perturbations in the coefficient of the designed filter may result in the serious deterioration of the system performance, including instability. Therefore, the filter should be designed to be insensitive to some amount of error with respect to its coefficients. On the other hand, *sensitivity analysis* allows the analyst to assess the effects of changes in the parameter values [12]. Hence, it is very useful to understand how changes in the parameter values influence the design [13]. After the hard work of many researchers in more than one decade, fundamental results have been obtained for the study of sensitivity analysis and performance limitations in automatic control systems [see, for example, 21, 88, 129, 132, 140, 141, and the references therein], and many different definitions of sensitivity have been used for *sensitivity analysis*. One of the effective synthesis methods is the *coefficient sensitivity method*, which describes the variations in performance due to variations in the parameters that affect the system dynamics [see 21, 60, 89, 92, 99, 121, 129, 130].

On the other hand, although the *structured vertex separator method* is proposed to deal with the numerical problem in Chapter 7, the number of linear matrix inequality (LMI) constraints involved in the design conditions is still large. What is worse, is that an *important drawback* of the design method in Chapter 7 is that the technique cannot be directly used to resolve the *interval multiplicative coefficient variation case* due to the complexity of the problem. Therefore, the *non-fragile* filtering problem for the interval multiplicative case remains challenging.

Motivated by the above points, this chapter is concerned with the *multi-objective problem* of designing *insensitive H_∞ filters* for *linear time-invariant continuous-time systems*, which *minimizes the filter coefficient sensitivity and meets the prescribed H_∞ norm constraint simultaneously*. First, coefficient sensitivity functions of transfer functions with respect to filter *additive/multiplicative coefficient variations* are defined, and the H_∞ norms of the sensitivity functions are used to measure the sensitivity of the transfer functions with respect to additive filter coefficient variations. Next, a new method for designing *insensitive* H_∞ *filters* subjected to *additive filter coefficient variations* is proposed in terms of solutions to a set of LMIs. Moreover, in addition to the *additive* case, the *multiplicative* case is also addressed where new measures are defined based on the average of the sensitivity functions, and LMI-based filter design conditions are presented. In addition, an indirect method is also presented to solve the multiplicative case. Then, in comparison to the method proposed in Chapter 7, two advantages of the proposed method lie in the fact that the proposed method can resolve the multiplicative coefficient variation case well and the number of LMI constraints involved in the design conditions is significantly reduced.

8.2   Problem Statement

Consider a class of continuous-time systems described as follows:

$$
\begin{aligned}
\dot{x}(t) &= Ax(t) + Bw(t) \\
y(t) &= Cx(t) + Dw(t) \\
z(t) &= Lx(t)
\end{aligned}
\tag{8.1}
$$

where $x(t) \in R^n$ is the state vector, $w(t) \in R^r$ is the disturbance input which is assumed to belong to $L_2[0, \infty)$, $z(t) \in R^q$ is the regulated output, and $y(t) \in R^p$ is the measured output, respectively. The system matrices A, B, C, D, and L are known real constant matrices with appropriate dimensions.

Considering the filter coefficient variations caused by inaccuracies of the filter implementation, the actual n_fth-order linear time-invariant filter is assumed to be

$$
\begin{aligned}
\dot{\bar{x}}(t) &= A_F\bar{x}(t) + B_Fy(t) \\
\bar{z}(t) &= C_F\bar{x}(t)
\end{aligned}
\tag{8.2}
$$

where $\bar{x}(t) \in R^{n_f}$ is the filter state, and $\bar{z}(t) \in R^q$ is the estimation of $z(t)$. The filter (8.2) is referred to as the full-order filter and reduced-order filter when $n_f = n$ and $n_f < n$, respectively. In addition, $A_F \in R^{n_f \times n_f}$, $B_F \in R^{n_f \times p}$, and $C_F \in R^{q \times n_f}$ represent the filter matrices with uncertainties. In this chapter, the following two classes of the filter coefficient matrices are considered.

• The additive form [102, 103, 142, 143]:

$$
A_F = \left[\ a_{f_{ij}} + \theta_{a_{f_{ij}}}\ \right]_{n_f \times n_f}, B_F = \left[\ b_{f_{ik}} + \theta_{b_{f_{ik}}}\ \right]_{n_f \times p}
$$
$$
C_F = \left[\ c_{f_{lj}} + \theta_{c_{f_{lj}}}\ \right]_{q \times n_f}
\tag{8.3}
$$

where $i, j = 1, \cdots, n_f; k = 1, \cdots, p; l = 1, \cdots, q$.

• The multiplicative form [147, 148]:

$$A_F = \left[(1 + \theta_{a_{f_{ij}}}) a_{f_{ij}} \right]_{n_f \times n_f}, B_F = \left[(1 + \theta_{b_{f_{ik}}}) b_{f_{ik}} \right]_{n_f \times p}$$

$$C_F = \left[(1 + \theta_{c_{f_{lj}}}) c_{f_{lj}} \right]_{q \times n_f} \tag{8.4}$$

where $i, j = 1, \cdots, n_f; k = 1, \cdots, p; l = 1, \cdots, q.$

Remark 8.1 *The models of multiplicative uncertainty used to describe the controller and the filter coefficient variations have been investigated by a lot of work [147, 148] since the model of multiplicative uncertainty is frequently used in many practical systems such as product lines, environmental risk assessments, and economic systems [see for example, 9, 118]. However, an important drawback of the design method in Chapter 7 is that the technique cannot be directly used to resolve the interval multiplicative coefficient variation case due to the complexity of the problem. Therefore, the non-fragile filtering problem for the interval multiplicative case remains challenging.*

To facilitate the following sections, denote

$$A_{Fe} = \left[a_{f_{ij}} \right]_{n_f \times n_f}, B_{Fe} = \left[b_{f_{ik}} \right]_{n_f \times p}, C_{Fe} = \left[c_{f_{lj}} \right]_{q \times n_f}$$
$$i = 1, \ldots, n_f; j = 1, \ldots, n_f; k = 1, \ldots, p; l = 1, \ldots, q$$

where $A_{Fe} \in R^{n_f \times n_f}$, $B_{Fe} \in R^{n_f \times p}$, and $C_{Fe} \in R^{q \times n_f}$ are filter matrices to be designed. $a_{f_{ij}}$, $b_{f_{ik}}$, and $c_{f_{lj}}$ are the elements of the filter matrices A_{Fe}, B_{Fe}, and C_{Fe}, respectively. $\theta_{a_{f_{ij}}}$, $\theta_{b_{f_{ik}}}$, and $\theta_{c_{f_{lj}}}$ represent the magnitudes of the deviation of the filter gains.

Applying the filter (8.2) to the system (8.1), the *filtering error system* is obtained

$$\begin{aligned} \dot{\xi}(t) &= \bar{A}_F \xi(t) + \bar{B}_F w(t) \\ e(t) &= \bar{C}_F \xi(t) \end{aligned} \tag{8.5}$$

where $\xi(t)^T = \left[x^T(t) \quad \bar{x}^T(t) \right]$, $e(t) = z(t) - \bar{z}(t)$ is the estimation error, and

$$\bar{A}_F = \left[\begin{array}{cc} A & 0 \\ B_F C & A_F \end{array} \right], \bar{B}_F = \left[\begin{array}{c} B \\ B_F D \end{array} \right], \bar{C}_F = \left[\begin{array}{cc} L & -C_F \end{array} \right].$$

The transfer function matrix of the *filtering error system* (8.5) from $w(t)$ to $e(t)$ is given by

$$T(s) = \bar{C}_F (sI - \bar{A}_F)^{-1} \bar{B}_F.$$

For convenience in the following sections, denote

$$T_e(s, [a_{f_{ij}}]_{n_f \times n_f}, [b_{f_{ik}}]_{n_f \times p}, [c_{f_{lj}}]_{q \times n_f}) = \bar{C}_{Fe}(sI - \bar{A}_{Fe})^{-1} \bar{B}_{Fe} \tag{8.6}$$

where

$$\bar{A}_{Fe} = \left[\begin{array}{cc} A & 0 \\ B_{Fe} C & A_{Fe} \end{array} \right], \bar{B}_{Fe} = \left[\begin{array}{c} B \\ B_{Fe} D \end{array} \right], \bar{C}_{Fe} = \left[\begin{array}{cc} L & -C_{Fe} \end{array} \right]. \tag{8.7}$$

In the following, coefficient sensitivity functions are defined in a similar way by Lutz and Hakimi [99].

Definition 8.1 *Let $q_{f_{uv}}$ denote the (u,v)th element of the matrix Q_F with Q_F being an $m \times n$ real matrix and let $f(\rho, [q_{f_{uv}}]_{m \times n})$ be a matrix function of Q_F. The coefficient sensitivity function of $f(\rho, [q_{f_{uv}}]_{m \times n})$ with respect to $q_{f_{uv}}$, that is, the (u,v)th element of Q_F, is given by*

$$S_{q_{f_{uv}}}(f(\rho, [q_{f_{uv}}]_{m \times n})) = \frac{df(\rho, [q_{f_{uv}}]_{m \times n})}{dq_{f_{uv}}} \qquad (8.8a)$$

for the additive form, while

$$S_{q_{f_{uv}}}(f(\rho, [q_{f_{uv}}]_{m \times n})) = \frac{df(\rho, [q_{f_{uv}}]_{m \times n} + \theta_{q_{f_{uv}}} q_{f_{uv}} e_u h_v^T)}{d\theta_{q_{f_{uv}}}} \Big|_{\theta_{q_{f_{uv}}}=0} \qquad (8.8b)$$

for the multiplicative form, respectively, where $\theta_{q_{f_{uv}}}$ is used to describe the magnitude of the deviation of the matrix coefficient $q_{f_{uv}}$, $e_k \in R^m$ and $h_k \in R^n$ denote the column vectors in which the kth element equals 1 and the others equal 0.

Remark 8.2 *In this chapter, the filter is designed to tolerate the coefficient variations in filter matrices which are caused by inaccuracies of the filter implementation. The sensitivity functions (8.8a) and (8.8b) are used to describe the sensitive properties of matrix coefficients with respect to additive coefficient variations and multiplicative coefficient variations as in (8.3) and (8.4), respectively.*

Then, based on Definition 8.1 and by means of the techniques developed in Gevers and Li [43] and Hilaire, Chevrel, and Trinquet [60], the following lemma is presented.

Lemma 8.1 *Let $T_e(s, [a_{f_{ij}}]_{n_f \times n_f}, [b_{f_{ik}}]_{n_f \times p}, [c_{f_{lj}}]_{q \times n_f})$ be defined in (8.6), and $a_{f_{ij}}$, $b_{f_{ik}}$, and $c_{f_{lj}}$ are the elements of the filter matrices A_{Fe}, B_{Fe}, and C_{Fe}, respectively. Then, the sensitivity functions of the transfer function with respect to the elements of the filter matrices are given as follows:*

$$\begin{aligned}
S_{a_{f_{ij}}}&(T_e(s, [a_{f_{ij}}]_{n_f \times n_f}, [b_{f_{ik}}]_{n_f \times p}, [c_{f_{lj}}]_{q \times n_f})) \\
&= \bar{C}_{Fe}(sI - \bar{A}_{Fe})^{-1} \mathcal{N}_{ij}^a (sI - \bar{A}_{Fe})^{-1} \bar{B}_{Fe} \\
&= \hat{C}_{Fa}(sI - \hat{A}_{Fa})^{-1} \hat{B}_{Fa}, i = 1, \cdots, n_f; j = 1, \cdots, n_f
\end{aligned} \qquad (8.9)$$

$$\begin{aligned}
S_{b_{f_{ik}}}&(T_e(s, [a_{f_{ij}}]_{n_f \times n_f}, [b_{f_{ik}}]_{n_f \times p}, [c_{f_{lj}}]_{q \times n_f})) \\
&= \bar{C}_{Fe}(sI - \bar{A}_{Fe})^{-1} \mathcal{N}_{ik}^b (sI - \bar{A}_{Fe})^{-1} \bar{B}_{Fe} \\
&\quad + \bar{C}_{Fe}(sI - \bar{A}_{Fe})^{-1} \mathcal{M}_{ik}^b \\
&= \hat{C}_{Fb}(sI - \hat{A}_{Fb})^{-1} \hat{B}_{Fb}, i = 1, \cdots, n_f; k = 1, \cdots, p
\end{aligned} \qquad (8.10)$$

$$\begin{aligned}
S_{c_{f_{lj}}}&(T_e(s, [a_{f_{ij}}]_{n_f \times n_f}, [b_{f_{ik}}]_{n_f \times p}, [c_{f_{lj}}]_{q \times n_f})) \\
&= \mathcal{N}_{lj}^c (sI - \bar{A}_{Fe})^{-1} \bar{B}_{Fe} \\
&= \hat{C}_{Fc}(sI - \hat{A}_{Fc})^{-1} \hat{B}_{Fc}, l = 1, \cdots, q; j = 1, \cdots, n_f
\end{aligned} \qquad (8.11)$$

where

$$
\mathcal{N}_{ij}^a = \begin{bmatrix} 0 & 0 \\ 0 & \Re_{ij}^a \end{bmatrix}, \hat{A}_{Fa} = \begin{bmatrix} \bar{A}_{Fe} & \mathcal{N}_{ij}^a \\ 0 & \bar{A}_{Fe} \end{bmatrix}, \hat{B}_{Fa} = \begin{bmatrix} 0 \\ \bar{B}_{Fe} \end{bmatrix}
$$

$$
\hat{C}_{Fa} = \begin{bmatrix} \bar{C}_{Fe} & 0 \end{bmatrix}; \mathcal{N}_{ik}^b = \begin{bmatrix} 0 & 0 \\ \Re_{ik}^b C & 0 \end{bmatrix}, \mathcal{M}_{ik}^b = \begin{bmatrix} 0 \\ \Re_{ik}^b D \end{bmatrix} \quad (8.12)
$$

$$
\hat{A}_{Fb} = \begin{bmatrix} \bar{A}_{Fe} & \mathcal{N}_{ik}^b & 0 \\ 0 & \bar{A}_{Fe} & 0 \\ 0 & 0 & \bar{A}_{Fe} \end{bmatrix}, \hat{B}_{Fb} = \begin{bmatrix} 0 \\ \bar{B}_{Fe} \\ \mathcal{M}_{ik}^b \end{bmatrix}
$$

$$
\hat{C}_{Fb} = \begin{bmatrix} \bar{C}_{Fe} & 0 & \bar{C}_{Fe} \end{bmatrix}; \mathcal{N}_{lj}^c = \begin{bmatrix} 0 & -\Re_{lj}^c \end{bmatrix}, \hat{C}_{Fc} = \mathcal{N}_{lj}^c
$$

$$
\hat{A}_{Fc} = \bar{A}_{Fe}, \hat{B}_{Fc} = \bar{B}_{Fe}
$$

and

$$
\Re_{ij}^a = e_i e_j^T, \Re_{ik}^b = e_i h_k^T, \Re_{lj}^c = g_l e_j^T \quad (8.13)
$$

for the additive form, and

$$
\Re_{ij}^a = e_i e_i^T A_{Fe} e_j e_j^T, \Re_{ik}^b = e_i e_i^T B_{Fe} h_k h_k^T, \Re_{lj}^c = g_l g_l^T C_{Fe} e_j e_j^T \quad (8.14)
$$

for the multiplicative form, respectively. In addition, $e_k \in R^{n_f}$, $h_k \in R^p$, and $g_k \in R^q$ denote the column vectors in which the kth element equals 1 and the others equal 0.

Proof 8.1 *Based on Definition 2.3, Definition 2.4, and by means of the property of system operations, it is routine to compute that*

$$
S_{b_{f_{ik}}}(T_e(s, [a_{f_{ij}}]_{n_f \times n_f}, [b_{f_{ik}}]_{n_f \times p}, [c_{f_{lj}}]_{q \times n_f}))
$$

$$
= \bar{C}_{Fe}(sI - \bar{A}_{Fe})^{-1}\mathcal{N}_{ik}^b(sI - \bar{A}_{Fe})^{-1}\bar{B}_{Fe} + \bar{C}_{Fe}(sI - \bar{A}_{Fe})^{-1}\mathcal{M}_{ik}^b
$$

$$
= \begin{bmatrix} \bar{C}_{Fe} & 0 \end{bmatrix} \left(zI - \begin{bmatrix} \bar{A}_{Fe} & \mathcal{N}_{ik}^b \\ 0 & \bar{A}_{Fe} \end{bmatrix} \right)^{-1} \begin{bmatrix} 0 \\ \bar{B}_{Fe} \end{bmatrix}
$$

$$
+ \bar{C}_{Fe}(zI - \bar{A}_{Fe})^{-1}\mathcal{M}_{ik}^b
$$

$$
= \begin{bmatrix} \bar{C}_{Fe} & 0 & \bar{C}_{Fe} \end{bmatrix} \left(zI - \begin{bmatrix} \bar{A}_{Fe} & \mathcal{N}_{ik}^b & 0 \\ 0 & \bar{A}_{Fe} & 0 \\ 0 & 0 & \bar{A}_{Fe} \end{bmatrix} \right)^{-1} \begin{bmatrix} 0 \\ \bar{B}_{Fe} \\ \mathcal{M}_{ik}^b \end{bmatrix}
$$

$$
= \hat{C}_{Fbik}(sI - \hat{A}_{Fbik})^{-1}\hat{B}_{Fbik}
$$

where $\mathcal{N}_{ik}^b$ and $\mathcal{M}_{ik}^b$ are defined in (8.12). Similarly, it can be proved that (8.9) and (8.11) hold. This completes the proof.

In this chapter, the H_∞ norm of the sensitivity functions of the filtering error system's transfer function with respect to the perturbations in the filter's coefficients is used to act as the *coefficient sensitivity measure*. Therefore, the coefficient sensitivity measures $M_{a_{f_{ij}}}$, $M_{b_{f_{ik}}}$, and $M_{c_{f_{lj}}}$ will be taken as

$$
\begin{aligned}
M_{a_{f_{ij}}} &= \| S_{a_{f_{ij}}}(T_e(s, [a_{f_{ij}}]_{n_f \times n_f}, [b_{f_{ik}}]_{n_f \times p}, [c_{f_{lj}}]_{q \times n_f})) \|_\infty \\
M_{b_{f_{ik}}} &= \| S_{b_{f_{ik}}}(T_e(s, [a_{f_{ij}}]_{n_f \times n_f}, [b_{f_{ik}}]_{n_f \times p}, [c_{f_{lj}}]_{q \times n_f})) \|_\infty \quad (8.15) \\
M_{c_{f_{lj}}} &= \| S_{c_{f_{lj}}}(T_e(s, [a_{f_{ij}}]_{n_f \times n_f}, [b_{f_{ik}}]_{n_f \times p}, [c_{f_{lj}}]_{q \times n_f})) \|_\infty
\end{aligned}
$$

$$
\text{for} \quad i, \quad j = 1, \cdots, n_f; k = 1, \cdots, p; l = 1, \cdots, q.
$$

Remark 8.3 *Improved coefficient sensitivity measures can be given as follows:*

$$
\begin{aligned}
M_{a_{f_{ij}}} &= \| W_{a_{f_{ij}}}(\delta) S_{a_{f_{ij}}}(T(\delta, [a_{f_{ij}}]_{n_f \times n_f}, [b_{f_{ik}}]_{n_f \times p}, [c_{f_{lj}}]_{q \times n_f})) \|_\infty, \\
M_{b_{f_{ik}}} &= \| W_{b_{f_{ik}}}(\delta) S_{b_{f_{ik}}}(T(\delta, [a_{f_{ij}}]_{n_f \times n_f}, [b_{f_{ik}}]_{n_f \times p}, [c_{f_{lj}}]_{q \times n_f})) \|_\infty, \\
M_{c_{f_{lj}}} &= \| W_{c_{f_{lj}}}(\delta) S_{c_{f_{lj}}}(T(\delta, [a_{f_{ij}}]_{n_f \times n_f}, [b_{f_{ik}}]_{n_f \times p}, [c_{f_{lj}}]_{q \times n_f})) \|_\infty, \\
&\text{for } i, j = 1, \cdots, n_f, l = 1, \cdots, q; k = 1, \cdots, p
\end{aligned}
$$

where $W_{a_{f_{ij}}}(\delta)$, $W_{b_{f_{ik}}}(\delta)$, and $W_{c_{f_{lj}}}(\delta)$ are the weighting functions, and the above type of sensitivity measure can be called the weighted sensitivity measure, which has been used extensively to sensitivity optimization problems [see, e.g., 63, 64, 91, and the references therein]. Obviously, the weighted sensitivity measure would be of practical use as this sensitivity measure allows to emphasize or de-emphasize the filter's sensitivity in certain coefficients. However, for the sake of simplicity, the usual measures in (8.15) are used in this chapter.

Given positive scalars γ and β, the design problem under consideration is to find an insensitive H_∞ filter such that the filtering error system is *asymptotically stable* and keeps

$$
\| T_e(s, [a_{f_{ij}}]_{n_f \times n_f}, [b_{f_{ik}}]_{n_f \times p}, [c_{f_{lj}}]_{q \times n_f}) \|_\infty < \gamma
$$

in the meantime, and satisfies $M_{a_{f_{ij}}} < \beta$, $M_{b_{f_{ik}}} < \beta$, and $M_{c_{f_{lj}}} < \beta$ for $i, j = 1, \cdots, n_f; k = 1, \cdots, p; l = 1, \cdots, q$.

8.3   Insensitive H_∞ Filter Design

From the definition of the sensitivity measures and Lemma 2.11, the following lemma is obtained. The lemma guarantees that the filtering error system (8.5) is asymptotically stable and satisfies an H_∞ norm performance requirement with the insensitive constraint simultaneously.

Lemma 8.2 *Consider the system in (8.1). Let scalars $\gamma > 0$ and $\beta > 0$. Then, the filtering error system (8.5) is asymptotically stable and*

$$
\begin{aligned}
\| T_e(s, [a_{f_{ij}}]_{n_f \times n_f}, [b_{f_{ik}}]_{n_f \times p}, [c_{f_{lj}}]_{q \times n_f}) \|_\infty &< \gamma \\
M_{a_{f_{ij}}} < \beta, M_{b_{f_{ik}}} < \beta, M_{c_{f_{lj}}} &< \beta \quad (8.16) \\
\text{for } i, j = 1, \cdots, n_f; k = 1, \cdots, p; l = 1, \cdots, q &
\end{aligned}
$$

hold, if there exist positive definite symmetric matrices $P_s \in R^{(n+n_f) \times (n+n_f)}$, $P_a \in R^{2(n+n_f) \times 2(n+n_f)}$, $P_b \in R^{3(n+n_f) \times 3(n+n_f)}$, and $P_c \in R^{(n+n_f) \times (n+n_f)}$ such that

$$\begin{bmatrix} He\{P_s \bar{A}_{Fe}\} & P_s \bar{B}_{Fe} & \bar{C}_{Fe}^T \\ * & -\gamma^2 I & 0 \\ * & * & -I \end{bmatrix} < 0, \tag{8.17}$$

$$\begin{bmatrix} He\{P_a \hat{A}_{Fa}\} & P_a \hat{B}_{Fa} & \hat{C}_{Fa}^T \\ * & -\beta^2 I & 0 \\ * & * & -I \end{bmatrix} < 0, \tag{8.18}$$

$$\begin{bmatrix} He\{P_b \hat{A}_{Fb}\} & P_b \hat{B}_{Fb} & \hat{C}_{Fb}^T \\ * & -\beta^2 I & 0 \\ * & * & -I \end{bmatrix} < 0, \tag{8.19}$$

$$\begin{bmatrix} He\{P_c \hat{A}_{Fc}\} & P_c \hat{B}_{Fc} & \hat{C}_{Fc}^T \\ * & -\beta^2 I & 0 \\ * & * & -I \end{bmatrix} < 0 \tag{8.20}$$

hold for $i, j = 1, \cdots, n_f; k = 1, \cdots, p; l = 1, \cdots, q$, where $\hat{A}_{F\chi}, \hat{B}_{F\chi}, \hat{C}_{F\chi}$, and $\hat{A}_{F\chi}(\chi = a, b, c)$ are defined in (8.9), (8.10), and (8.11), respectively.

8.3.1 Additive Filter Coefficient Variation Case

In this section, an LMI-based method for designing insensitive H_∞ filters is presented for the additive filter coefficient variations. The main result is given as follows.

Theorem 8.1 *Consider the system in (8.1). Let scalars $\gamma > 0$, $\beta > 0$. Then, the filtering error system (8.5) is asymptotically stable and the conditions in (8.16) hold if, for some positive scalars λ_a, λ_b, and λ_c, there exist matrices $F_A \in R^{n_f \times n_f}$, $F_B \in R^{n_f \times p}$, $F_C \in R^{q \times n_f}$, $0 < S \in R^{n \times n}$, and $0 < N \in R^{n_f \times n_f}$ such that the following LMIs hold for $i, j = 1, \cdots, n_f; k = 1, \cdots, p; l = 1, \cdots, q$:*

$$\begin{bmatrix} He\{M_{as}\} & M_{bs} & M_{cs}^T \\ * & -\gamma^2 I & 0 \\ * & * & -I \end{bmatrix} < 0, \tag{8.21}$$

$$\begin{bmatrix} He\{ \begin{bmatrix} \lambda_a M_{as} & \lambda_a M_{af} \\ 0 & \lambda_a M_{as} \end{bmatrix} \} & \begin{bmatrix} 0 \\ \lambda_a M_{bs} \end{bmatrix} & \begin{bmatrix} M_{cs}^T \\ 0 \end{bmatrix} \\ * & -\beta^2 I & 0 \\ * & * & -I \end{bmatrix} < 0, \tag{8.22}$$

$$
\left[He\left\{ \begin{bmatrix} \lambda_b M_{as} & \lambda_b M_{b1f} & 0 \\ 0 & \lambda_b M_{as} & 0 \\ 0 & 0 & \lambda_b M_{as} \\ * & & \\ * & & \end{bmatrix} \right\} \begin{bmatrix} 0 \\ \lambda_b M_{bs} \\ \lambda_b M_{b2f} \\ -\beta^2 I \\ * \end{bmatrix} \begin{bmatrix} M_{cs}^T \\ 0 \\ M_{cs}^T \\ 0 \\ -I \end{bmatrix} \right] < 0,
$$

$$(8.23)$$

$$
\begin{bmatrix} \lambda_c He\{M_{as}\} & \lambda_c M_{bs} & M_{cf}^T \\ * & -\beta^2 I & 0 \\ * & * & -I \end{bmatrix} < 0 \tag{8.24}
$$

where

$$
M_{as} = \begin{bmatrix} SA + \mathcal{H}F_B C & \mathcal{H}F_A \\ N\mathcal{H}^T A + F_B C & F_A \end{bmatrix}, M_{bs} = \begin{bmatrix} SB + \mathcal{H}F_B D \\ N\mathcal{H}^T B + F_B D \end{bmatrix}
$$

$$
M_{cs} = \begin{bmatrix} L & -F_C \end{bmatrix}, M_{cf} = \begin{bmatrix} 0 & -\Re_{lj}^c \end{bmatrix}, M_{af} = \begin{bmatrix} 0 & \mathcal{H}N\Re_{ij}^a \\ 0 & N\Re_{ij}^a \end{bmatrix}
$$

$$
M_{b1f} = \begin{bmatrix} \mathcal{H}N\Re_{ik}^b C & 0 \\ N\Re_{ik}^b C & 0 \end{bmatrix}, M_{b2f} = \begin{bmatrix} \mathcal{H}N\Re_{ik}^b D \\ N\Re_{ik}^b D \end{bmatrix}
$$

with $\Re_{aij}$, $\Re_{bik}$, and $\Re_{clj}$ satisfying (8.13) and $\mathcal{H} = \begin{bmatrix} I_{(n_f \times n_f)} \\ 0_{(n-n_f) \times n_f} \end{bmatrix}$. More-over, from the solutions of the above inequalities, the n_fth-order insensitive filter can be given by

$$
A_{Fe} = N^{-1} F_A, B_{Fe} = N^{-1} F_B, C_{Fe} = F_C. \tag{8.25}
$$

Proof 8.2 *Lemma 8.2 shows that the filtering error system (8.5) is asymptotically stable and satisfies the H_∞ norm performance requirement with the insensitive constraint simultaneously, if there exist positive matrices P_s, P_a, P_b, and P_c satisfying (8.17), (8.18), (8.19), and (8.20), respectively. In the following proof, we will show that Theorem 8.1 is sufficient to guarantee that (8.17), (8.18), (8.19), and (8.20) hold.*

Let the Lyapunov matrices P_s, P_a, P_b, and P_c satisfy the following constraints

$$
P_s = P, P_a = \lambda_a(I_2 \otimes P), P_b = \lambda_b(I_3 \otimes P), P_c = \lambda_c P \tag{8.26}
$$

where λ_a, λ_b, and λ_c are positive scalar parameters to be searched.

Further, it follows from the proof of Theorem 3 in Wu and Ho [134], that we partition the matrix P as

$$
P = \begin{bmatrix} P_1 & P_2 \\ * & P_3 \end{bmatrix}, P_2 = \begin{bmatrix} P_4 \\ 0_{(n-n_f) \times n_f} \end{bmatrix}. \tag{8.27}
$$

Because the n_fth-order case is being considered, the filter order n_f will be less than or equal to the plant order n. The partitioned submatrices in (8.27) will be dimensionalized as $0 < P_1 \in R^{n \times n}, P_2 \in R^{n \times n_f}, 0 < P_3 \in R^{n_f \times n_f}$, *and $P_4 \in R^{n_f \times n_f}$. By recalling the assumption in Wu and Ho [134], we may assume, without loss of generality, that P_4 is nonsingular.*

From (8.27), the following equation can be implied:

$$P = \begin{bmatrix} P_1 & P_2 \\ * & P_3 \end{bmatrix} = \begin{bmatrix} P_1 & \mathcal{H}P_4 \\ * & P_3 \end{bmatrix}$$

where $\mathcal{H}$ is defined in Theorem 8.1.

Construct the following matrices

$$J = \begin{bmatrix} I & 0 \\ 0 & P_3^{-1}P_4^T \end{bmatrix},$$

and

$$\begin{bmatrix} F_A & F_B \\ F_C & 0 \end{bmatrix} = (P_4 \oplus I) \begin{bmatrix} A_{Fe} & B_{Fe} \\ C_{Fe} & 0 \end{bmatrix} (P_3^{-1}P_4^T \oplus I). \qquad (8.28)$$

For brevity, we only prove that the inequality (8.22) is equivalent to (8.18) with the constraint $P_a = \lambda_a(I_2 \otimes P)$. Performing a congruence transformation to (8.18) by $\mathrm{diag}\{(I_2 \otimes P)^T, I, I\}$, we obtain

$$\begin{bmatrix} He\{\Xi\} & \begin{bmatrix} 0 \\ \lambda_a J^T P \bar{B}_{Fe} \end{bmatrix} & \begin{bmatrix} J^T \bar{C}_{Fe}^T \\ 0 \end{bmatrix} \\ * & -\beta^2 I & 0 \\ * & * & -I \end{bmatrix} < 0$$

where

$$\Xi = \begin{bmatrix} \lambda_a J^T P \bar{A}_{Fe}J & \lambda_a J^T P \mathcal{N}_{ij}^a J \\ 0 & \lambda_a J^T P \bar{A}_{Fe}J \end{bmatrix}$$

$$J^T P \bar{A}_{Fe}J = \begin{bmatrix} P_1 A + \mathcal{H}P_4 B_{Fe}C & \mathcal{H}P_4 A_{Fe}P_3^{-1}P_4^T \\ \Xi_1 & P_4 A_{Fe}P_3^{-1}P_4^T \end{bmatrix}$$

$$J^T P \mathcal{N}_{ij}^a J = \begin{bmatrix} 0 & \mathcal{H}P_4 \Re_{ij}^a P_3^{-1}P_4^T \\ 0 & P_4 \Re_{ij}^a P_3^{-1}P_4^T \end{bmatrix}$$

$$J^T P \bar{B}_{Fe} = \begin{bmatrix} P_1 B + \mathcal{H}P_4 B_{Fe}D \\ P_4 P_3^{-1}P_4^T \mathcal{H}^T B + P_4 B_{Fe}D \end{bmatrix}$$

$$\bar{C}_{Fe}J = \begin{bmatrix} L & -C_{Fe}P_3^{-1}P_4^T \end{bmatrix}$$

$$\Xi_1 = P_4 P_3^{-1}P_4^T \mathcal{H}^T A + P_4 B_{Fe}C$$

with $\Re_{ij}^a (i, j = 1, \cdots, n_f)$ being defined in (8.13).

In addition, note that (8.28) is equivalent to

$$\begin{bmatrix} A_{Fe} & B_{Fe} \\ C_{Fe} & 0 \end{bmatrix} = \begin{bmatrix} P_4^{-1} & 0 \\ 0 & I \end{bmatrix} \mathcal{F} \begin{bmatrix} (P_3^{-1}P_4^T)^{-1} & 0 \\ 0 & I \end{bmatrix} \qquad (8.29)$$

$$= \begin{bmatrix} (P_3^{-1}P_4^T)N^{-1} & 0 \\ 0 & I \end{bmatrix} \mathcal{F} \begin{bmatrix} (P_3^{-1}P_4^T)^{-1} & 0 \\ 0 & I \end{bmatrix}$$

with $N = P_4 P_3^{-1} P_4^T$, $\mathcal{F} = \begin{bmatrix} F_A & F_B \\ F_C & 0 \end{bmatrix}$. *Note that the filter matrices* A_{Fe}, B_{Fe}, *and* C_{Fe} *can be written as (8.29), which implies that* $(P_3^{-1} P_4^T)^{-1}$ *can be viewed as a similarity transformation on the state-space realization of the filter and, as such, has no effect on the filter mapping from* y *to* $\bar{z}(t)$. *Without loss of generality, we may set* $P_3^{-1} P_4^T = I$, *thus the filter in (8.2) can be constructed by (8.25).*

Moreover, following from (8.29) and $S = P_1$, *we have*

$$J^T P \bar{A}_{Fe} J = \begin{bmatrix} SA + \mathcal{H}F_B C & \mathcal{H}F_A \\ N\mathcal{H}^T A + F_B C & F_A \end{bmatrix}$$

$$J^T P \bar{B}_{Fe} = \begin{bmatrix} SB + \mathcal{H}F_B D \\ N\mathcal{H}^T B + F_B D \end{bmatrix}$$

$$J^T P \mathcal{N}_{ij}^a J = \begin{bmatrix} 0 & \mathcal{H}N\Re_{ij}^a \\ 0 & N\Re_{ij}^a \end{bmatrix}$$

$$\bar{C}_{Fe} J = \begin{bmatrix} L & -F_C \end{bmatrix}.$$

Then (8.22) can be easily obtained. Note that because of the invertibility of the matrices involved in the above proof process, the above proof process is a reversible process, therefore, (8.22) is equivalent to (8.18) with the constraint $P_a = \lambda_a (I_2 \otimes P)$.

Moreover, performing congruence transformations to (8.17) and (8.20) by $diag\{J, I, I\}$ *and to (8.19) by* $diag\{diag\{J, J, J\}, I, I\}$, *respectively, and following similar steps as the proof process of (8.22), the inequalities (8.21), (8.23) and (8.24) are equivalent to (8.17), (8.19), and (8.20) with the constraints in (8.26), respectively. This completes the proof.*

Remark 8.4 *It is obvious that the conditions given in Lemma 8.2 are not LMIs due to the product of the variables* P_s, P_a, P_b, *and* P_c *with the filter matrices* A_{Fe}, B_{Fe}, *and* C_{Fe}, *respectively. On the other hand, the sensitive coefficient matrices are given in almost block diagonal form of the filtering error system matrices. A common method, which can be found in the existing results concerning the multi-objective problems [see, for example, 29–31], is to let the constrains in (8.26) hold before converting all the inequalities to LMIs.*

Remark 8.5 *It is interesting to note that the* n_f*th-order filter (8.2) is connected with the full-order filter by choosing* $\mathcal{H}$. *For the case of* $n_f = n$, *the matrix* $\mathcal{H}$ *defined in Theorem 8.1 will become an identity matrix with dimension* n, *which means that the* n_f*th-order filter design will recover the full-order filter design.*

Remark 8.6 *Theorem 8.1 offers an LMI-based condition for the existence of insensitive* H_∞ *filters with respect to additive coefficient variations. The result may be conservative mainly due to the introduction of the constraints in (8.26). Fortunately, the scalars* λ_a, λ_b, *and* λ_c *offer the flexibility of setting different performance requirements for the normal* H_∞ *constraint and the* *insensitive constraint which can be exploited to reduce such conservativeness by appropriately selecting the design parameters λ_a, λ_b, and λ_c. The relevant discussion and corresponding numerical algorithm can be found in Xie et al. [138, and references therein].*

Remark 8.7 *When the scalars λ_a, λ_b, and λ_c are set to be fixed parameters, the conditions (8.21), (8.22), and (8.23) of Theorem 8.1 become linear. For the prescribed H_∞ performance γ, the problem of H_∞ filter design under insensitive constraints can be converted to the following optimization problem:*

$$\min_{S,N,F_A,F_B,F_C} \beta,$$
$$s.t.\ S > 0, N > 0, (8.21), (8.22), (8.23), (8.24). \tag{8.30}$$

The minimal sensitivity β^* is the optimization value of β, and the designed filter's gains can be obtained by (8.25).

Remark 8.8 *When the insensitive constraint is unconsidered, Theorem 8.1 reduces to the standard H_∞ filtering design method [42].*

Remark 8.9 *In some cases, the magnitude of the designed filter matrices A_{Fe} (B_{Fe} or C_{Fe}) may be too large to be applied in practice. For solving the problem, the following constraints are needed:*

$$N > \mu I,\ F_A F_A^T < \nu I, \tag{8.31}$$

then, the magnitude of A_{Fe} can be reduced. In fact, by $A_{Fe} = N^{-1}F_A$ and (8.31), we can obtain

$$\| A_{Fe} \| < \sqrt{\nu}/\mu.$$

The similar method can be used for the gains B_{Fe} and C_{Fe}, namely,

$$F_B F_B^T < \nu I, F_C^T F_C < \nu I. \tag{8.32}$$

8.3.2   Multiplicative Filter Coefficient Variation Case

In this section, due to the complexity problem of the multiplicative coefficient variations, new coefficient measures are defined to obtain convex conditions for the insensitive filtering problem. In addition, an indirect approach is also provided.

At first, based on Lemma 8.2 and the coefficient measures defined in (8.15), the following theorem can be obtained by using the similar techniques in Theorem 8.1.

Theorem 8.2 *Consider the system in (8.1). Let scalars $\gamma > 0$, $\beta > 0$. Then, the filtering error system (8.5) is asymptotically stable and the conditions in (8.16) hold if, for some positive scalars λ_a, λ_b, and λ_c, there exist matrices $F_A \in R^{n_f \times n_f}$, $F_B \in R^{n_f \times p}$, $F_C \in R^{q \times n_f}$, $0 < S \in R^{n \times n}$, and $0 < N \in R^{n_f \times n_f}$, such that (8.21), (8.22), (8.23), and (8.24) hold with $\Re_{aij}$, $\Re_{bik}$, and $\Re_{clj}$ satisfying (8.14) and $\mathcal{H}$ being defined in Theorem 8.1.*

Because the matrices $\Re_{aij}$, $\Re_{bik}$, and $\Re_{clj}$ defined in (8.14) contain the filter matrix elements, and what is worse, there exist cross-product terms between the variable N and the matrices $\Re_{aij}$ and $\Re_{bik}$, the conditions given in Theorem 8.2 are non-convex and cannot be solved directly. In order to solve the non-convex problem, the following coefficient sensitivity measures based on the average of the sensitivity functions are introduced. The coefficient sensitivity measure $M_{A_{Fe}}$ is taken as

$$
\begin{aligned}
M_{A_{Fe}} &= \|\frac{1}{n_f \times n_f} \sum_{i=1}^{n_f} \sum_{j=1}^{n_f} S_{a_{f_{ij}}}(T_e(s, [a_{f_{ij}}]_{n_f \times n_f}, [b_{f_{ik}}]_{n_f \times p}, [c_{f_{lj}}]_{q \times n_f}))\|_\infty \\
&= \|\frac{1}{n_f \times n_f} \sum_{i=1}^{n_f} \sum_{j=1}^{n_f} \bar{C}_{Fe}(sI - \bar{A}_{Fe})^{-1} \mathcal{N}_{ij}^a (sI - \bar{A}_{Fe})^{-1} \bar{B}_{Fe}\|_\infty \\
&= \|\bar{C}_{Fe}(sI - \bar{A}_{Fe})^{-1} \mathcal{N}_{a_f}(sI - \bar{A}_{Fe})^{-1} \bar{B}_{Fe}\|_\infty \\
&= \|\hat{C}_{Fa}(sI - \hat{A}_{Fa})^{-1} \hat{B}_{Fa}\|_\infty.
\end{aligned}
$$
$$(8.33)$$

Similarly, we can obtain the coefficient sensitivity measures $M_{B_{Fe}}$ and $M_{C_{Fe}}$ as follows:

$$
\begin{aligned}
M_{B_{Fe}} &= \|\frac{1}{n_f \times p} \sum_{i=1}^{n_f} \sum_{k=1}^{p} S_{b_{f_{ik}}}(T_e(s, [a_{f_{ij}}]_{n_f \times n_f}, [b_{f_{ik}}]_{n_f \times p}, [c_{f_{lj}}]_{q \times n_f}))\|_\infty \\
&= \|\hat{C}_{Fb}(sI - \hat{A}_{Fb})^{-1} \hat{B}_{Fb}\|_\infty
\end{aligned}
$$
$$(8.34)$$

$$
\begin{aligned}
M_{C_{Fe}} &= \|\frac{1}{q \times n_f} \sum_{l=1}^{q} \sum_{j=1}^{n_f} S_{c_{f_{lj}}}(T_e(s, [a_{f_{ij}}]_{n_f \times n_f}, [b_{f_{ik}}]_{n_f \times p}, [c_{f_{lj}}]_{q \times n_f}))\|_\infty \\
&= \|\hat{C}_{Fc}(sI - \hat{A}_{Fc})^{-1} \hat{B}_{Fc}\|_\infty
\end{aligned}
$$
$$(8.35)$$

where

$$
\mathcal{N}_{a_f} = \begin{bmatrix} 0 & 0 \\ 0 & \frac{1}{n_f \times n_f}\Re_{af} \end{bmatrix}, \hat{A}_{Fa} = \begin{bmatrix} \bar{A}_{Fe} & \mathcal{N}_{a_f} \\ 0 & \bar{A}_{Fe} \end{bmatrix}, \hat{B}_{Fa} = \begin{bmatrix} 0 \\ \bar{B}_{Fe} \end{bmatrix}
$$

$$
\hat{C}_{Fa} = \begin{bmatrix} \bar{C}_{Fe} & 0 \end{bmatrix}; \mathcal{N}_{b_f} = \begin{bmatrix} 0 & 0 \\ \frac{1}{n_f \times p}\Re_{b_f} C & 0 \end{bmatrix}, \mathcal{M}_{b_f} = \begin{bmatrix} 0 \\ \frac{1}{n_f \times p}\Re_{b_f} D \end{bmatrix}
$$

$$
\hat{A}_{Fb} = \begin{bmatrix} \bar{A}_{Fe} & \mathcal{N}_{b_f} & 0 \\ 0 & \bar{A}_{Fe} & 0 \\ 0 & 0 & \bar{A}_{Fe} \end{bmatrix}, \hat{B}_{Fb} = \begin{bmatrix} 0 \\ \bar{B}_{Fe} \\ \mathcal{M}_{b_f} \end{bmatrix}, \hat{C}_{Fb} = \begin{bmatrix} \bar{C}_{Fe} & 0 & \bar{C}_{Fe} \end{bmatrix}
$$

$$
\mathcal{N}_{c_f} = \begin{bmatrix} 0 & -\frac{1}{q \times n_f}\Re_{cf} \end{bmatrix}, \hat{C}_{Fc} = \mathcal{N}_{c_f}, \hat{A}_{Fc} = \bar{A}_{Fe}, \hat{B}_{Fc} = \bar{B}_{Fe}
$$

and

$$
\Re_{af} = f_{n_f} f_{n_f}^T, \Re_{bf} = f_{n_f} f_p^T, \Re_{cf} = f_q f_{n_f}^T,
$$
$$(8.36)$$

for the additive form, while

$$
\Re_{af} = A_{Fe}, \Re_{bf} = B_{Fe}, \Re_{cf} = C_{Fe},
$$

for the multiplicative form, where $f_{n_f} \in R^{n_f}$, $f_p \in R^p$, and $f_q \in R^q$ denote the column vectors in which all the elements equal 1.

Compared with the coefficient sensitivity measures defined in (8.15), the advantage of the new measures is that $\Re_{af}, \Re_{bf}$, and $\Re_{cf}$ are compatible with the filter matrices. The feature is crucial in dealing with the filtering problem with respect to multiplicative coefficient variations.

Based on the new sensitivity measures and Lemma 2.11, employing similar technologies of Theorem 8.1, the following theorem presents a sufficient condition for the insensitive H_∞ filtering problems with respect to the multiplicative filter coefficient variations and additive filter coefficient variations, respectively.

Theorem 8.3 *Consider the system in (8.1). Let scalars $\gamma > 0$, $\beta > 0$. Then, the filtering error system (8.5) is asymptotically stable and the conditions*

$$\|T_e(s, [a_{f_{ij}}]_{n_f \times n_f}, [b_{f_{ik}}]_{n_f \times p}, [c_{f_{lj}}]_{q \times n_f})\|_\infty < \gamma,$$
$$M_{A_{Fe}} < \beta, M_{B_{Fe}} < \beta, M_{C_{Fe}} < \beta$$

hold if, for some positive scalars λ_a, λ_b, and λ_c, there exist matrices $F_A \in R^{n_f \times n_f}$, $F_B \in R^{n_f \times p}$, $F_C \in R^{q \times n_f}$, $0 < S \in R^{n \times n}$, and $0 < N \in R^{n_f \times n_f}$, such that (8.21), (8.22), (8.23), and (8.24) hold, where

$$M_{cf} = \begin{bmatrix} 0 & -\frac{1}{q \times n_f} F_C \end{bmatrix}, \quad M_{af} = \begin{bmatrix} 0 & \frac{1}{n_f \times n_f} \mathcal{H} F_A \\ 0 & \frac{1}{n_f \times n_f} F_A \end{bmatrix}$$

$$M_{b1f} = \begin{bmatrix} \frac{1}{n_f \times p} \mathcal{H} F_B C & 0 \\ \frac{1}{n_f \times p} F_B C & 0 \end{bmatrix}, \quad M_{b2f} = \begin{bmatrix} \frac{1}{n_f \times p} \mathcal{H} F_B D \\ \frac{1}{n_f \times p} F_B D \end{bmatrix}$$

for the multiplicative case, while

$$M_{cf} = \begin{bmatrix} 0 & -\frac{1}{q \times n_f} \Re_{cf} \end{bmatrix}, \quad M_{af} = \begin{bmatrix} 0 & \frac{1}{n_f \times n_f} \mathcal{H} N \Re_{af} \\ 0 & \frac{1}{n_f \times n_f} N \Re_{af} \end{bmatrix}$$

$$M_{b1f} = \begin{bmatrix} \frac{1}{n_f \times p} \mathcal{H} N \Re_{bf} C & 0 \\ \frac{1}{n_f \times p} N \Re_{bf} C & 0 \end{bmatrix}, \quad M_{b2f} = \begin{bmatrix} \frac{1}{n_f \times p} \mathcal{H} N \Re_{bf} D \\ \frac{1}{n_f \times p} N \Re_{bf} D \end{bmatrix}$$

for the additive case, respectively, where $\Re_{af}$, $\Re_{bf}$, and $\Re_{cf}$ are defined in (8.36). In addition, M_{as}, M_{bs}, M_{cs}, and $\mathcal{H}$ are defined in Theorem 8.1. Moreover, the n_fth-order insensitive filter can also be given by (8.25).

Remark 8.10 *From the theory point of view, the new coefficient sensitivity measures defined in (8.33), (8.34), and (8.35) cannot guarantee the all concerning individual sensitivity. However, the concerning individual sensitivity can be practically reduced in many examples by using these new sensitivity measures, see the examples in Section 8.6.*

In what follows, we provide an alternative approach to obtain the insensitivity H_∞ filter which can tolerate the multiplicative coefficient variations and can guarantee the individual insensitivity to the multiplicative coefficient variations. But, the approach may be more conservative than the one of minimizing average sensitivity measures.

In consideration of the fact that the filter coefficient variations are trivial, therefore, if the magnitudes of the designed filter gains A_{Fe}, B_{Fe}, and C_{Fe} obtained by Theorem 8.1 are small enough, then the obtained filter is also insensitive to the multiplicative coefficient variations. Fortunately, we can obtain the insensitive filter with small magnitude by solution of the conditions in Theorem 8.1 with the constraints (8.31) and (8.32). In addition, the effectiveness of this approach is illustrated by numerical examples. Then, by using the following algorithm, this approach can be realized.

Algorithm 8.1

Step 1. *Fix $\gamma, \lambda_a, \lambda_b$, and λ_c, and select appropriate μ and ν.*

Step 2. *β is minimized if the following optimization problem is solvable*

$$\min_{S,N,F_A,F_B,F_C} \beta,$$
$$\text{s.t. } S > 0, N > 0, (8.21), (8.22), (8.23), (8.24), (8.31), (8.32). \tag{8.37}$$

Step 3. *The designed filter's parameters finally can be obtained by (8.25).*

Remark 8.11 *The LMI-based conditions for designing insensitive H_∞ filters given in Theorem 8.1 and Theorem 8.3 can be easily extended to deal with the robust insensitive H_∞ filtering problem for systems with polytopic uncertainties because the system matrices are affinely involved in the proposed design conditions.*

8.4 Computation of Robust H_∞ Performance Index

In order to illustrate the effectiveness of the proposed method, in this section we evaluate the robust H_∞ performances of the filtering error system with filter coefficient variations.

In this chapter, $\theta_{a_{f_{ij}}}$, $\theta_{b_{f_{ik}}}$, and $\theta_{c_{f_{lj}}}$ represent the interval type of coefficient variations with the following form:

$$
\begin{aligned}
|\theta_{a_{f_{ij}}}| &\leq \theta, i, j = 1, \cdots, n_f \\
|\theta_{b_{f_{ik}}}| &\leq \theta, i = 1, \cdots, n_f, k = 1, \cdots, p \\
|\theta_{c_{f_{lj}}}| &\leq \theta, l = 1, \cdots, q, j = 1, \cdots, n_f.
\end{aligned} \tag{8.38}
$$

In the following, for a given filter, we will establish LMI-based sufficient conditions of evaluating the H_∞ performances for the filter subjected to the additive coefficient variations and multiplicative coefficient variations, respectively.

Additive Coefficient Variation Case

When the filter is subject to the additive filter coefficient variations described by (8.3), the result can be obtained by a slight modification of Lemma 10 given in Yang and Che [143], that is, the order of filter n is changed to n_f. Consequently, it is omitted for brevity.

Multiplicative Coefficient Variation Case

For the case of multiplicative filter coefficient variations described by (8.4), denote

$$G_{m1} = \begin{bmatrix} g_{m11} & g_{m12} & \cdots & g_{m1l_a} \end{bmatrix},$$
$$G_{m2} = \begin{bmatrix} g_{m21}^T & g_{m22}^T & \cdots & g_{m2l_a}^T \end{bmatrix}^T$$

where $l_a = n_f^2 + n_f p + n_f q$, and

$$g_{m1k} = \begin{bmatrix} (0_{1\times n} & e_i^T)P & 0_{1\times r} & 0_{1\times q} \end{bmatrix}^T$$
$$g_{m2k} = \begin{bmatrix} 0_{1\times n} & e_i^T A_{Fe}e_j e_j^T & 0_{1\times r} & 0_{1\times q} \end{bmatrix}$$
for $k = (i-1)n_f + j$, $i, j = 1, \cdots, n_f$

$$g_{m1k} = \begin{bmatrix} (0_{1\times n} & e_i^T)P & 0_{1\times r} & 0_{1\times q} \end{bmatrix}^T$$
$$g_{m2k} = \begin{bmatrix} e_i^T B_{Fe}h_j h_j^T C & 0_{1\times n_f} & e_i^T B_{Fe}h_j h_j^T D & 0_{1\times q} \end{bmatrix}$$
for $k = n_f^2 + (i-1)p + j$, $i = 1, \cdots, n_f$, $j = 1, \cdots, p$

$$g_{m1k} = \begin{bmatrix} 0_{1\times n} & 0_{1\times n_f} & 0_{1\times r} & -g_i^T \end{bmatrix}^T$$
$$g_{m2k} = \begin{bmatrix} 0_{1\times n} & g_i^T C_{Fe}e_j e_j^T & 0_{1\times r} & 0_{1\times q} \end{bmatrix}$$
for $k = n_f^2 + n_f p + (i-1)n_f + j$, $i = 1, \cdots, q, j = 1, \cdots, n_f$.

At the same time, by using the same techniques as those in the proof of Lemma 10 in Yang and Che [143], the following lemma can be presented.

Lemma 8.3 *Consider the system in (8.1). Let $\gamma > 0$, $\theta > 0$ be given constants and filter gain matrices A_{Fe}, B_{Fe}, and C_{Fe} be given. Then, $\|T(s)\|_\infty < \gamma$ holds for all $\theta_{a_{f_{ij}}}$, $\theta_{b_{f_{ik}}}$, and $\theta_{c_{f_{lj}}}$ satisfying (8.4) and (8.38) if there exist a positive-definite symmetric matrix $P > 0$ and a symmetric matrix Σ with the structure described by (7.13) and $s_a = 1, \cdots, n_f^2 + n_f p + n_f q$ such that the* *following LMIs hold:*

$$\begin{bmatrix} Q_s & G_{m1} \\ G_{m1}^T & 0 \end{bmatrix} + \begin{bmatrix} G_{m2} & 0 \\ 0 & I \end{bmatrix}^T \Sigma \begin{bmatrix} G_{m2} & 0 \\ 0 & I \end{bmatrix} < 0$$

$$\begin{bmatrix} I \\ \Delta_1 \end{bmatrix}^T \begin{bmatrix} \sigma_{11}^i & \sigma_{12}^i \\ \sigma_{12}^i & \sigma_{22}^i \end{bmatrix} \begin{bmatrix} I \\ \Delta_1 \end{bmatrix} \geq 0$$

$$\Delta_1 = diag\{\theta_{k_{i-1}+j} \cdots \theta_{k_i}\}, for\ all\ \theta_{k_{i-1}+j} \in \{-\theta, \theta\}$$

$$j = 1, \cdots, k_i - k_{i-1}, i = 1, \cdots, s_a$$

where

$$Q_s = \begin{bmatrix} He\{P\bar{A}_{Fe}\} & P\bar{B}_{Fe} & \bar{C}_{Fe}^T \\ * & -\gamma^2 I & 0 \\ * & * & -I \end{bmatrix},$$

with $\bar{A}_{Fe}$, $\bar{B}_{Fe}$, and $\bar{C}_{Fe}$ being defined in (8.7).

8.5 Comparison with the Existing Design Method

The non-fragile H_∞ filter design for continuous-time systems was investigated in Yang and Che [143], in which the designed filter was assumed to be with interval additive filter coefficient variations. In comparison with the existing method, the new proposed method in this chapter has two main advantages:

(a) It is difficult to use the existing method in Yang and Che [143] to obtain convex conditions for the filter design problem with respect to interval multiplicative coefficient variations, while this problem can be resolved well by using the new proposed method.

(b) The number of the LMIs involved in the design conditions by the new proposed method is significantly reduced. The fact can be found in Table 8.1. Let s_a denote the number of structured vertex separators [143], and $s_a = 1, \cdots, s_{amax}(s_{amax} = n_f^2 + n_f p + n_f q)$, where s_{amax} denotes the number of filter uncertain parameters. Let $\tilde{s}_a = \sum_{i=1}^{s_a} 2^{k_i - k_{i-1}}$ where $k_i(i = 0, \cdots, s_a)$ are integers satisfying $k_0 = 0 < k_1 < \cdots < k_{s_a} = s_{amax}$. The following illustration is intended to make the table clear.

Besides (8.21), the number of LMIs involved in sensitive constraint is s_{amax} in Theorem 8.1, while the number of LMIs involved in sensitive constraint is three in Theorem 8.3. On the other hand, it should be noted that the number of LMI constraints involved in the design conditions given by Theorem 5 in Yang and Che [143] grows exponentially with the

TABLE 8.1

Number of LMIs Involved

| Present Chapter | | Yang and Che [143] | | |
Theorem 8.1	Theorem 8.3	Theorem 5	Theorem 7	
—	—	—	$s_a = 1$	$s_a \neq 1$
$1 + s_{amax}$	4	$2^{s_{amax}}$	$2^{s_{amax}}$	$1 + \tilde{s}_a$

number of uncertain parameters, which may result in a numerical problem for systems with high dimensions. Although the structured vertex separator method is proposed in Theorem 7 in Yang and Che [143] to deal with this problem, the number of LMI constraints involved in the design condition may still be large.

8.6   Example

In this section, two numerical examples are presented to illustrate the effectiveness of the proposed insensitive H_∞ filter design method, and comparisons between the novel proposed method and the existing method [143] are provided. In addition, in order to do the comparison, in this section, the full-order filtering problem (i.e., $n_f = n$) is considered.

Also, in this section, SF denotes the standard H_∞ filter, IF denotes the insensitive H_∞ filter, and NF denotes the non-fragile H_∞ filter for brevity. In addition, we always set $\theta = 0.02$ and $k_i = i, i = 1, \cdots, 15$, that is, $s_a = 15$ as well as $k_i = 3i, i = 1, \cdots, 5$, that is, $s_a = 5$ in this section.

Example 8.1 *Consider a linear continuous-time system of the form (8.1) with the following matrices (borrowed from Yang and Che [143]):*

$$A = \begin{bmatrix} 0 & 1 & 0 \\ 0 & 0 & 1 \\ -1 & -2.5 & -1 \end{bmatrix}, B = \begin{bmatrix} -1.5 & 0 \\ 1 & 0 \\ 2 & 0 \end{bmatrix}$$

$$L = \begin{bmatrix} 3 & -2 & 1 \end{bmatrix}, C = \begin{bmatrix} -2 & 1 & 2 \end{bmatrix}, D = \begin{bmatrix} 0 & 0.9 \end{bmatrix}$$

By using Remark 8.8, the standard H_∞ filter is obtained and accordingly the optimal H_∞ performance is $\gamma = 4.6536$.

Additive Coefficient Variations Case

First, we design an insensitive H_∞ filter by using Theorem 8.1. By solving the optimization problem (8.30) with $\gamma^2 = 33.8$, $\lambda_a = \lambda_b = 7$, and $\lambda_c = 4$, we obtain the sensitivity performance $\beta = 20.2568$ and the insensitive filter F_{ain}

with gain matrices as

$$A_{Fe} = 10^9 * \begin{bmatrix} 0.2888 & -0.2150 & -0.4479 \\ -0.4592 & 0.3418 & 0.7122 \\ 0.7335 & -0.5459 & -1.1376 \end{bmatrix}, B_{Fe} = 10^8 * \begin{bmatrix} -1.9846 \\ 3.1555 \\ -5.0404 \end{bmatrix}$$

$$C_{Fe} = \begin{bmatrix} -4.0747 & 2.0948 & -0.8936 \end{bmatrix}.$$

Observing that the gains of F_{ain} are too large and are not desired, as indicated in Remark 8.9, applying Theorem 8.1 with (8.31) and (8.32) and with $\mu = 0.3, \nu = 100000$, we obtain the sensitivity performance $\beta = 23.2299$ and the insensitive filter $F_{ain\mu\nu}$ with gain matrices as

$$A_{Fe} = \begin{bmatrix} 18.8902 & -32.5433 & -70.1680 \\ -46.1768 & 17.9570 & 38.0298 \\ 78.6196 & -61.9701 & -123.9871 \end{bmatrix}, B_{Fe} = \begin{bmatrix} -26.8187 \\ 19.9925 \\ -54.7735 \end{bmatrix}$$

$$C_{Fe} = \begin{bmatrix} -2.5305 & 2.3930 & -0.1173 \end{bmatrix}.$$

On the other hand, by solving the conditions in Theorem 8.3 with $\gamma^2 = 25.58$, $\lambda_a = \lambda_b = 2$, and $\lambda_c = 0.05$, we obtain the sensitivity performance $\beta = 6.4716$ and the insensitive filter F_{ain} with gain matrices as

$$A_{Fe} = 10^6 * \begin{bmatrix} -2.2208 & -0.3176 & 0.3949 \\ -1.8159 & -0.2597 & 0.3229 \\ -1.1495 & -0.1644 & 0.2044 \end{bmatrix}, B_{Fe} = 10^5 * \begin{bmatrix} 5.0338 \\ 4.1160 \\ 2.6053 \end{bmatrix}$$

$$C_{Fe} = \begin{bmatrix} -2.5478 & 2.0930 & -0.9612 \end{bmatrix}.$$

Since the gains of F_{ain} are also too large, applying Theorem 8.3 with (8.31) and (8.32) and with $\mu = 0.5, \nu = 10000$, we obtain the sensitivity performance $\beta = 6.5653$ and the insensitive filter $F_{ain\mu\nu}$ with gain matrices as

$$A_{Fe} = \begin{bmatrix} -7.8807 & -2.8624 & -2.5114 \\ -5.4775 & -3.5230 & -1.8065 \\ 0.7756 & -6.8086 & -7.3332 \end{bmatrix}, B_{Fe} = \begin{bmatrix} 0.7164 \\ -0.0704 \\ -2.5422 \end{bmatrix}$$

$$C_{Fe} = \begin{bmatrix} -2.2181 & 2.3058 & -0.9080 \end{bmatrix}.$$

Comparing the above design without (8.31) and (8.32), the magnitudes of the filter gains have been reduced.

By using Remark 8.8, Theorem 8.1/Theorem 8.3, and Theorem 7 in Yang and Che [143], we can obtain the gains of the standard filter, insensitive filter, and non-fragile filter, respectively. Lemma 10 given in Yang and Che [143] is also applicable to evaluate the H_∞ performances of the above three designed filters. Then, the H_∞ performance indexes for the two different cases can be obtained and are listed in Table 8.2 and Table 8.3, respectively. Obviously, from Table 8.2 and Table 8.3, comparing with the optimal H_∞ performance $\gamma = 4.6536$, the coefficient variations of the standard filter destabilize the filtering error system, that is, the standard filter is sensitive to the filter coefficient

TABLE 8.2
Comparison of H_∞ Performance Indexes without (8.31)
and (8.32)

s_a	SF	IF		NF
		Theorem 8.1	Theorem 8.3	
$s_a = 15$	13.3500	5.8666	5.5363	5.2996
$s_a = 5$	13.1341	5.8658	5.5149	5.2832

TABLE 8.3
Comparison of H_∞ Performance Indexes with (8.31) and
(8.32)

s_a	SF	IF		NF
		Theorem 8.1	Theorem 8.3	
$s_a = 15$	13.7707	5.8474	5.5640	5.3036
$s_a = 5$	13.5661	5.8462	5.5414	5.2993

variations, while the filters designed by the new proposed method and the existing method are insensitive to the filter coefficient variations; moreover, the two methods achieve the same level of performance.

The numbers of the LMI constraints involved in the designs are given in Table 8.4. From Table 8.4, it is easy to see that the new proposed method can overcome the difficulty of implementing the LMI constraints in the existing method, which shows the superiority of the new method.

Multiplicative Coefficient Variations Case

In what follows, we design an insensitive H_∞ filter by using Theorem 8.3. By solving the conditions in Theorem 8.3 with $\gamma^2 = 26.5$, $\lambda_a = \lambda_c = 2$, and $\lambda_b = 2.01$, we obtain the sensitivity performance index $\beta = 14.6595$ and the insensitive filter F_{min} with gain matrices as

$$A_{Fe} = \begin{bmatrix} -13.9798 & -1.9508 & -3.6563 \\ -5.5296 & -1.7229 & -1.5088 \\ -12.7851 & -5.7467 & -5.6845 \end{bmatrix}, B_{Fe} = \begin{bmatrix} 0.8174 \\ -0.1150 \\ -0.0229 \end{bmatrix}$$

$$C_{Fe} = \begin{bmatrix} -1.7652 & 2.4598 & -0.1395 \end{bmatrix}.$$

TABLE 8.4
Number of LMI Constraints Involved

Present Chapter		Yang and Che [143]			
Theorem 8.1	Theorem 8.3	Theorem 5	Theorem 7		
			$s_a = 1$	$s_a = 5$	$s_a = 15$
—	—	—			
16	4	32768	32768	41	31

TABLE 8.5

Comparison of H_∞ Performance Indexes

s_a	Standard H_∞ Filter	Insensitive H_∞ Filter	
—	—	Theorem 8.3	Algorithm 8.1
$s_a = 15$	Infeasible	6.0103	7.7347
$s_a = 5$	Infeasible	5.9674	7.7196

Observing that the gains of F_{min} are not large, the constraints (8.31) and (8.32) are not needed.

In addition, using Algorithm 8.1, by solving the optimal problem (8.37) with $\gamma^2 = 40$, $\lambda_a = \lambda_c = 7$, $\lambda_b = 4$, and with $\mu = 0.7$, $\nu = 1000$, we obtain the sensitivity performance index $\beta = 63.4126$ and the insensitive filter F_{min} with gain matrices as

$$A_{Fe} = \begin{bmatrix} -20.0114 & -2.7054 & -7.3982 \\ -10.9043 & -0.5825 & 1.1313 \\ -12.0229 & -9.5280 & -16.8272 \end{bmatrix}, B_{Fe} = \begin{bmatrix} 0.5364 \\ 1.6693 \\ -3.9470 \end{bmatrix}$$

$$C_{Fe} = \begin{bmatrix} -3.4239 & 2.2002 & -0.5467 \end{bmatrix}.$$

Then, Lemma 8.3 is applicable for evaluating the H_∞ performance of the designed filter, and the H_∞ performances of the standard filter and insensitive filter are listed in Table 8.5.

Obviously, from Table 8.5, comparing with the optimal H_∞ performance $\gamma = 4.6536$, the multiplicative filter coefficient variations result in the serious deterioration of the performance of the standard filter, that is, the standard filter is sensitive to the multiplicative filter coefficient variations, while the filter designed by using the new proposed method is insensitive to the filter coefficient variations. Furthermore, from Table 8.5, it can be seen that the proposed method given by Theorem 8.3 gives smaller performance indexes, which shows that Theorem 8.3 can provide less conservative designs than Algorithm 8.1.

In order to show the effectiveness of our method more clearly, a simulation is also performed. In the following simulation, let the system initial state be $x_0^T = \begin{bmatrix} 0 & 0 & 0 \end{bmatrix}$ and the filter initial state be $\bar{x}_0^T = \begin{bmatrix} 0 & 0 & 0 \end{bmatrix}$. In addition, we assume the disturbance input $w^T(t) = \begin{bmatrix} w_1^T(t) & w_2^T(t) \end{bmatrix}$ as follows:

$$w_1(t) = w_2(t) = \begin{cases} 0.1\sin(0.3t) & 10 \le t \le 20, \\ 0 & otherwise. \end{cases}$$

Figure 8.1 and Figure 8.2 show the response of the estimation error $e(t)$ for the insensitive filter and the standard filter with $\theta = 0$ and $\theta = 0.02$, respectively. From Figure 8.1, one can see that the standard filter performs well; however, the performance of the standard filter is seriously deteriorated for the filter coefficient variation case in Figure 8.2. These phenomena show the effectiveness of the new proposed design method.

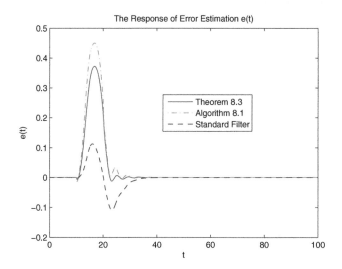

FIGURE 8.1

Comparison between the insensitive H_∞ filter and the standard H_∞ filter ($\theta = 0$).

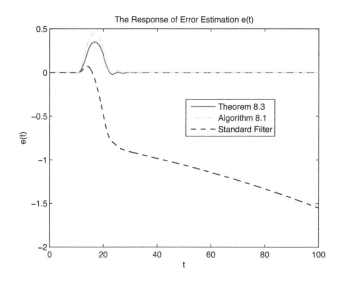

FIGURE 8.2

Comparison between the insensitive H_∞ filter and the standard H_∞ filter ($\theta = 0$).

TABLE 8.6

Comparison of H_∞ Performance Indexes

s_a	SF	IF		NF
		Theorem 8.1	Theorem 8.3	
$s_a = 15$	Infeasible	3.3027	3.3469	3.2746
$s_a = 5$	Infeasible	3.3005	3.3466	3.2745

TABLE 8.7

Comparison of H_∞ Performance Indexes

s_a	Standard H_∞ Filter	Insensitive H_∞ Filter	
—	—	Theorem 8.3	Algorithm 8.1
$s_a = 15$	32.7909	3.7829	4.3360
$s_a = 5$	32.7909	3.7783	4.2834

Example 8.2 *In this section, to check how the proposed insensitive H_∞ filter works in a real system, a numerical example based on a linearized model of an F-404 engine from Eustace et al. [33], which is also studied in Ahn, Han, and Kwon [2], Che and Yang [56], and Han and Kwon, [18], is taken. The corresponding system parameters are*

$$A = \begin{bmatrix} -1.4600 & 0 & 2.4280 \\ 0.1643 + 0.5\sigma & -0.4 + \sigma & -0.3788 \\ 0.3107 & 0 & -2.2300 \end{bmatrix}, B = \begin{bmatrix} 0.35 & 0 \\ 0.55 & 0 \\ -3.86 & 0 \end{bmatrix}$$

$$L = \begin{bmatrix} 0 & 0 & 4 \end{bmatrix}, C = \begin{bmatrix} 1 & 0 & 0 \end{bmatrix}, D = \begin{bmatrix} -0.8 & 0.9 \end{bmatrix}$$

with $\sigma = 0.39$. Then, by using Remark 8.8, the standard H_∞ filter is obtained and accordingly the optimal H_∞ performance is $\gamma = 3.2479$. In this example, we omitted the gain parameters of the designed filters due to the limited space.

For the additive coefficient variations case, by solving the conditions in Theorem 8.1 with $\gamma = 3.3$, $\lambda_a = \lambda_b = 1.7$, and $\lambda_c = 20$, and solving the conditions in Theorem 8.3 with $\gamma = 3.37$, $\lambda_a = \lambda_b = 1.95$, and $\lambda_c = 2.05$, at the same time, Lemma 7.4 given in Chapter 7 is applicable to evaluate the H_∞ performances of the designed filters, and the H_∞ performance indexes can be obtained and are listed in Table 8.6.

For the multiplicative coefficient variations case, by solving the conditions in Theorem 8.3 with $\gamma = 3.7$, $\lambda_a = \lambda_c = 1.95$, and $\lambda_b = 2.05$, and solving the optimal problem (8.37) with $\gamma = 3.5$, $\lambda_a = \lambda_c = 1.7$, $\lambda_b = 20$, and with $\mu = 0.07, \nu = 100$, respectively, at the same time, Lemma 8.3 is applicable for evaluating the H_∞ performance of the designed filter. Then, Table 8.7 is obtained.

Obviously, from Table 8.6 and Table 8.7, the filter coefficient variations result in the serious deterioration of the performance of the standard filter, that is, the standard filter is sensitive to the filter coefficient variations, while the filters designed by the new proposed method and the existing method are insensitive to the filter coefficient variations, and the two methods achieve the same level of performance.

8.7 Conclusion

In this chapter, the problem of designing insensitive H_∞ filters for linear continuous-time systems has been addressed. Coefficient sensitivity functions of transfer functions with respect to filter additive/multiplicative coefficient variations are defined, and the H_∞ norms of the sensitivity functions are used to measure the sensitivity of the transfer functions with respect to filter coefficient variations. A method for designing insensitive H_∞ filters subjected to filter coefficient variations is given in terms of the *linear matrix inequality (LMI)* optimization techniques. The new proposed method has two main advantages. The first one is to resolve the problem of designing the insensitive filters subjected to the multiplicative filter coefficient variations in a new framework, while this problem cannot be resolved using the existing method. The second one is that the number of the LMIs involved in the design conditions does not grow exponentially, and is significantly reduced comparing with the existing method.

9

Insensitive H_∞ Filtering of Delta Operator Systems

9.1 Introduction

In Chapter 8, we have successfully designed the insensitive H_∞ filters for linear continuous-time systems based on minimizing the coefficient sensitivity. However, the artificial constraints imposed on the Lyapunov matrices in Chapter 8 might lead to considerable conservativeness, where a common *Lyapunov matrix* was adopted for the same class of individual sensitivity functions. Thus, in this chapter, some efforts will be made in the direction of reducing the conservativeness of the analysis and design methods for improving the system's performance.

On the other hand, it is well known that the usual *shift operator* approach suffers from numerical ill-conditioning at a sufficiently small sampling period. Therefore, in order to solve this problem, the *delta operator* instead of the traditional shift operator was constructed to study sampling continuous-time systems [44,45]. It is worth mentioning that the delta form gives a less sensitive fixed point coefficient representation than the shift operator form [60,84,119]. Hence, considerable attention has focused on designing low parametric sensitivity finite word length (FWL) realizations of controllers or filters based on the delta operator formulation [see 20, 55, 65, 119, 133, and the references therein]. This research shows that the advantages of the delta operator include superior round-off noise performance, more accurate coefficient representation, and less sensitive control law design. Therefore, the problem of designing insensitive H_∞ filters with respect to filter *coefficient variations* based on the delta operator formulation is a comparable worthy research issue.

Motivated by the above points, this chapter is concerned with the *multiobjective problem* of designing *insensitive H_∞ filters* for *delta operator systems*, which minimizes the *filter coefficient sensitivity* and meets the prescribed H_∞ *norm* constraint simultaneously. The contribution of this chapter is twofold. First, the filter design problem is solved based on *delta operator* models, which establishes an interesting link between continuous-time and discrete-time systems analysis. Second, *Finsler's lemma* is used to derive novel conditions which are adapted to treat multiple objective optimization problems in a potentially *less conservative* framework. The chapter is organized as follows.

First, *coefficient sensitivity functions* of transfer functions with respect to *additive/multiplicative filter coefficient variations* are defined, and the H_∞ norms of the sensitivity functions are used to measure the sensitivity of the transfer functions with respect to *additive/multiplicative filter coefficient variations*. Next, sufficient conditions for designing *insensitive H_∞ filters* with respect to *additive/multiplicative filter coefficient variations* based on two different sensitivity measures are presented in terms of a set of linear matrix inequalities (LMIs). Finally, a numerical example is given to illustrate the effectiveness of the developed techniques, where a comparison is also made between the delta operator approach (or δ-domain) and the forward-shift operator approach (or z-domain), which verifies that the δ-domain is less sensitive than the z-domain at high sampling rates.

9.2 Problem Statement

Let h be the *sample period* and q be the standard forward-shift operator, that is, $qx(k) = x(k+1)$. Then, by using the definition of delta operator in Definition 2.1, the corresponding discrete-time system of (8.1) is given as follows:

$$\begin{cases} \delta x_\delta(k) = A_\delta x_\delta(k) + B_\delta w(k) \\ y(k) = C x_\delta(k) + D w(k) \\ z(k) = L x_\delta(k) \\ x_\delta(k) = 0, \quad k \leq 0 \\ A_\delta = (A_z - I)/h, A_q = e^{Ah}, B_\delta = B_q/h, B_q = \int_0^h e^{A\tau} B d\tau \end{cases} \tag{9.1}$$

where A_δ and B_δ are the corresponding delta operator system matrices, A_q and B_q are the z-domain discrete system matrices. C, D, and L are the same as the z-domain discrete system matrices. It is obvious that

$$\lim_{h \to 0} A_\delta = \lim_{h \to 0} (e^{Ah} - I)/h = A, \quad \lim_{h \to 0} B_\delta = B,$$

which implies that the delta representation converges to the continuous-time representation as $h \to 0$.

We are interested in designing the following n_fth-order delta operator filter for the estimation of $z(k)$:

$$\begin{aligned} \delta \bar{x}(k) &= A_F \bar{x}(k) + B_F y(k) \\ \bar{z}(k) &= C_F \bar{x}(k) \end{aligned} \tag{9.2}$$

where $\bar{x}(k) \in R^n$ is the filter state and $\bar{z}(k) \in R^q$ is the estimation of $z(k)$. The filter (9.2) is referred to as the full-order filter and reduced-order filter when $n_f = n$ and $n_f < n$, respectively. $A_F \in R^{n_f \times n_f}$, $B_F \in R^{n_f \times p}$, and $C_F \in$

$R^{q \times n_f}$ are filter matrices to be designed. Moreover, it should be mentioned that $a_{f_{ij}}, b_{f_{ik}}$, and $c_{f_{lj}}$ ($i, j = 1, \cdots, n_f; k = 1, \cdots, p; l = 1, \cdots, q$) are used to denote the elements of the filter matrices A_F, B_F, and C_F, respectively.

Applying the filter (9.2) to the system (9.1), we obtain the *filtering error system*

$$
\begin{aligned}
\delta \xi(k) &= A_{cl}\xi(k) + B_{cl}w(k) \\
e(k) &= C_{cl}\xi(k)
\end{aligned}
\tag{9.3}
$$

where $\xi(k) = \begin{bmatrix} x_\delta(k) \\ \bar{x}(k) \end{bmatrix}$, $e(k) = z(k) - \bar{z}(k)$ is the estimation error, and

$$
A_{cl} = \begin{bmatrix} A_\delta & 0 \\ B_F C & A_F \end{bmatrix}, B_{cl} = \begin{bmatrix} B_\delta \\ B_F D \end{bmatrix}, C_{cl} = \begin{bmatrix} L & -C_F \end{bmatrix}.
$$

The transfer function matrix of the filtering error system (9.3) from $w(k)$ to $e(k)$ is given by

$$
T(\delta, [a_{f_{ij}}]_{n_f \times n_f}, [b_{f_{ik}}]_{n_f \times p}, [c_{f_{lj}}]_{q \times n_f}) = C_{cl}(\delta I - A_{cl})^{-1}B_{cl}.
\tag{9.4}
$$

Then, based on Definition 8.1 and by means of the techniques developed in Gevers and Li [43] and Hilaire, Chevrel, and Trinquet [60], the following lemma is presented.

Lemma 9.1 *Let $T(\delta, a_{f_{ij}}, b_{f_{ik}}, c_{f_{lj}})$ be defined in (9.4), and $a_{f_{ij}}, b_{f_{ik}}$, and $c_{f_{lj}}$ are the elements of the filter matrices A_F, B_F, and C_F, respectively. Then, the sensitivity functions of the transfer function with respect to the elements of the filter matrices are given as follows:*

$$
\begin{aligned}
&S_{a_{f_{ij}}}(T(\delta, [a_{f_{ij}}]_{n_f \times n_f}, [b_{f_{ik}}]_{n_f \times p}, [c_{f_{lj}}]_{q \times n_f})) \\
&= C_{cl}(\delta I - A_{cl})^{-1}N_{a_{f_{ij}}}(\delta I - A_{cl})^{-1}B_{cl} \\
&= \hat{C}_{a_{f_{ij}}}(\delta I - \hat{A}_{a_{f_{ij}}})^{-1}\hat{B}_{a_{f_{ij}}}, i = 1, \cdots, n_f; j = 1, \cdots, n_f \\
&S_{b_{f_{ik}}}(T(\delta, [a_{f_{ij}}]_{n_f \times n_f}, [b_{f_{ik}}]_{n_f \times p}, [c_{f_{lj}}]_{q \times n_f})) \\
&= C_{cl}(\delta I - A_{cl})^{-1}N_{b_{f_{ik}}}(\delta I - A_{cl})^{-1}B_{cl} + C_{cl}(\delta I - A_{cl})^{-1}\tilde{N}_{b_{f_{ik}}} \\
&= \hat{C}_{b_{f_{ik}}}(\delta I - \hat{A}_{b_{f_{ik}}})^{-1}\hat{B}_{b_{f_{ik}}}, i = 1, \cdots, n_f; k = 1, \cdots, p \\
&S_{c_{f_{lj}}}(T(\delta, [a_{f_{ij}}]_{n_f \times n_f}, [b_{f_{ik}}]_{n_f \times p}, [c_{f_{lj}}]_{q \times n_f})) \\
&= N_{c_{f_{lj}}}(\delta I - A_{cl})^{-1}B_{cl} \\
&= \hat{C}_{c_{f_{lj}}}(\delta I - \hat{A}_{c_{f_{lj}}})^{-1}\hat{B}_{c_{f_{lj}}}, l = 1, \cdots, q; j = 1, \cdots, n_f
\end{aligned}
$$

where

$$N_{a_{f_{ij}}} = \begin{bmatrix} 0 & 0 \\ 0 & R_{a_{f_{ij}}} \end{bmatrix}, \hat{A}_{a_{f_{ij}}} = \begin{bmatrix} A_{cl} & N_{a_{f_{ij}}} \\ 0 & A_{cl} \end{bmatrix}, \hat{B}_{a_{f_{ij}}} = \begin{bmatrix} 0 \\ B_{cl} \end{bmatrix}$$

$$\hat{C}_{a_{f_{ij}}} = \begin{bmatrix} C_{cl} & 0 \end{bmatrix}; N_{b_{f_{ik}}} = \begin{bmatrix} 0 & 0 \\ R_{b_{f_{ik}}} C & 0 \end{bmatrix}, \tilde{N}_{b_{f_{ik}}} = \begin{bmatrix} 0 \\ R_{b_{f_{ik}}} D \end{bmatrix} \quad (9.5)$$

$$\hat{A}_{b_{f_{ik}}} = \begin{bmatrix} A_{cl} & N_{b_{f_{ik}}} & 0 \\ 0 & A_{cl} & 0 \\ 0 & 0 & A_{cl} \end{bmatrix}, \hat{B}_{b_{f_{ik}}} = \begin{bmatrix} 0 \\ B_{cl} \\ \tilde{N}_{b_{f_{ik}}} \end{bmatrix}$$

$$\hat{C}_{b_{f_{ik}}} = \begin{bmatrix} C_{cl} & 0 & C_{cl} \end{bmatrix}; N_{c_{f_{lj}}} = \begin{bmatrix} 0 & -R_{c_{f_{lj}}} \end{bmatrix}, \hat{C}_{c_{f_{lj}}} = N_{c_{f_{lj}}}$$

$$\hat{A}_{c_{f_{lj}}} = A_{cl}, \hat{B}_{c_{f_{lj}}} = B_{cl}$$

and

$$R_{a_{f_{ij}}} = e_i e_j^T, R_{b_{f_{ik}}} = e_i h_k^T, R_{c_{f_{lj}}} = g_l e_j^T \quad (9.6a)$$

for the additive form, and

$$R_{a_{f_{ij}}} = e_i e_i^T A_F e_j e_j^T, R_{b_{f_{ik}}} = e_i e_i^T B_F h_k h_k^T, R_{c_{f_{lj}}} = g_l g_l^T C_F e_j e_j^T \quad (9.6b)$$

for the multiplicative form, respectively. In addition, $e_k \in R^{n_f}$, $h_k \in R^p$, and $g_k \in R^q$ denote the column vectors in which the kth element equals 1 and the others equal 0.

   Then, the similar *coefficient sensitivity measures* $M_{a_{f_{ij}}}$, $M_{b_{f_{ik}}}$, and $M_{f_{c_{lj}}}$, as in the last chapter, are taken as follows:

$$\begin{aligned} M_{a_{f_{ij}}} &= \|S_{a_{f_{ij}}}(T(\delta, [a_{f_{ij}}]_{n_f \times n_f}, [b_{f_{ik}}]_{n_f \times p}, [c_{f_{lj}}]_{q \times n_f}))\|_\infty, \\ M_{b_{f_{ik}}} &= \|S_{b_{f_{ik}}}(T(\delta, [a_{f_{ij}}]_{n_f \times n_f}, [b_{f_{ik}}]_{n_f \times p}, [c_{f_{lj}}]_{q \times n_f}))\|_\infty, \\ M_{c_{f_{lj}}} &= \|S_{c_{f_{lj}}}(T(\delta, [a_{f_{ij}}]_{n_f \times n_f}, [b_{f_{ik}}]_{n_f \times p}, [c_{f_{lj}}]_{q \times n_f}))\|_\infty, \\ &\quad \text{for } i, j = 1, \cdots, n_f, l = 1, \cdots, q; k = 1, \cdots, p. \end{aligned} \quad (9.7)$$

   The insensitive H_∞ filtering problem to be solved is addressed as follows. Design a filter in (9.2) such that the filtering error system (9.3) satisfies the following requirements:

1. While there is no exogenous disturbance, that is $w(k) \equiv 0$, the filtering error system (9.3) is asymptotically stable.

2. Given positive scalars γ and β, the filtering error system (9.3) keeps

$$\|T(\delta, [a_{f_{ij}}]_{n_f \times n_f}, [b_{f_{ik}}]_{n_f \times p}, [c_{f_{lj}}]_{q \times n_f})\|_\infty < \gamma,$$

and in the meantime satisfies $M_{a_{f_{ij}}} < \beta, M_{b_{f_{ik}}} < \beta$, and $M_{c_{f_{lj}}} < \beta$ for $i, j = 1, \cdots, n_f; k = 1, \cdots, p; l = 1, \cdots, q$.

The following theorem presents a sufficient condition to make the delta operator discrete-time system *asymptotically* stable and satisfies the H_∞ performance requirement.

Theorem 9.1 *The filtering error system (9.3) is asymptotically stable with an H_∞ disturbance attention level γ, if the following two equivalent conditions hold:*

(a) There exist matrices $P = P^T > 0$, A_{cl}, B_{cl}, and C_{cl} such that

$$
\begin{bmatrix}
-\frac{1}{h}P & \frac{1}{h}P + PA_{cl} & 0 & PB_{cl} \\
* & -\frac{1}{h}P & C_{cl}^T & 0 \\
* & * & -I & 0 \\
* & * & * & -\gamma^2 I
\end{bmatrix} < 0.
\tag{9.8}
$$

(b) There exist matrices $P = P^T > 0$, G, A_{cl}, B_{cl}, and C_{cl} such that

$$
\begin{bmatrix}
\frac{1}{h}(P - He\{G\}) & \frac{1}{h}G + GA_{cl} & 0 & GB_{cl} \\
* & -\frac{1}{h}P & C_{cl}^T & 0 \\
* & * & -I & 0 \\
* & * & * & -\gamma^2 I
\end{bmatrix} < 0.
\tag{9.9}
$$

Proof 9.1 *Define the Lyapunov function as*

$$
V(k) = \xi(k)^T P \xi(k)
$$

where $P > 0$ is a Lyapunov weighting matrix to be determined.
The matrix inequality (9.9) can be equivalently rewritten as

$$
\Sigma + He\{\mathcal{H}\mathcal{M}^T\} < 0
\tag{9.10}
$$

where

$$
\Sigma = \begin{bmatrix}
\frac{1}{h}P & 0 & 0 & 0 \\
0 & -\frac{1}{h}P & 0 & 0 \\
0 & 0 & I & 0 \\
0 & 0 & 0 & -\gamma^2 I
\end{bmatrix}, \mathcal{H} = \begin{bmatrix}
\frac{1}{h}G & 0 \\
0 & 0 \\
0 & I \\
0 & 0
\end{bmatrix}
\tag{9.11}
$$

$$
\mathcal{M}^T = \begin{bmatrix}
-I & I + hA_{cl} & 0 & hB_{cl} \\
0 & C_{cl} & -I & 0
\end{bmatrix}.
$$

From the definition of delta operator, we have

$$
\delta\xi(k) = (\xi(k+1) - \xi(k))/h.
\tag{9.12}
$$

Combining (9.3) and (9.12), the filtering error system (9.3) can be rewritten as

$$
\begin{aligned}
\xi(k+1) &= (hA_{cl} + I)\xi(k) + hB_{cl}w(k) \\
e(k) &= C_{cl}\xi(k).
\end{aligned}
\tag{9.13}
$$

It follows from (9.13) that $\mathcal{M}^T\hat{\xi}(k) = 0$ for all non-zero $\hat{\xi}(k) \neq 0$. Here, $\hat{\xi}(k)$ is defined as $\hat{\xi}(k)^T = \left[\begin{array}{cccc} \xi(k+1) & \xi(k) & e(k) & w(k) \end{array} \right]$. By using Lemma 2.9, the condition (9.10) is equivalent to $\hat{\xi}(k)^T \Sigma \hat{\xi}(k) < 0, \forall \mathcal{M}^T\hat{\xi}(k) = 0, \hat{\xi}(k) \neq 0$.

It is obvious that $\hat{\xi}(k)^T \Sigma \hat{\xi}(k) < 0$ is equivalent to

$$\frac{1}{h}(\xi(k+1)^T P\xi(k+1) - \xi(k)^T P\xi(k)) + e(k)^T e(k) - \gamma^2 w(k)^T w(k) < 0. \quad (9.14)$$

In order to determine a sufficient condition for the filtering error system (9.3) to satisfy the H_∞ constraint, the following index is considered:

$$J(e, w) = \sum_{k=0}^{\infty}[e^T(k)e(k) - \gamma^2 w^T(k)w(k)], \quad w(k) \in l_2[0, \infty), \quad w(k) \neq 0.$$

Under zero-initial condition $\xi(0) = 0$, we have $V(0) = 0$ and $V(\infty) \geq 0$, which leads to

$$
\begin{aligned}
J(e, w) &= J(e, w) + \sum_{k=0}^{\infty}\frac{1}{h}(V(k+1) - V(k)) + \frac{1}{h}(V(0) - V(\infty)) \\
&\leq J(e, w) + \sum_{k=0}^{\infty}\delta V(k) \\
&= \sum_{k=0}^{\infty}(e^T(k)e(k) - \gamma^2 w^T(k)w(k) + \delta V(k)) \qquad (9.15) \\
&= \sum_{k=0}^{\infty}\left\{ e^T(k)e(k) - \gamma^2 w^T(k)w(k) + \frac{1}{h}\xi^T(k+1)P\xi^T(k+1) \right. \\
&\qquad\qquad \left. -\frac{1}{h}\xi^T(k)P\xi(k) \right\} \\
&= \sum_{k=0}^{\infty}\hat{\xi}(k)^T \Sigma \hat{\xi}(k).
\end{aligned}
$$

Therefore, combining (9.15) with (9.14), we have

$$J(e, w) < 0.$$

Consequently, we can conclude from Definition 2.5 that the condition (9.9) guarantees the filtering error system (9.3) to be asymptotically stable and satisfies the H_∞ performance constraint.

Similar to the above statements, one can obtain the condition (9.8) just by letting the matrix $\mathcal{H}$ of (9.11) as $\mathcal{H}^T = \left[\begin{array}{cccc} \frac{1}{h}P & 0 & 0 & 0 \\ 0 & 0 & I & 0 \end{array} \right]$. This completes the proof.

Remark 9.1 *The introduction of the extra variable G in condition (b) enables us to associate different Lyapunov functions to different objectives or even systems in the case of the multi-model approach. Obviously, the condition (b) is less conservative than the conditions (a) and (2.29) because the slack variable G provides extra free dimensions in the solution space. Of course, more slack variables can be introduced by using Lemma 2.9; however, the notations will get more complicated in the following filter design problem. Therefore, we use the condition (b) to design the insensitive H_∞ filter in this chapter.*

The following lemma can be obtained based on Theorem 9.1 and the definitions of the sensitivity measures.

Lemma 9.2 *Consider the system of the form (9.1). For scalars $\gamma > 0$ and $\beta > 0$, the filtering error system (9.3) is asymptotically stable and the conditions in (9.20) hold, if there exist symmetric positive-definite matrices $P^s \in R^{(n+n_f) \times (n+n_f)}$, $P_{ij}^a \in R^{2(n+n_f) \times 2(n+n_f)}$, $P_{ik}^b \in R^{3(n+n_f) \times 3(n+n_f)}$, and $P_{lj}^c \in R^{(n+n_f) \times (n+n_f)}$, and matrices $G^s \in R^{(n+n_f) \times (n+n_f)}$, $G_{ij}^a \in R^{2(n+n_f) \times 2(n+n_f)}$, $G_{ik}^b \in R^{3(n+n_f) \times 3(n+n_f)}$, and $G_{lj}^c \in R^{(n+n_f) \times (n+n_f)}$ such that*

$$
\begin{bmatrix}
\frac{1}{h}(P^s - He\{G^s\}) & \frac{1}{h}G^s + G^s A_{cl} & 0 & G^s B_{cl} \\
* & -\frac{1}{h}P^s & C_{cl}^T & 0 \\
* & * & -I & 0 \\
* & * & * & -\gamma^2 I
\end{bmatrix} < 0 \qquad (9.16)
$$

$$
\begin{bmatrix}
\frac{1}{h}(P_{ij}^a - He\{G_{ij}^a\}) & \frac{1}{h}G_{ij}^a + G_{ij}^a \hat{A}_{a_{f_{ij}}} & 0 & G_{ij}^a \hat{B}_{a_{f_{ij}}} \\
* & -\frac{1}{h}P_{ij}^a & \hat{C}_{a_{f_{ij}}}^T & 0 \\
* & * & -I & 0 \\
* & * & * & -\beta^2 I
\end{bmatrix} < 0 \qquad (9.17)
$$

$$
\begin{bmatrix}
\frac{1}{h}(P_{ik}^b - He\{G_{ik}^b\}) & \frac{1}{h}G_{ik}^b + G_{ik}^b \hat{A}_{b_{f_{ik}}} & 0 & G_{ik}^b \hat{B}_{b_{f_{ik}}} \\
* & -\frac{1}{h}P_{ik}^b & \hat{C}_{b_{f_{ik}}}^T & 0 \\
* & * & -I & 0 \\
* & * & * & -\beta^2 I
\end{bmatrix} < 0 \qquad (9.18)
$$

$$
\begin{bmatrix}
\frac{1}{h}(P_{lj}^c - He\{G_{lj}^c\}) & \frac{1}{h}G_{lj}^c + G_{lj}^c \hat{A}_{c_{f_{lj}}} & 0 & G_{lj}^c \hat{B}_{c_{f_{lj}}} \\
* & -\frac{1}{h}P_{lj}^c & \hat{C}_{c_{f_{lj}}}^T & 0 \\
* & * & -I & 0 \\
* & * & * & -\beta^2 I
\end{bmatrix} < 0 \qquad (9.19)
$$

hold for $i, j = 1, \cdots, n_f; k = 1, \cdots, p; l = 1, \cdots, q$, where $\hat{A}_\chi, \hat{B}_\chi$, and $\hat{C}_\chi (\chi = a_{f_{ij}}, b_{f_{ik}},$ or $c_{f_{lj}})$ are defined in (9.5).

9.3 Insensitive H_∞ Filter Design

In this section, the problem of insensitive H_∞ filter design for the system in (9.1) is considered. At first, a sufficient condition for the existence of insensitive filters with respect to additive coefficient variations is presented. Then, a new type of sensitivity measure is defined and a sufficient condition is also provided to solve the problem of insensitive H_∞ filters with respect to multiplicative coefficient variations.

9.3.1 Additive Coefficient Variation Case

The following theorem based on Lemma 9.2 presents a sufficient condition under which the insensitive H_∞ filter problem with respect to additive coefficient variations is solvable.

Theorem 9.2 *Consider the system in (9.1) and let scalars $\gamma > 0$ and $\beta > 0$ be given constants. Then, the filtering error system (9.3) is asymptotically stable and*

$$\|T(\delta, [a_{f_{ij}}]_{n_f \times n_f}, [b_{f_{ik}}]_{n_f \times p}, [c_{f_{lj}}]_{q \times n_f})\|_\infty < \gamma, M_{a_{f_{ij}}} < \beta, M_{b_{f_{ik}}} < \beta,$$
$$M_{c_{f_{lj}}} < \beta, \text{for } i,j = 1, \cdots, n_f; k = 1, \cdots, p; l = 1, \cdots, q$$

$$(9.20)$$

hold if, for some positive scalars $\lambda_a, \lambda_b,$ and λ_c, there exist symmetric positive-definite matrices $Q^s \in R^{(n+n_f)\times(n+n_f)}$, $Q_{ij}^a \in R^{2(n+n_f)\times 2(n+n_f)}$, $Q_{ik}^b \in R^{3(n+n_f)\times 3(n+n_f)}$, $Q_{lj}^c \in R^{(n+n_f)\times(n+n_f)}$, and the matrices $G_{11} \in R^{n\times n}$, $G_{21} \in R^{n_f \times n}$, $\tilde{G}_{12} \in R^{n_f \times n_f}$, $F_A \in R^{n_f \times n_f}$, $F_B \in R^{n_f \times p}$, and $F_C \in R^{q \times n_f}$ such that the following LMIs hold for $i,j = 1, \cdots, n_f; k = 1, \cdots, p; l = 1, \cdots, q$:

$$\begin{bmatrix} \frac{1}{h}(Q_s - He\{M_g\}) & \frac{1}{h}M_g + M_{as} & 0 & M_{bs} \\ * & -\frac{1}{h}Q_s & M_{cs}^T & 0 \\ * & * & -I & 0 \\ * & * & * & -\gamma^2 I \end{bmatrix} < 0 \qquad (9.21)$$

$$\begin{bmatrix} \frac{1}{h}(Q_{ij}^a - \lambda_a He\{I_2 \otimes M_g\}) & \Xi_{a_{f_{ij}}} & 0 & \begin{bmatrix} 0 \\ \lambda_a M_{bs} \end{bmatrix} \\ * & -\frac{1}{h}Q_{ij}^a & \begin{bmatrix} M_{cs}^T \\ 0 \end{bmatrix} & 0 \\ * & * & -I & 0 \\ * & * & * & -\beta^2 I \end{bmatrix} < 0 \quad (9.22)$$

$$
\left[
\begin{array}{cccc}
\frac{1}{h}(Q_{ik}^b - \lambda_b He\{I_3 \otimes M_g\}) & \Xi_{b_{f_{ik}}} & 0 & \lambda_b \begin{bmatrix} 0 \\ M_{bs} \\ M_{2b_{f_{ik}}} \end{bmatrix} \\
* & -\frac{1}{h}Q_{ik}^b & \begin{bmatrix} M_{cs}^T \\ 0 \\ M_{cs}^T \end{bmatrix} & 0 \\
* & * & -I & 0 \\
* & * & * & -\beta^2 I
\end{array}
\right] < 0
$$

(9.23)

$$
\left[
\begin{array}{ccccc}
\frac{1}{h}(Q_{lj}^c - \lambda_c He\{M_g\}) & \frac{\lambda_c}{h}M_g + \lambda_c M_{as} & 0 & \lambda_c M_{bs} \\
* & -\frac{1}{h}Q_{lj}^c & M_{c_{f_{lj}}}^T & 0 \\
* & * & -I & 0 \\
* & * & * & -\beta^2 I
\end{array}
\right] < 0 \quad (9.24)
$$

where

$$
M_{as} = \begin{bmatrix} G_{11}A_\delta + HF_BC & HF_A \\ G_{21}A_\delta + F_BC & F_A \end{bmatrix}, M_{bs} = \begin{bmatrix} G_{11}B_\delta + HF_BD \\ G_{21}B_\delta + F_BD \end{bmatrix}
$$

$$
\Xi_{a_{f_{ij}}} = \frac{\lambda_a}{h}(I_2 \otimes M_g) + \lambda_a \begin{bmatrix} M_{as} & M_{a_{f_{ij}}} \\ 0 & M_{as} \end{bmatrix}, M_{cs} = \begin{bmatrix} L & -F_C \end{bmatrix}
$$

$$
\Xi_{b_{f_{ik}}} = \frac{\lambda_b}{h}(I_3 \otimes M_g) + \lambda_b \begin{bmatrix} M_{as} & M_{1b_{f_{ik}}} & 0 \\ 0 & M_{as} & 0 \\ 0 & 0 & M_{as} \end{bmatrix}
$$

$$
M_{2b_{f_{ik}}} = \begin{bmatrix} H\tilde{G}_{12}R_{b_{f_{ik}}}D \\ \tilde{G}_{12}R_{b_{f_{ik}}}D \end{bmatrix}, M_g = \begin{bmatrix} G_{11} & H\tilde{G}_{12} \\ G_{21} & \tilde{G}_{12} \end{bmatrix}
$$

$$
M_{1b_{f_{ik}}} = \begin{bmatrix} H\tilde{G}_{12}R_{b_{f_{ik}}}C & 0 \\ \tilde{G}_{12}R_{b_{f_{ik}}}C & 0 \end{bmatrix}, M_{a_{f_{ij}}} = \begin{bmatrix} 0 & H\tilde{G}_{12}R_{a_{f_{ij}}} \\ 0 & \tilde{G}_{12}R_{a_{f_{ij}}} \end{bmatrix}
$$

$$
M_{c_{f_{lj}}} = \begin{bmatrix} 0 & -R_{c_{f_{lj}}} \end{bmatrix}
$$

with $R_{a_{f_{ij}}}$, $R_{b_{f_{ik}}}$, and $R_{c_{f_{lj}}}$ satisfying (9.6a) and $\mathcal{H} = \begin{bmatrix} I_{(n_f \times n_f)} \\ 0_{(n-n_f) \times n_f} \end{bmatrix}$. Moreover, from the solutions of the above inequalities, the n_fth-order insensitive filter can be given by

$$
A_F = \tilde{G}_{12}^{-1}F_A, B_F = \tilde{G}_{12}^{-1}F_B, C_F = F_C. \tag{9.25}
$$

Proof 9.2 *Lemma 9.2 shows that the filtering error system (9.3) is asymptotically stable and satisfies the H_∞ norm performance requirement with the insensitive constraint simultaneously, if there exist symmetric positive-definite matrices P^s, P_{ij}^a, P_{ik}^b, and P_{lj}^c and matrices G^s, G_{ij}^a, G_{ik}^b, and G_{lj}^c $(i, j = 1, \cdots, n_f; k = 1, \cdots, p; l = 1, \cdots, q)$ satisfying (9.16), (9.17), (9.18), and (9.19), respectively. In the following proof, we will show that Theorem 9.2 is sufficient to guarantee that (9.16), (9.17), (9.18), and (9.19) hold.*

From (9.5), one can observe that the sensitive coefficient matrices are given in almost block diagonal form of the filtering error system matrices. Therefore, in order to solve the problem, the following additional constraints are introduced to the matrices G^s, G_{ij}^a, G_{ik}^b, and G_{lj}^c, which may lead to some conservativeness:

$$G^s = G, G_{ij}^a = \lambda_a I_2 \otimes G, G_{ik}^b = \lambda_b I_3 \otimes G, G_{lj}^c = \lambda_c G \qquad (9.26)$$

where λ_a, λ_b, and λ_c are positive scalar parameters to be searched.

Moreover, motivated by Tuan, Apkarian, and Nguyen [123] and Wu and Ho [134], we partition the matrix G as

$$G = \begin{bmatrix} G_{11} & G_{12} \\ G_{21} & G_{22} \end{bmatrix}, G_{12} = \begin{bmatrix} \tilde{G}_{12} \\ 0_{(n-n_f) \times n_f} \end{bmatrix} \qquad (9.27)$$

where the partitioned submatrices will be dimensionalized as $G_{11} \in R^{n \times n}, G_{12} \in R^{n \times n_f}, G_{21} \in R^{n_f \times n}, G_{22} \in R^{n_f \times n_f}, \tilde{G}_{12} \in R^{n_f \times n_f}$. Because the n_fth-order case is being considered, the filter order n_f will be less than or equal to the plant order n. By recalling the assumption [123, 134], we may assume, without loss of generality, that $\tilde{G}_{12}$ is nonsingular.

From (9.27), the following equation can be implied:

$$G_{12} = H\tilde{G}_{12}$$

where $\mathcal{H}$ is defined in Theorem 9.2.

Construct the following matrices

$$J = diag\{I, G_{22}^{-T}\tilde{G}_{12}^T\}$$

and

$$\begin{bmatrix} F_A & F_B \\ F_C & 0 \end{bmatrix} = diag\{\tilde{G}_{12}, I\} \begin{bmatrix} A_F & B_F \\ C_F & 0 \end{bmatrix} diag\{G_{22}^{-T}\tilde{G}_{12}^T, I\}. \qquad (9.28)$$

For brevity, we only prove that the inequality (9.22) is equivalent to (9.17) with the constraint $G_{ij}^a = \lambda_a I_2 \otimes G$. Performing a congruence transformation to (9.17) by $diag\{I_2 \otimes J, I_2 \otimes J, I, I\}$, we obtain

$$\begin{bmatrix} \Xi_{p1} & \frac{\lambda_a}{h}I_2 \otimes (J^T GJ) + \Xi_{p2} & 0 & \begin{bmatrix} 0 \\ \lambda_a J^T GB_{cl} \end{bmatrix} \\ * & -\frac{1}{h}Q_{ij}^a & \begin{bmatrix} J^T C_{cl}^T \\ 0 \end{bmatrix} & 0 \\ * & * & -I & 0 \\ * & * & * & -\beta^2 I \end{bmatrix} < 0$$

where

$$Q_{ij}^a = (I_2 \otimes J^T)P_{ij}^a(I_2 \otimes J), \Xi_{p1} = \frac{1}{h}(Q_{ij}^a - He\{\lambda_a I_2 \otimes (J^T GJ)\})$$

$$J^T GA_{cl}J = \begin{bmatrix} G_{11}A_\delta + H\tilde{G}_{12}B_F C & H\tilde{G}_{12}A_F G_{22}^{-T}\tilde{G}_{12}^T \\ \tilde{G}_{12}G_{22}^{-1}G_{21}A_\delta + \tilde{G}_{12}B_F C & \tilde{G}_{12}A_F G_{22}^{-T}\tilde{G}_{12}^T \end{bmatrix}$$

$$\Xi_{p2} = \begin{bmatrix} \lambda_a J^T GA_{cl}J & \lambda_a J^T GN_{a_{f_{ij}}}J \\ 0 & \lambda_a J^T GA_{cl}J \end{bmatrix}$$

$$J^T GJ = \begin{bmatrix} G_{11} & H\tilde{G}_{12}G_{22}^{-T}\tilde{G}_{12}^T \\ \tilde{G}_{12}G_{22}^{-1}G_{21} & \tilde{G}_{12}G_{22}^{-T}\tilde{G}_{12}^T \end{bmatrix}$$

$$J^T GN_{a_{f_{ij}}}J = \begin{bmatrix} 0 & H\tilde{G}_{12}R_{a_{f_{ij}}}G_{22}^{-T}\tilde{G}_{12}^T \\ 0 & \tilde{G}_{12}R_{a_{f_{ij}}}G_{22}^{-T}\tilde{G}_{12}^T \end{bmatrix}$$

$$J^T GB_{cl} = \begin{bmatrix} G_{11}B_\delta + H\tilde{G}_{12}B_F D \\ \tilde{G}_{12}G_{22}^{-1}G_{21}B_\delta + \tilde{G}_{12}B_F D \end{bmatrix}$$

$$C_{cl}J = \begin{bmatrix} L & -C_F G_{22}^{-T}\tilde{G}_{12}^T \end{bmatrix}$$

with $R_{a_{f_{ij}}}(i,j = 1, \cdots, n_f)$ being defined in (9.6a).

It is of interest to note that (9.28) is equivalent to

$$\begin{bmatrix} A_F & B_F \\ C_F & 0 \end{bmatrix} = \begin{bmatrix} \tilde{G}_{12}^{-1} & 0 \\ 0 & I \end{bmatrix} \mathcal{F} \begin{bmatrix} (G_{22}^{-T}\tilde{G}_{12}^T)^{-1} & 0 \\ 0 & I \end{bmatrix} \tag{9.29}$$

$$= \begin{bmatrix} (G_{22}^{-T}\tilde{G}_{12}^T)N^{-1} & 0 \\ 0 & I \end{bmatrix} \mathcal{F} \begin{bmatrix} (G_{22}^{-T}\tilde{G}_{12}^T)^{-1} & 0 \\ 0 & I \end{bmatrix}$$

with $N = \tilde{G}_{12}G_{22}^{-T}\tilde{G}_{12}^T$, $\mathcal{F} = \begin{bmatrix} F_A & F_B \\ F_C & 0 \end{bmatrix}$. Note that the filter matrices $A_F, B_F,$ and C_F can be written as (9.29), which implies that $(G_{22}^{-T}\tilde{G}_{12}^T)^{-1}$ can be viewed as a similarity transformation on the state-space realization of the filter and, as such, has no effect on the filter mapping from y to $\bar{z}(t)$. Without loss of generality, we may set $\tilde{G}_{12}G_{22}^{-1} = I$, thus the filter in (9.2) can be constructed by (9.25).

Moreover, following from (9.29), we have

$$J^T GA_{cl}J = \begin{bmatrix} G_{11}A_\delta + H\tilde{G}_{12}B_F C & H\tilde{G}_{12}A_F \\ G_{21}A_\delta + \tilde{G}_{12}B_F C & \tilde{G}_{12}A_F \end{bmatrix}$$

$$J^T GN_{a_{f_{ij}}}J = \begin{bmatrix} 0 & H\tilde{G}_{12}R_{a_{f_{ij}}} \\ 0 & \tilde{G}_{12}R_{a_{f_{ij}}} \end{bmatrix}$$

$$J^T GB_{cl} = \begin{bmatrix} G_{11}B_\delta + H\tilde{G}_{12}B_F D \\ G_{21}B_\delta + \tilde{G}_{12}B_F D \end{bmatrix}$$

$$C_{cl}J = \begin{bmatrix} L & -C_F \end{bmatrix}.$$

Then, (9.22) can be easily obtained. Note that because of the invertibility of the matrices involved in the above proof process, the above proof process is a reversible process, and therefore (9.22) is equivalent to (9.17) with the constraint $G_{ij}^a = \lambda_a I_2 \otimes G$.

Moreover, performing congruence transformations to (9.16) and (9.19) by $diag\{J, J, I, I\}$ and to (9.18) by $diag\{I_3 \otimes J, I_3 \otimes J, I, I\}$, respectively, and following similar steps as the proof process of (9.22), the inequalities (9.21), (9.23), and (9.24) are equivalent to (9.16), (9.18), and (9.19) with the constraints in (9.26), respectively. This completes the proof.

Remark 9.2 *In order to obtain the LMI conditions in Theorem 9.2, the artificial constraints in (9.26) are imposed, which are different from those of Chapter 8 in which the constraints are imposed on the Lyapunov matrix. Furthermore, the Lyapunov matrices of Theorem 9.2 are dependent on the indices i, j, k, and l for different sensitivity functions which are also different from Chapter 8, where a single Lyapunov matrix is adapted. Thus, the proposed approach is adapted to treat multiple objective problems in a potentially less conservative framework than methods that associate a single Lyapunov function with the different control objectives such as in Chapter 8.*

Remark 9.3 *The scalars λ_a, λ_b, and λ_c are introduced, aiming at reducing the conservatism caused by introducing the artificial constraints in (9.26). When the scalars λ_a, λ_b, and λ_c are set to be fixed parameters, the conditions (9.21), (9.22), and (9.23) of Theorem 9.2 become linear. For the prescribed H_∞ performance γ, the problem of H_∞ filter design under insensitive constraints can be converted into the following optimization problem:*

$$\min_{Q^s, Q_{ij}^a, Q_{ik}^b, Q_{lj}^c, G_{11}, G_{21}, \tilde{G}_{12}, F_A, F_B, F_C} \beta^2,$$
$$s.t. \ Q^s > 0, Q_{ij}^a > 0, Q_{ik}^b > 0, Q_{lj}^c > 0, (9.21), (9.22), (9.23), (9.24). \quad (9.30)$$

The minimal sensitivity β^* is the optimization value of β, and the designed filter's gains can be obtained by (9.25).

Remark 9.4 *When the insensitive constraint is unconsidered, Theorem 9.2 reduces to the standard H_∞ filtering design method.*

9.3.2   Multiplicative Filter Coefficient Variation Case

In this section, due to the complexity problem of the multiplicative coefficient variations, similar to Chapter 8, new coefficient sensitivity measures based on average values of coefficient sensitivity functions are used to obtain convex conditions for the insensitive filtering problem.

Because the matrices $R_{a_{f_{ij}}}$, $R_{b_{f_{ik}}}$, and $R_{c_{f_{lj}}}$ defined in (9.6b) contain the filter matrix elements, and what is worse, is that there exist cross-product terms between the variable $\tilde{G}_{12}$ and the matrices $R_{a_{f_{ij}}}$ and $R_{b_{f_{ik}}}$, the conditions will be non-convex and cannot be solved directly if the coefficient sensitivity measures defined in (9.7) are used directly. In order to solve the non-convex problem, the coefficient sensitivity measures based on the average of the sensitivity functions as in Chapter 8 are introduced. The coefficient sensitivity measures M_{A_F}, M_{B_F}, and M_{C_F} are given as follows:

$$
\begin{aligned}
M_{A_F} &= \|\tfrac{1}{n_f \times n_f} \sum_{i=1}^{n_f} \sum_{j=1}^{n_f} S_{a_{f_{ij}}}(T(\delta, [a_{f_{ij}}]_{n_f \times n_f}, [b_{f_{ik}}]_{n_f \times p}, [c_{f_{lj}}]_{q \times n_f}))\|_\infty \\
&= \|\hat{C}_{a_f}(\delta I - \hat{A}_{a_f})^{-1} \hat{B}_{a_f}\|_\infty
\end{aligned}
\tag{9.31a}
$$

$$
\begin{aligned}
M_{B_F} &= \|\tfrac{1}{n_f \times p} \sum_{i=1}^{n_f} \sum_{k=1}^{p} S_{b_{f_{ik}}}(T(\delta, [a_{f_{ij}}]_{n_f \times n_f}, [b_{f_{ik}}]_{n_f \times p}, [c_{f_{lj}}]_{q \times n_f}))\|_\infty \\
&= \|\hat{C}_{b_f}(\delta I - \hat{A}_{b_f})^{-1} \hat{B}_{b_f}\|_\infty
\end{aligned}
\tag{9.31b}
$$

$$
\begin{aligned}
M_{C_F} &= \|\tfrac{1}{q \times n_f} \sum_{l=1}^{q} \sum_{j=1}^{n_f} S_{c_{f_{lj}}}(T(\delta, [a_{f_{ij}}]_{n_f \times n_f}, [b_{f_{ik}}]_{n_f \times p}, [c_{f_{lj}}]_{q \times n_f}))\|_\infty \\
&= \|\hat{C}_{c_f}(\delta I - \hat{A}_{c_f})^{-1} \hat{B}_{c_f}\|_\infty
\end{aligned}
\tag{9.31c}
$$

where

$$
N_{a_f} = \begin{bmatrix} 0 & 0 \\ 0 & \frac{1}{n_f \times n_f} R_{a_f} \end{bmatrix}, \hat{A}_{a_f} = \begin{bmatrix} A_{cl} & N_{a_f} \\ 0 & A_{cl} \end{bmatrix}, \hat{B}_{a_f} = \begin{bmatrix} 0 \\ B_{cl} \end{bmatrix}
$$

$$
\hat{C}_{a_f} = \begin{bmatrix} C_{cl} & 0 \end{bmatrix}, N_{b_f} = \begin{bmatrix} 0 & 0 \\ \frac{1}{n_f \times p} R_{b_f} C & 0 \end{bmatrix}, \tilde{N}_{b_f} = \begin{bmatrix} 0 \\ \frac{1}{n_f \times p} R_{b_f} D \end{bmatrix}
$$

$$
\hat{A}_{b_f} = \begin{bmatrix} A_{cl} & N_{b_f} & 0 \\ 0 & A_{cl} & 0 \\ 0 & 0 & A_{cl} \end{bmatrix}, \hat{B}_{b_f} = \begin{bmatrix} 0 \\ B_{cl} \\ \tilde{N}_{b_f} \end{bmatrix}, \hat{C}_{b_f} = \begin{bmatrix} C_{cl} & 0 & C_{cl} \end{bmatrix}
$$

$$
N_{c_f} = \begin{bmatrix} 0 & -\frac{1}{q \times n_f} R_{c_f} \end{bmatrix}, \hat{C}_{c_f} = N_{c_f}, \hat{A}_{c_f} = A_{cl}, \hat{B}_{c_f} = B_{cl}
\tag{9.32}
$$

and

$$
R_{a_f} = A_F, R_{b_f} = B_F, R_{c_f} = C_F
$$

for the multiplicative form.

Compared with the coefficient sensitivity measures defined in (9.7), the advantage of the new measures is that R_{a_f}, R_{b_f}, and R_{c_f} are compatible with the filter matrices. The feature is crucial in dealing with the filtering problem with respect to multiplicative coefficient variations.

At the same time, the following lemma can be obtained.

Lemma 9.3 *Consider the system of the form (9.1). For scalars $\gamma > 0$ and $\beta > 0$, the filtering error system (9.3) is asymptotically stable and the conditions in (9.33) hold, if there exist symmetric positive-definite matrices $P^s \in R^{(n+n_f) \times (n+n_f)}$, $P^a \in R^{2(n+n_f) \times 2(n+n_f)}$, $P^b \in R^{3(n+n_f) \times 3(n+n_f)}$, and $P^c \in R^{(n+n_f) \times (n+n_f)}$, and matrices $G^s \in R^{(n+n_f) \times (n+n_f)}$, $G^a \in R^{2(n+n_f) \times 2(n+n_f)}$,*

$G^b \in R^{3(n+n_f) \times 3(n+n_f)}$, and $G^c \in R^{(n+n_f) \times (n+n_f)}$ such that (9.16) and

$$
\begin{bmatrix}
\frac{1}{h}(P^a - He\{G^a\}) & \frac{1}{h}G^a + G^a \hat{A}_{a_f} & 0 & G^a \hat{B}_{a_f} \\
* & -\frac{1}{h}P^a & \hat{C}_{a_f}^T & 0 \\
* & * & -I & 0 \\
* & * & * & -\beta^2 I
\end{bmatrix} < 0
$$

$$
\begin{bmatrix}
\frac{1}{h}(P^b - He\{G^b\}) & \frac{1}{h}G^b + G^b \hat{A}_{b_f} & 0 & G^b \hat{B}_{b_f} \\
* & -\frac{1}{h}P^b & \hat{C}_{b_f}^T & 0 \\
* & * & -I & 0 \\
* & * & * & -\beta^2 I
\end{bmatrix} < 0
$$

$$
\begin{bmatrix}
\frac{1}{h}(P^c - He\{G^c\}) & \frac{1}{h}G^c + G^c \hat{A}_{c_f} & 0 & G^c \hat{B}_{c_f} \\
* & -\frac{1}{h}P^c & \hat{C}_{c_f}^T & 0 \\
* & * & -I & 0 \\
* & * & * & -\beta^2 I
\end{bmatrix} < 0
$$

hold, where $\hat{A}_\chi, \hat{B}_\chi$, and $\hat{C}_\chi (\chi = a_f, b_f,$ or $c_f)$ are defined in (9.32).

Based on the new sensitivity measures and Lemma 9.3, employing similar technologies of Theorem 9.2, the following theorem presents a sufficient condition for the insensitive H_∞ filtering problem with respect to the multiplicative filter coefficient variations.

Theorem 9.3 *For system (9.1), let scalars $\gamma > 0$, $\beta > 0$ then the filtering error system (9.3) is asymptotically stable and the conditions*

$$
\|T(\delta, [a_{f_{ij}}]_{n_f \times n_f}, [b_{f_{ik}}]_{n_f \times p}, [c_{f_{lj}}]_{q \times n_f})\|_\infty < \gamma, M_{A_F} < \beta, M_{B_F} < \beta, M_{C_F} < \beta
\tag{9.33}
$$

hold if, for some positive scalars λ_a, λ_b, and λ_c, there exist symmetric positive-definite matrices $Q^s \in R^{(n+n_f) \times (n+n_f)}$, $Q^a \in R^{2(n+n_f) \times 2(n+n_f)}$, $Q^b \in R^{3(n+n_f) \times 3(n+n_f)}$, $Q^c \in R^{(n+n_f) \times (n+n_f)}$, and the matrices $G_{11} \in R^{n \times n}$, $G_{21} \in R^{n_f \times n}$, $\tilde{G}_{12} \in R^{n_f \times n_f}$, $F_A \in R^{n_f \times n_f}$, $F_B \in R^{n_f \times p}$, and $F_C \in R^{q \times n_f}$ such that (9.21) and

$$
\begin{bmatrix}
\frac{1}{h}(Q^a - \lambda_a He\{I_2 \otimes M_g\}) & \Xi_{a_f} & 0 & \begin{bmatrix} 0 \\ \lambda_a M_{bs} \end{bmatrix} \\
* & -\frac{1}{h}Q^a & \begin{bmatrix} M_{cs}^T \\ 0 \end{bmatrix} & 0 \\
* & * & -I & 0 \\
* & * & * & -\beta^2 I
\end{bmatrix} < 0
$$

$$\begin{bmatrix} \frac{1}{h}(Q^b - \lambda_b He\{I_3 \otimes M_g\}) & \Xi_{b_f} & 0 & \lambda_b \begin{bmatrix} 0 \\ M_{bs} \\ M_{2b_f} \end{bmatrix} \\ * & -\frac{1}{h}Q^b & \begin{bmatrix} M_{cs}^T \\ 0 \\ M_{cs}^T \end{bmatrix} & 0 \\ * & * & -I & 0 \\ * & * & * & -\beta^2 I \end{bmatrix} < 0$$

$$\begin{bmatrix} \frac{1}{h}(Q^c - \lambda_c He\{M_g\}) & \frac{\lambda_c}{h} M_g + \lambda_c M_{as} & 0 & \lambda_c M_{bs} \\ * & -\frac{1}{h}Q^c & M_{c_f}^T & 0 \\ * & * & -I & 0 \\ * & * & * & -\beta^2 I \end{bmatrix} < 0$$

hold, where

$$\Xi_{a_f} = \frac{\lambda_a}{h}(I_2 \otimes M_g) + \lambda_a \begin{bmatrix} M_{as} & M_{a_f} \\ 0 & M_{as} \end{bmatrix}, M_{a_f} = \begin{bmatrix} 0 & \frac{1}{n_f \times n_f}\mathcal{H}F_A \\ 0 & \frac{1}{n_f \times n_f}F_A \end{bmatrix}$$

$$\Xi_{b_f} = \frac{\lambda_b}{h}(I_3 \otimes M_g) + \lambda_b \begin{bmatrix} M_{as} & M_{1b_f} & 0 \\ 0 & M_{as} & 0 \\ 0 & 0 & M_{as} \end{bmatrix}$$

$$M_{1b_f} = \begin{bmatrix} \frac{1}{n_f \times p}\mathcal{H}F_B C & 0 \\ \frac{1}{n_f \times p}F_B C & 0 \end{bmatrix}, M_{2b_f} = \begin{bmatrix} \frac{1}{n_f \times p}\mathcal{H}F_B D \\ \frac{1}{n_f \times p}F_B D \end{bmatrix}$$

$$M_{c_f} = \begin{bmatrix} 0 & -\frac{1}{q \times n_f}F_C \end{bmatrix}$$

for the multiplicative case. In addition, $M_{as}, M_{bs}, M_{cs}, M_g$, and $\mathcal{H}$ are defined in Theorem 9.2. Moreover, the n_fth-order insensitive filter can also be given by (9.25).

Remark 9.5 *Generally speaking, the new coefficient sensitivity measures defined in (9.31) may fail to ensure (9.17), (9.18), and (9.19) hold for $i, j = 1, \cdots, n_f; k = 1, \cdots, p; l = 1, \cdots, q$. However, they may practically reduce the individual sensitivity in many cases, see the examples in Section 9.4. It is worth mentioning that the convex conditions for the filter design problem with respect to multiplicative coefficient variations can be obtained by using the new measures. Furthermore, using the same algorithm as Algorithm 8.1 in Chapter 8, an alternative approach can be provided to obtain the insensitive H_∞ filter which can guarantee the individual insensitive to the multiplicative coefficient variations.*

Remark 9.6 *Obviously, the problem of designing insensitive filters is equivalent to minimizing β by using (9.30). However, it is necessary to take into account that the artificial constraints (9.26) and the partition of G as in (9.27)* *may introduce some conservativeness into the value of β. Fortunately, the actual measures in (9.7) can be computed directly when the filter's matrices A_F, B_F, and C_F are given, where the largest one is taken as the actual sensitivity measure of the designed filter.*

9.4 Example

Example 9.1 *In this section, a numerical example based on a linearized model of an F-404 engine from Eustace et al. [33], which is also studied in Ahn, Han, and Kwon [2] and Han and Kwon [56], is taken. By using the delta operator in sampling the continuous-time system in Chapter 8 with $h = 0.001s$, the relevant discrete-time system in a δ-domain can be obtained as follows:*

$$\delta x_\delta(k) = \begin{bmatrix} -1.4586 & 0 & 2.4235 \\ 0.3590 & -0.01 & -0.3779 \\ 0.3101 & 0 & -2.2271 \end{bmatrix} x_\delta(k) + \begin{bmatrix} 0.3451 & 0 \\ 0.5508 & 0 \\ -3.8556 & 0 \end{bmatrix} w(k)$$

$$y(k) = \begin{bmatrix} 1 & 0 & 0 \end{bmatrix} x_\delta(k) + \begin{bmatrix} -0.8 & 0.9 \end{bmatrix} w(k)$$

$$z(k) = \begin{bmatrix} 0 & 0 & 4 \end{bmatrix} x_\delta(k).$$

Furthermore, it should be mentioned that the reduced-order filtering problem (i.e., $n_f = 2$) is considered in this section.

Solving the LMI given in (9.21), the standard second-order filter with the optimal H_∞ performance index $\gamma = 4.3636$ is given as follows:

$$A_F = \begin{bmatrix} -8.6030 & 0.9726 \\ 10.0265 & -1.2158 \end{bmatrix}, B_F = \begin{bmatrix} -9.3140 \\ 10.7591 \end{bmatrix}$$

$$C_F = \begin{bmatrix} -2.1592 & 0.3787 \end{bmatrix}.$$

Additive Coefficient Variation Case

In the following, the case of additive coefficient variations is considered. By solving the optimization problem (9.30) with $\gamma = 4.4$, $\lambda_a = \lambda_c = 10$, and $\lambda_b = 8$, we obtain the sensitivity performance $\beta = 16.1966$ and the insensitive second-order filter with gain matrices as

$$A_F = \begin{bmatrix} -11.6548 & 0.0897 \\ 6.5832 & -8.4102 \end{bmatrix}, B_F = \begin{bmatrix} -12.1646 \\ 9.0847 \end{bmatrix}$$

$$C_F = \begin{bmatrix} -2.2094 & 0.2483 \end{bmatrix}.$$

In addition, the actual coefficient sensitivity measures for the insensitive filter designed by Theorem 9.2 and the standard filter, respectively, can be computed by using the method given in Remark 9.6. The results are given in Table 9.1.

TABLE 9.1

Comparison of Coefficient Sensitivity
Measures β

Standard H_∞ Filter	Insensitive H_∞ Filter
Remark 9.4	Theorem 9.2
8.7106	4.5095

It can be seen that the sensitivity of the insensitive filter is much lower than the standard one. On the other hand, it is shown that insensitivity to filter coefficient changes can be obtained with very little loss in standard performance.

To show the effectiveness of the proposed method more clearly, a simulation is also performed. $\theta_{a_{f_{ij}}}$, $\theta_{b_{f_{ik}}}$, and $\theta_{c_{f_{lj}}}$ represent the interval type of coefficient variations with the following form:

$$|\theta_{a_{f_{ij}}}| \le \theta, i, j = 1, \cdots, n_f,$$
$$|\theta_{b_{f_{ik}}}| \le \theta, i = 1, \cdots, n_f, k = 1, \cdots, p,$$
$$|\theta_{c_{f_{lj}}}| \le \theta, l = 1, \cdots, q, j = 1, \cdots, n_f.$$

Moreover, in the following simulation, let the system's initial state be $x_0 = \begin{bmatrix} 0 & 0 & 0 \end{bmatrix}$, the filter's initial state be $\bar{x}_0 = \begin{bmatrix} 0 & 0 & 0 \end{bmatrix}$, and the disturbance input $w^T(k) = \begin{bmatrix} w_1^T(k) & w_2^T(k) \end{bmatrix}$ be

$$w_1(k) = w_2(k) = \begin{cases} 0.2 & 5 \le k \le 10, \\ 0 & otherwise. \end{cases}$$

Figure 9.1 shows the response of the estimation error $e(k)$ for the insensitive filter and the standard filter with $\theta = 0$ and $\theta = 0.04$, respectively. From this figure, it can be observed that the filter designed by Theorem 9.2 is insensitive to the additive coefficient variations while, on the contrary, the standard filter is unstable under the same filter coefficient variations.

In the following, the results in Table 9.2 are given to provide a comparison between the actual sensitivity measures of filters designed in the formulation of delta and shift operator models. It can be seen from the table that the delta operator is less sensitive than the traditional shift operator at high sampling rates. Further, the problem of system instability by using the usual shift operator model when $h = 0.001s$ is successfully solved by using a delta operator model.

Multiplicative Coefficient Variation Case

In the following, the case of multiplicative coefficient variations is considered. By solving the conditions given in Theorem 9.3 with $\gamma = 5.4$, $\lambda_a = \lambda_b = 1.3$,

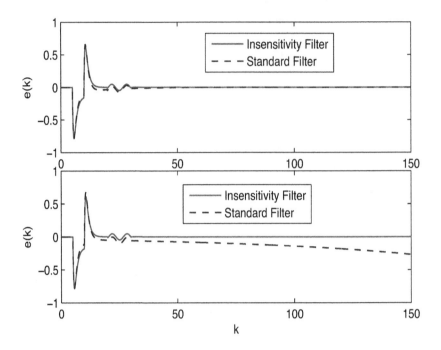

FIGURE 9.1
Comparison of the response of the estimation error $e(k)$ (the upper one for $\theta = 0$ and the lower one for $\theta = 0.04$).

and $\lambda_c = 1.5$, the insensitive second-order filter with gain matrices as

$$A_F = \begin{bmatrix} -628.6188 & -0.0012 \\ 287.4084 & -0.0623 \end{bmatrix}, B_F = \begin{bmatrix} -482.0449 \\ 220.5115 \end{bmatrix}$$
$$C_F = \begin{bmatrix} -1.5197 & 0.0000 \end{bmatrix}, \beta = 5.1896.$$

Moreover, the actual coefficient sensitivity measures for the insensitive filter designed by Theorem 9.3 and the standard filter, respectively, can be computed and are given in Table 9.3.

It also can be seen that the sensitivity of the insensitive filter is much lower than the standard one.

The responses of the estimation error $e(k)$ for the insensitive filter designed by Theorem 9.3 and the standard filter with $\delta = 0$ and $\delta = 0.04$, respectively, are shown in Figure 9.2. From Figure 9.2, one can see that the insensitive filter designed by Theorem 9.3 is superior in its ability to tolerate multiplicative coefficient variations as compared to the standard filter. The above example clearly demonstrates the obvious superiority of the proposed approach.

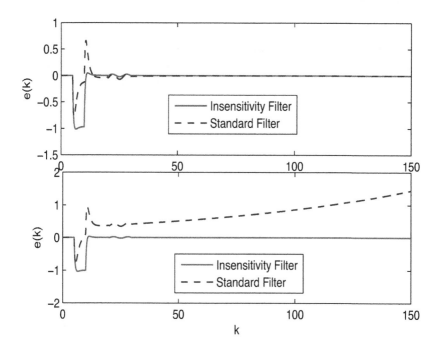

FIGURE 9.2
Comparison of the response of the estimation error $e(k)$ (the upper one for $\theta = 0$ and the lower one for $\theta = 0.04$).

TABLE 9.2
Comparison of Coefficient Sensitivity Measures β for Various Sample Periods

Sample Period	Delta Operator	Shift Operator [145]
h=0.001s	4.5095	Infeasible
	$\gamma = 4.4,\ \lambda_a = \lambda_c = 10,\ \lambda_b = 8$	
h=0.1s	4.2423	9.1694
	$\gamma = 4.4,\ \lambda_a = \lambda_c = 10,\ \lambda_b = 8$	$\gamma = 9,\ \lambda_a = \lambda_b = \lambda_c = 10$

TABLE 9.3
Comparison of Coefficient Sensitivity
Measures β

Standard H_∞ Filter	Insensitive H_∞ Filter
Remark 9.4	Theorem 9.3
86.2928	5.0450

9.5   Conclusion

The coefficient sensitivity approach is employed to investigate the problem of designing insensitive H_∞ filters for discrete-time systems with respect to filter coefficient variations based on delta operator models. The filter design problem is formulated as a multi-objective optimization problem. Finsler's lemma is used to derive potential less conservative sufficient conditions for this multi-objective problem. Furthermore, *linear matrix inequality (LMI)* conditions are obtained for the existence of admissible filters with respect to additive/multiplicative coefficient variations based on two different types of sensitivity measures. Finally, a numerical example is also provided to demonstrate the effectiveness of the proposed methods.

10

Insensitive H_∞ Output Tracking Control

10.1 Introduction

The *output tracking control* problem is an important problem. Compared with the stabilization problem, tracking control is more difficult [70]. Over the past decades, this problem has attracted considerable experimental and theoretical attention due to the demands from practical dynamic processes in industry [70,94,104,127,128]. The fuzzy H_∞ tracking control problem for nonlinear networked control systems with a prescribed H_∞ tracking performance was investigated in Jia et al. [70], and Lin, Wang, and Lee [94]. Martin, Devasia, and Paden [104] developed a control scheme for output tracking of nonminimum phase flat systems and applied the scheme to a Vertical Take Off and Landing aircraft example. The problem of H_∞ output tracking control for linear network-based control systems was investigated in Wang and Yang [127]. However, the above results are all based on a common assumption that the output tracking controller will be implemented exactly, which may result in large tracking errors and even lose track of the input reference signal [122], which will further be illustrated via a numerical example in Section 10.4. Therefore, how to design an insensitive output tracking controller is important and challenging in both theory and practice, which motivated us for this study.

In this chapter, the problem of designing an insensitive H_∞ output tracking controller for discrete-time systems by using *delta operator* systems is investigated. The delta operator has the advantage of better numerical properties at high sampling rates and provides a theoretically unified formulation of continuous-time and discrete-time systems [44,45,151]. It is important to note that *a novel Bounded Real Lemma (BRL) (Theorem 10.1) for delta operator systems*, which is more suitable for systems with uncertainties, is presented using the techniques proposed in Jia [71]. Then, based on this new lemma, a linear matrix inequality (LMI)-based sufficient condition for designing insensitive H_∞ output tracking controllers is presented. Finally, a numerical example based on a linearized model of the *F-18 aircraft* is provided to prove the capability of the proposed approach to handle the problem considered in this chapter.

10.2 Problem Statement

Consider a linear time-invariant continuous-time system described by

$$
\begin{aligned}
\dot{x}(t) &= Ax(t) + B_1 u(t) + B_2 w(t) \\
y(t) &= Cx(t) + Dw(t)
\end{aligned}
\tag{10.1}
$$

where $x(t) \in R^n$ is the state with $x(0) = 0$; $y(t) \in R^p$ is the output; $u(t) \in R^m$ is the control input; $w(t) \in R^q$ is the disturbance input, which is assumed to belong to $L_2[0, \infty)$; and A, B_1, B_2, C, and D are known constant matrices of appropriate dimensions.

The control objective is to design a controller so as to make the output $y(t)$ of the plant track a reference signal to meet the required tracking performance. Suppose the reference signal is generated by

$$
\begin{aligned}
\dot{x}_r(t) &= A_r x_r(t) + B_r r(t) \\
y_r(t) &= C_r x_r(t)
\end{aligned}
\tag{10.2}
$$

where $x_r(t) \in R^{n_1}$, $r(t) \in R^r$, and $y_r(t) \in R^p$ are the reference state, the energy bounded reference input, and the reference output, respectively.

These days most control systems contain regulators implemented with a microprocessor; moreover, it is known that, for digital controllers implemented using *z-transform*, the effects of rounding become more pronounced as the *sampling rate* increases. Therefore, in order to overcome this problem, the *delta operator* approach [44, 45, 151] is used. Then, the *δ-transform* of (10.1) and (10.2) are thus given by

$$
\left\{
\begin{aligned}
&\delta x(k) = A_\delta x(k) + B_{1\delta} u(k) + B_{2\delta} w(k) \\
&y(k) = Cx(k) + Dw(k), \\
&x(k) = 0, \quad k \leq 0
\end{aligned}
\right.
\tag{10.3}
$$

and

$$
\left\{
\begin{aligned}
&\delta x_r(k) = A_{r\delta} x_r(k) + B_{r\delta} r(k) \\
&y_r(k) = C_{r\delta} x_r(k), \\
&x_r(k) = 0, \quad k \leq 0,
\end{aligned}
\right.
\tag{10.4}
$$

where

$$
A_\delta = (A_z - I)/h, B_{1\delta} = B_{1z}/h, B_{2\delta} = B_{2z}/h,
$$
$$
A_{r\delta} = (A_{rz} - I)/h, B_{r\delta} = B_{rz}/h
$$

with the parameters of (10.1) and (10.2) by using *z-transform*

$$
A_z = e^{Ah}, B_{1z} = \int_0^h e^{A\tau} B_1 d\tau, B_{2z} = \int_0^h e^{A\tau} B_2 d\tau,
$$
$$
A_{rz} = e^{A_r h}, B_{rz} = \int_0^h e^{A_r \tau} B_r d\tau.
$$

Remark 10.1 *Letting $h \to 0$, one can see that*

$$\lim_{h\to 0} A_z = \lim_{h\to 0} e^{Ah} = I, \ \lim_{h\to 0} B_{1z} = 0, \ \lim_{h\to 0} B_{2z} = 0,$$

which means that all eigenvalues of A_z converge to 1, which is clearly undesirable. As a result, a discrete-time system sampled by the usual z-transform becomes extremely sensitive to roundoff errors [84, 119]. In addition, one can observe that

$$\lim_{h\to 0} A_\delta = \lim_{h\to 0} (e^{Ah} - I)/h = A,$$
$$\lim_{h\to 0} B_{1\delta} = B_1, \ \lim_{h\to 0} B_{2\delta} = B_2,$$

which results in a natural convergence to the continuous-time system as $h \to 0$. Then, many numerical problems for high sampling rates are avoided by using the delta operator (see more details in Goodwin et al. [44,45]). The above facts all demonstrate the necessity of using the delta operator.

The state feedback controller is taken as the following form:

$$u(k) \quad = \quad Kx(k) + \tilde{K}x_r(k) \tag{10.5}$$

where K and $\tilde{K}$ are the state feedback controller gains which will be designed. k_{ij} and $\tilde{k}_{il}$ $(i = 1, \cdots, m; j = 1, \cdots, n; l = 1, \cdots, n_1)$ are used to denote the elements of the controller matrices K and $\tilde{K}$, respectively.

Substituting (10.5) into (10.3) and (10.4), we obtain the augmented closed-loop system

$$\delta\zeta(k) \quad = \quad A_{cl}\zeta(k) + B_{cl}v(k)$$
$$e(k) \quad = \quad C_{cl} \tag{10.6}$$
$$zeta(k) + D_{cl}v(k)$$

where $e(k) = y(k) - y_r(k)$, and

$$\zeta(k) = \begin{bmatrix} x(k) \\ x_r(k) \end{bmatrix}, v(k) = \begin{bmatrix} w(k) \\ r(k) \end{bmatrix},$$

$$A_{cl} = \begin{bmatrix} A_\delta + B_{1\delta}K & B_{1\delta}\tilde{K} \\ 0 & A_{r\delta} \end{bmatrix} = \bar{A} + \bar{B}\bar{K},$$

$$B_{cl} = \begin{bmatrix} B_{2\delta} & 0 \\ 0 & B_{r\delta} \end{bmatrix}, C_{cl} = \begin{bmatrix} C & -C_r \end{bmatrix},$$

$$D_{cl} = \begin{bmatrix} D & 0 \end{bmatrix}$$

with $\bar{A} = \begin{bmatrix} A_\delta & 0 \\ 0 & A_{r\delta} \end{bmatrix}, \bar{B} = \begin{bmatrix} B_{1\delta} \\ 0 \end{bmatrix}, \bar{K} = \begin{bmatrix} K & \tilde{K} \end{bmatrix}$.

Then, the closed-loop transfer function from $v(k)$ to $y(k)$ is given by

$$T(\delta, [k_{ij}]_{m\times n}, [\tilde{k}_{il}]_{m\times n_1}) = C_{cl}(\delta I - A_{cl})^{-1}B_{cl} + D_{cl}. \tag{10.7}$$

Remark 10.2 *It is well known that the limitations in available computer memory and word-length capabilities of the digital processor and the A/D and D/A converters may result in roundoff errors in numerical computations leading to controller implementation imprecision [66]. Hence, our goal is to design a controller which is insensitive to some amount of error with respect to its coefficients. The sensitivity function defined in (8.8a) is used to describe the sensitive properties of matrix functions with respect to additive coefficient variations. While for the designed controller, it should be insensitive to the following interval additive controller coefficient variations which have been extensively researched in the existing literature [66, 142, 143]:*

$$k_{ij} + \theta_{k_{ij}}, \tilde{k}_{il} + \theta_{\tilde{k}_{il}}$$
$$for\ i = 1, \ldots, m; j = 1, \cdots, n; l = 1, \cdots, n_1$$

with

$$|\theta_{k_{ij}}| \le \theta, i = 1, \cdots, m, j = 1, \cdots, n,$$
$$|\theta_{\tilde{k}_{il}}| \le \theta, i = 1, \cdots, m, l = 1, \cdots, n_1,$$

where $\theta_{k_{ij}}$ and $\theta_{\tilde{k}_{il}}$ (for $i, j = 1, \cdots, n; k = 1, \cdots, p; l = 1, \cdots, q$) are used to denote the magnitudes of the deviation of the controller's coefficients of K and $\tilde{K}$, respectively. θ denotes the upper bound of $\theta_{k_{ij}}$ and $\theta_{\tilde{k}_{il}}$. The above interval additive coefficient variation model is from Li [89], which has been extensively used to describe the finite word length (FWL) effects.

Then, based on Definition 8.1 and by means of the techniques developed in Chapter 8 and Chapter 9, Lemma 10.1 is presented.

Lemma 10.1 *Let $T(\delta, [k_{ij}]_{m \times n}, [\tilde{k}_{il}]_{m \times n_1})$ be defined in (10.7), k_{ij} and $\tilde{k}_{il}$ are the elements of the controller's matrices K and $\tilde{K}$, respectively. Then, the sensitivity functions of the transfer function with respect to the elements of the controller's matrices are given as follows:*

$$S_{k_{ij}}(T(\delta, [k_{ij}]_{m \times n}, [\tilde{k}_{il}]_{m \times n_1}))$$
$$= C_{cl}(\delta I - A_{cl})^{-1} N_{ij}^{\tilde{a}}(\delta I - A_{cl})^{-1} B_{cl}$$
$$= \begin{bmatrix} C_{cl} & 0 \end{bmatrix} \left(\delta I - \begin{bmatrix} A_{cl} & N_{ij}^k \\ 0 & A_{cl} \end{bmatrix} \right)^{-1} \begin{bmatrix} 0 \\ B_{cl} \end{bmatrix}$$
$$= C_{ij}^k(\delta I - A_{ij}^k)^{-1} B_{ij}^k, i = 1, \ldots, m; j = 1, \cdots, n$$
$$S_{\tilde{k}_{il}}(T(\delta, [k_{ij}]_{m \times n}, [\tilde{k}_{il}]_{m \times n_1}))$$
$$= C_{cl}(\delta I - A_{cl})^{-1} N_{il}^k(\delta I - A_{cl})^{-1} B_{cl}$$
$$= \begin{bmatrix} C_{cl} & 0 \end{bmatrix} \left(\delta I - \begin{bmatrix} A_{cl} & N_{il}^{\tilde{k}} \\ 0 & A_{cl} \end{bmatrix} \right)^{-1} \begin{bmatrix} 0 \\ B_{cl} \end{bmatrix}$$
$$= C_{il}^{\tilde{k}}(\delta I - A_{il}^{\tilde{k}})^{-1} B_{il}^{\tilde{k}}, i = 1, \ldots, m; l = 1, \cdots, n_1$$

where

$$A_{ij}^k = \begin{bmatrix} A_{cl} & N_{ij}^k \\ 0 & A_{cl} \end{bmatrix}, B_{ij}^k = \begin{bmatrix} 0 & B_{cl}^T \end{bmatrix}^T,$$

$$C_{ij}^k = \begin{bmatrix} C_{cl} & 0 \end{bmatrix}, A_{il}^{\tilde{k}} = \begin{bmatrix} A_{cl} & N_{il}^{\tilde{k}} \\ 0 & A_{cl} \end{bmatrix},$$

$$B_{il}^{\tilde{k}} = \begin{bmatrix} 0 & B_{cl}^T \end{bmatrix}^T, C_{il}^{\tilde{k}} = \begin{bmatrix} C_{cl} & 0 \end{bmatrix}, \quad (10.8)$$

$$N_{ij}^k = \begin{bmatrix} B_{1\delta}R_{ij}^k & 0 \\ 0 & 0 \end{bmatrix}, N_{il}^{\tilde{k}} = \begin{bmatrix} 0 & B_{1\delta}R_{il}^{\tilde{k}} \\ 0 & 0 \end{bmatrix},$$

and

$$R_{ij}^k = e_i e_j^T, \quad R_{il}^{\tilde{k}} = e_i h_l^T. \quad (10.9)$$

In addition, e_i and h_i denote the column vectors in which the ith element equals 1 and the others equal 0.

In this chapter, the *coefficient sensitivity measures* $M_{k_{ij}}$ and $M_{\tilde{k}_{il}}$ are the same as those in the last two chapters, that is,

$$M_{k_{ij}} = \|S_{k_{ij}}(T(\delta, [k_{ij}]_{m \times n}, [\tilde{k}_{il}]_{m \times n_1})\|_\infty,$$
$$i = 1, \cdots, m; j = 1, \cdots, n,$$
$$M_{\tilde{k}_{il}} = \|S_{\tilde{k}_{il}}(T(\delta, [k_{ij}]_{m \times n}, [\tilde{k}_{il}]_{m \times n_1}))\|_\infty, \quad (10.10)$$
$$i = 1, \cdots, m; l = 1, \cdots, n_1.$$

A standard bounded real lemma for the delta operator systems is presented in Lemma 2.13. However, an obvious disadvantage is that one unique *Lyapunov matrix* will be adopted for multi-objective control problems. Moreover, the construct products between the Lyapunov matrix and the closed-loop system matrices also introduce conservativeness. For the sake of overcoming the above disadvantages, a novel bounded real lemma for delta operator systems is obtained by introducing additional matrices and eliminating the product coupling of the system matrices and Lyapunov matrix.

Then, on the basis of the techniques developed in Jia [71], the following result can be derived.

Theorem 10.1 *For a prescribed scalar $\gamma > 0$, the closed-loop system in (10.6) is asymptotically stable, and satisfies the norm constraint $\|T(\delta, [k_{ij}]_{m \times n}, [\tilde{k}_{il}]_{m \times n_1})\|_\infty < \gamma$ if and only if there exist matrices $Q = Q^T > 0$, Z, and a sufficiently small positive scalar ε such that*

$$\begin{bmatrix} Q - He\{Z\} & * & * & * & * \\ (I + \varepsilon A_{cl})Z & -Q & * & * & * \\ \sqrt{\varepsilon h} A_{cl} Z & 0 & -Q & * & * \\ 0 & \sqrt{\varepsilon} B_{cl}^T & \sqrt{h} B_{cl}^T & -\gamma^2 I & * \\ \sqrt{\varepsilon} C_{cl} Z & 0 & 0 & D_{cl} & -I \end{bmatrix} < 0. \quad (10.11)$$

Proof 10.1 *For convenience and compactness, set*

$$
J_1 = \begin{bmatrix} I & 0 & 0 & 0 \\ 0 & 0 & 0 & I \\ 0 & I & 0 & 0 \\ 0 & 0 & I & 0 \end{bmatrix}.
$$

Then, pre- and post-multiplying (2.29) by J_1 and J_1^T, respectively, yields

$$
\begin{bmatrix} A_{cl}^T P + P A_{cl} & * & * & * \\ \sqrt{h} P A_{cl} & -P & * & * \\ B_{cl}^T P & \sqrt{h} B_{cl}^T P & -\gamma^2 I & * \\ C_{cl} & 0 & D_{cl} & -I \end{bmatrix} < 0. \tag{10.12}
$$

From Lemma 2.10, we readily find that (10.12) is equivalent to

$$
\begin{bmatrix} -\varepsilon P^{-1} & * & * & * & * \\ I + \varepsilon A_{cl}^T & -\varepsilon^{-1} P & * & * & * \\ 0 & \sqrt{h} P A_{cl} & -P & * & * \\ 0 & B_{cl}^T P & \sqrt{h} B_{cl}^T P & -\gamma^2 I & * \\ 0 & C_{cl} & 0 & D_{cl} & -I \end{bmatrix} < 0. \tag{10.13}
$$

Substituting $Q = P^{-1}$ into (10.13), after pre- and post-multiplying both sides of (10.13) by $\operatorname{diag}\{\varepsilon^{-\frac{1}{2}} I, \varepsilon^{\frac{1}{2}} Q, Q, I\}$, respectively, we have

$$
\begin{bmatrix} -Q & * & * & * & * \\ Q(I + \varepsilon A_{cl}^T) & -Q & * & * & * \\ 0 & \sqrt{\varepsilon h} A_{cl} Q & -Q & * & * \\ 0 & \sqrt{\varepsilon} B_{cl}^T & \sqrt{h} B_{cl}^T & -\gamma^2 I & * \\ 0 & \sqrt{\varepsilon} C_{cl} Q & 0 & D_{cl} & -I \end{bmatrix} < 0. \tag{10.14}
$$

After some suitable row–column exchanges, we can obtain the following equivalent form of (10.14) as follows:

$$
\begin{bmatrix} -I & * & * & * & * \\ 0 & -Q & * & * & * \\ \sqrt{\varepsilon} Q C_{cl}^T & (3,2) & -Q & * & * \\ 0 & 0 & \sqrt{\varepsilon h} A_{cl} Q & -Q & * \\ D_{cl}^T & 0 & \sqrt{\varepsilon} B_{cl}^T & \sqrt{h} B_{cl}^T & -\gamma^2 I \end{bmatrix} < 0 \tag{10.15}
$$

with $(3,2) = Q(I + \varepsilon A_{cl}^T)$.

The matrix inequality (10.15) shows that

$$
\begin{bmatrix} -I & * & * & * \\ 0 & -Q & * & * \\ \sqrt{\varepsilon} Q C_{cl}^T & Q(I + \varepsilon A_{cl}^T) & -Q & * \\ 0 & 0 & \sqrt{\varepsilon h} A_{cl} Q & -Q \end{bmatrix} < 0. \tag{10.16}
$$

Then, from Lemma 2.11, it is easy to obtain that

$$
\begin{bmatrix}
-I & * & * & * \\
0 & -Q & * & * \\
\sqrt{\varepsilon}Z^T C_{cl}^T & (3,2) & Q - He\{Z\} & * \\
0 & 0 & \sqrt{\varepsilon h}A_{cl}Z & -Q
\end{bmatrix} < 0 \qquad (10.17)
$$

with $(3,2) = Z^T(I + \varepsilon A_{cl}^T)$.

It follows from the simple row–column exchanges that (10.17) holds if and only if

$$
\begin{bmatrix}
-I & * & * & * \\
\sqrt{\varepsilon}Z^T C_{cl}^T & Q - He\{Z\} & * & * \\
0 & (I + \varepsilon A_{cl})Z & -Q & * \\
0 & \sqrt{\varepsilon h}A_{cl}Z & 0 & -Q
\end{bmatrix} < 0. \qquad (10.18)
$$

Then, by using the block replacement technique developed in Jia [71], that is, the replacement of the upper 4×4 block of (10.15) by the left side of (10.18) yields

$$
\begin{bmatrix}
-I & * & * & * & * \\
\sqrt{\varepsilon}Z^T C_{cl}^T & (2,2) & * & * & * \\
0 & (3,2) & -Q & * & * \\
0 & \sqrt{\varepsilon h}A_{cl}Z & 0 & -Q & * \\
D_{cl}^T & 0 & \sqrt{\varepsilon}B_{cl}^T & \sqrt{h}B_{cl}^T & -\gamma^2 I
\end{bmatrix} < 0 \qquad (10.19)
$$

where $(2,2) = Q - He\{Z\}$, $(3,2) = (I + \varepsilon A_{cl})Z$.

Again, carrying out simple block–row–column exchanges, (10.11) is reached. This completes the proof.

Remark 10.3 *From the proof of Theorem 10.1, one can see that (10.11) in Theorem 10.1 is actually equivalent to (2.29) in Lemma 2.13 for nominal systems. The advantage of (10.11) lies in the fact that it eliminates the couple terms between the Lyapunov matrix P and the system matrices by introducing an additional slack variable Z and a sufficiently small positive scalar ε [95]. This decoupling will enable Theorem 10.1 to be more flexible than Lemma 2.13 in solving multi-objective control problems and systems with polytopic uncertainties because different Lyapunov functions can be used. Therefore, Theorem 10.1 is more suitable for robust control problems for systems with polytopic uncertainties and the multi-objective optimization problems in a potentially less conservative framework.*

Then, the following lemma can be obtained based on Theorem 10.1 and the definitions of sensitivity measures as in (10.10).

Lemma 10.2 *Consider the system of the form (10.3). For scalars $\gamma > 0$ and $\beta > 0$, the closed-loop system in (10.6) is asymptotically stable and the conditions $(i = 1, \cdots, m; j = 1, \cdots, n; l = 1, \cdots, n_1)$*

$$
\|T(\delta, [k_{ij}]_{m \times n}, [\tilde{k}_{il}]_{m \times n_1})\|_\infty < \gamma, \ M_{k_{ij}} < \beta, \ M_{\tilde{k}_{il}} < \beta \qquad (10.20)
$$

hold, if there exist symmetric positive-definite matrices $Q \in R^{(n+n_1)\times(n+n_1)}$, $Q_{ij}^k \in R^{2(n+n_1)\times2(n+n_1)}$, $Q_{il}^{\tilde{k}} \in R^{2(n+n_1)\times2(n+n_1)}$, *matrices* $Z \in R^{(n+n_1)\times(n+n_1)}$, $Z_{ij}^k \in R^{2(n+n_1)\times2(n+n_1)}$, *and* $Z_{il}^{\tilde{k}} \in R^{2(n+n_1)\times2(n+n_1)}$, *and sufficiently small positive scalars* ε, ε_{ij}^k *and* $\varepsilon_{il}^{\tilde{k}}$ *such that (10.11) and the following inequalities hold for* $i = 1, \cdots, m; j = 1, \cdots, n; l = 1, \cdots, n_1$.

$$\begin{bmatrix} Q_{ij}^k - He\{Z_{ij}^k\} & * & * & * & * \\ (I + \varepsilon_{ij}^k A_{ij}^k)Z_{ij}^k & -Q_{ij}^k & * & * & * \\ \sqrt{\varepsilon_{ij}^k} h A_{ij}^k Z_{ij}^k & 0 & -Q_{ij}^k & * & * \\ 0 & \sqrt{\varepsilon_{ij}^k}(B_{ij}^k)^T & \sqrt{h}(B_{ij}^k)^T & -\beta^2 I & * \\ \sqrt{\varepsilon_{ij}^k} C_{ij}^k Z_{ij}^k & 0 & 0 & 0 & -I \end{bmatrix} < 0, \quad (10.21)$$

$$\begin{bmatrix} Q_{il}^{\tilde{k}} - He\{Z_{il}^{\tilde{k}}\} & * & * & * & * \\ (I + \varepsilon_{il}^{\tilde{k}} A_{il}^{\tilde{k}})Z_{il}^{\tilde{k}} & -Q_{il}^{\tilde{k}} & * & * & * \\ \sqrt{\varepsilon_{il}^{\tilde{k}}} h A_{il}^{\tilde{k}} Z_{il}^{\tilde{k}} & 0 & -Q_{il}^{\tilde{k}} & * & * \\ 0 & \sqrt{\varepsilon_{il}^{\tilde{k}}}(B_{il}^{\tilde{k}})^T & \sqrt{h}(B_{il}^{\tilde{k}})^T & -\beta^2 I & * \\ \sqrt{\varepsilon_{il}^{\tilde{k}}} C_{il}^{\tilde{k}} Z_{il}^{\tilde{k}} & 0 & 0 & 0 & -I \end{bmatrix} < 0. \quad (10.22)$$

10.3   Insensitive H_∞ Tracking Control Design

In this section, in order to ensure that the output $y(k)$ tracks the reference output signal $y_r(k)$, an insensitive H_∞ output tracking controller is designed such that the closed-loop system (10.6) is asymptotically stable and its H_∞ disturbance attenuation performance bound and the sensitivity performance index are minimized simultaneously.

From (10.8), one can observe that the sensitive coefficient matrices are given in almost block diagonal form of the closed-loop system matrices. Therefore, in order to solve the problem, the following additional constraints are introduced, which may lead to some conservativeness.

$$Z_{ij}^k = \lambda_1 I_2 \otimes Z, Z_{il}^{\tilde{k}} = \lambda_2 I_2 \otimes Z \tag{10.23}$$

where λ_1 and λ_2 are positive scalar parameters to be searched. Then, based on Theorem 10.1 and the definitions of sensitivity measures as in (10.10), after some simple matrix transformations, the main result is given as follows.

Theorem 10.2 *For scalars* $\gamma > 0$ *and* $\beta > 0$, *the closed-loop system in (10.6) is asymptotically stable and the conditions in (10.20) hold, if there exist symmetric positive-definite matrices* $Q \in R^{(n+n_1)\times(n+n_1)}$, $Q_{ij}^k =$

$$\begin{bmatrix} (Q_{ij}^k)_1 & (Q_{ij}^k)_2 \\ (Q_{ij}^k)_2^T & (Q_{ij}^k)_3 \end{bmatrix} \in R^{2(n+n_1)\times 2(n+n_1)}, \quad Q_{il}^{\tilde{k}} = \begin{bmatrix} (Q_{ij}^{\tilde{k}})_1 & (Q_{ij}^{\tilde{k}})_2 \\ (Q_{ij}^{\tilde{k}})_2^T & (Q_{ij}^{\tilde{k}})_3 \end{bmatrix} \in$$

$R^{2(n+n_1)\times 2(n+n_1)}$, *matrices* $Z \in R^{(n+n_1)\times (n+n_1)}$ *and* $M \in R^{p\times (n+n_1)}$, *and positive scalars* λ_1 *and* λ_2, *sufficiently small positive scalars* ε, ε_{ij}^k, *and* $\varepsilon_{il}^{\tilde{k}}$ *such that (10.11) and the following inequalities hold for* $i = 1,\cdots,m; j = 1,\cdots,n; l = 1,\cdots,n_1$:

$$\begin{bmatrix} Q - He\{Z\} & * & * & * & * \\ Z + \varepsilon(\bar{A}Z + \bar{B}M) & -Q & * & * & * \\ \sqrt{\varepsilon h}(\bar{A}Z + \bar{B}M) & 0 & -Q & * & * \\ 0 & \sqrt{\varepsilon}B_{cl}^T & \sqrt{h}B_{cl}^T & -\gamma^2 I & * \\ \sqrt{\varepsilon}C_{cl}Z & 0 & 0 & D_{cl} & -I \end{bmatrix} < 0, \qquad (10.24)$$

$$\begin{bmatrix} Q_{ij}^k - He\{\lambda_1 I_2 \otimes Z\} & * & * & * & * \\ \lambda_1 I_2 \otimes Z + \lambda_1 \varepsilon_{ij}^k \Omega_{ij}^k & -Q_{ij}^k & * & * & * \\ \lambda_1 \sqrt{\varepsilon_{ij}^k h}\Omega_{ij}^k & 0 & -Q_{ij}^k & * & * \\ 0 & \sqrt{\varepsilon_{ij}^k}\Omega_b & \sqrt{h}\Omega_b & -\beta^2 I & * \\ \lambda_1 \sqrt{\varepsilon_{ij}^k}\Omega_c & 0 & 0 & 0 & -I \end{bmatrix} < 0, \qquad (10.25)$$

$$\begin{bmatrix} Q_{il}^{\tilde{k}} - He\{\lambda_2 I_2 \otimes Z\} & * & * & * & * \\ \lambda_2 I_2 \otimes Z + \lambda_2 \varepsilon_{il}^{\tilde{k}} \Omega_{il}^{\tilde{k}} & -Q_{il}^{\tilde{k}} & * & * & * \\ \lambda_2 \sqrt{\varepsilon_{il}^{\tilde{k}} h}\Omega_{il}^{\tilde{k}} & 0 & -Q_{il}^{\tilde{k}} & * & * \\ 0 & \sqrt{\varepsilon_{il}^{\tilde{k}}}\Omega_b & \sqrt{h}\Omega_b & -\beta^2 I & * \\ \lambda_2 \sqrt{\varepsilon_{il}^{\tilde{k}}}\Omega_c & 0 & 0 & 0 & -I \end{bmatrix} < 0 \qquad (10.26)$$

where

$$\Omega_{ij}^k = \begin{bmatrix} \bar{A}Z + \bar{B}M & N_{ij}^k Z \\ 0 & \bar{A}Z + \bar{B}M \end{bmatrix},$$

$$\Omega_{il}^{\tilde{k}} = \begin{bmatrix} \bar{A}Z + \bar{B}M & N_{il}^{\tilde{k}} Z \\ 0 & \bar{A}Z + \bar{B}M \end{bmatrix},$$

$$\Omega_b = \begin{bmatrix} 0 & B_{cl}^T \end{bmatrix}, \Omega_c = \begin{bmatrix} C_{cl}Z & 0 \end{bmatrix}.$$

Moreover, if there exist solutions of the above inequalities, the standard output controller can be given by

$$\bar{K} = MZ^{-1}. \qquad (10.27)$$

Remark 10.4 *When the scalars* λ_1, λ_2, ε, ε_{ij}^k, *and* $\varepsilon_{il}^{\tilde{k}}$ *are set to be fixed parameters, the conditions in Theorem 10.2 become linear. The corresponding insensitive* H_∞ *output tracking controller realizes an optimal trade-off between* the standard H_∞ performance γ and the sensitivity performance of the transfer function with respect to controller coefficient variations β. Therefore, the problem of insensitive H_∞ output tracking controller design can be converted to the following optimization problem:

$$\min_{Z,M,Q,Q_{ij}^k,Q_{il}^{\bar{k}}} \gamma + \alpha\beta, \qquad (10.28)$$
$$s.t.\ (10.24),(10.25),(10.26)$$

where the weight α quantifies the relative importance of γ and β, and the designed controller's gains can be obtained by (10.27).

Remark 10.5 *When the insensitive constraint is unconsidered, which is called as standard output tracking control design, Theorem 10.2 reduces to the standard output tracking control design method, which (10.25) must be satisfied.*

Remark 10.6 *Obviously, the problem of designing insensitive controllers is equivalent to minimizing β by using (10.28). However, the artificial constraints in (10.23) may introduce some conservativeness. The actual measures in (10.10) can be computed directly when the controller's matrices K and $\tilde{K}$ are given, where the largest one is taken as the actual sensitivity measure of the designed controller.*

Remark 10.7 *It is well known that unavoidable discrepancies between mathematical models and real-world systems can result in the degradation of control system performance and even leading to instability [66]. On the other hand, it is of practical interest to design a controller such that it is insensitive to its own coefficient variations because the controller is implemented with finite precision arithmetic, quantized coefficients, and so forth. Therefore, how to design a controller for a given plant such that the controller is insensitive to its own coefficient variations and the plant model uncertainties should be the ultimate goal. Fortunately, the LMI-based conditions in Theorem 10.2 can make the proposed method be easily extended to deal with the robust insensitive H_∞ output tracking control problem for delta operator systems with polytopic uncertainties because the system matrices are affinely involved in the proposed design conditions. However, to make our idea more clear and to avoid complicated notation, we consider the case in which the designed controller is only insensitive to controller coefficient variations in this chapter.*

10.4   Example

In this section, to illustrate the effectiveness of the proposed insensitive H_∞ output tracking controller design method, a numerical example based on a linearized model of F-18 aircraft from Yang and Wang [146], is taken.

Example 10.1 *The corresponding system parameters are*

$$
A = \begin{bmatrix} -1.4 + 0.2\delta_1 & 1 + 0.1\delta_1 & -0.26 & -0.53 \\ -1.6 & -0.78 & -10 & -4.7 + 0.2\delta_2 \\ 0 & 0 & -8.2 & 0 \\ 0 & 0 & 0 & -7.3 \end{bmatrix},
$$

$$
B_1^T = \begin{bmatrix} 0 & 0 & 8.2 & 0 \\ 0 & 0 & 0 & 7.3 \end{bmatrix}, C = \begin{bmatrix} 1 & 0 & 0 & 0 \end{bmatrix}
$$

with $\delta_1 = 0.3$ and $\delta_2 = -0.4$, and the other matrices are assumed to be

$$
\begin{aligned}
B_2^T &= \begin{bmatrix} 0.8 & 0.3 & -1 & 0 \\ 1 & -0.5 & 0 & -0.2 \end{bmatrix}, \\
D &= \begin{bmatrix} -0.5 & -3 \end{bmatrix}.
\end{aligned}
$$

In addition, the reference model is given as follows:

$$
\begin{aligned}
\dot{x}_r(t) &= -0.85x_r(t) + 0.6r(t) \\
y_r(t) &= 0.5x_r(t).
\end{aligned}
$$

Then, by using the delta operator approach in sampling the continuous-time system and the reference model with $h = 0.1s$, the relevant discrete-time system's parameters in the δ-domain are given as follows:

$$
A_\delta = \begin{bmatrix} -1.3274 & 0.9240 & -0.5336 & -0.5296 \\ -1.4353 & -0.8251 & -6.4942 & -3.2067 \\ 0 & 0 & -5.5957 & 0 \\ 0 & 0 & 0 & -5.1809 \end{bmatrix},
$$

$$
B_{1\delta} = \begin{bmatrix} -0.1885 & -0.1943 \\ -3.0809 & -1.3400 \\ 5.5957 & 0 \\ 0 & 5.1809 \end{bmatrix}, A_{r\delta} = -0.8149,
$$

$$
B_{2\delta} = \begin{bmatrix} 0.7841 & 0.9147 \\ 0.6040 & -0.5175 \\ -0.6824 & 0 \\ 0 & -0.1419 \end{bmatrix}, B_{r\delta} = 0.5752.
$$

At first, based on Remark 10.5, the standard controller with $\varepsilon = 0.05$ is given as follows:

$$
K = \begin{bmatrix} 11.0830 & -7.4903 & 2.1785 & 3.9459 \\ -15.7117 & 12.0190 & -4.8587 & -6.5520 \end{bmatrix},
$$

$$
\tilde{K} = \begin{bmatrix} 3.7724 \\ -4.8212 \end{bmatrix}, \gamma = 3.1219.
$$

Then, by solving the optimization problem (10.28) with $\alpha = 10$, $\varepsilon = 0.05$,

TABLE 10.1

Comparison of Coefficient Sensitivity Measures β

Indexes	Standard Controller Remark 10.5	Insensitive Controller Theorem 10.2	Remark 10.6
γ	3.1219	3.4365	3.4365
β	1.7366	0.6181	0.1066

$\varepsilon_{ij}^{k} = \varepsilon_{il}^{\tilde{k}} = 0.04$, $\lambda_1 = 0.98$, and $\lambda_2 = 0.8$, we obtain the sensitivity performance $\beta = 0.6181$ and the insensitive controller with gain matrices as

$$K = \begin{bmatrix} 0.5549 & 0.5877 & -0.7059 & -0.2636 \\ 2.1531 & 0.9122 & -0.7305 & -0.7158 \end{bmatrix},$$

$$\tilde{K} = \begin{bmatrix} 0.4250 \\ 1.5344 \end{bmatrix}, \gamma = 3.4365, \beta = 0.6181.$$

In addition, the actual coefficient sensitivity measures for the insensitive controller designed by Theorem 10.2 and the standard controller designed by Remark 10.5, respectively, can be computed by using the method given in Remark 10.6. The results are given in Table 10.1.

It can be seen that the sensitivity of the insensitive controller is much lower than the standard one. On the other hand, from Table 10.1, it is clear that compared with the standard H_∞ performance index bound $\gamma = 3.1219$, the performance index obtained by Theorem 10.2 is degraded 7.79%, which shows that insensitivity to controller coefficient changes can be obtained with very little loss in standard performance by using the proposed insensitive design method.

Assume that $\theta_{k_{ij}}$ and $\theta_{\tilde{k}_{il}}$ represent the interval type of additive coefficient variations with the upper bound θ in Remark 10.2. Moreover, let the system's initial state be $x_0^T = \begin{bmatrix} 0 & 0 & 0 & 0 \end{bmatrix}$.

Suppose the input signals are given as follows

$$w_1(k) = w_2(k) = \begin{cases} -0.3\cos(k) & 10 \leq k \leq 12, \\ 0 & otherwise, \end{cases}$$

$$r(k) = \begin{cases} 0.15 & 10 \leq k \leq 12, \\ 0 & otherwise. \end{cases}$$

Figure 10.1 through Figure 10.3 show the responses of the tracking error $e(k)$ for the insensitive controller and the standard controller with $\theta = 0$, $\theta = 0.08$, and $\theta = 0.1$, respectively. From these figures, it can be observed that the controller designed by Theorem 10.2 is insensitive to the additive coefficient variations while, on the contrary, the standard controller is sensitive to the same controller coefficient variations. Particularly, the output $y(k)$ via the standard controller is losing track of the reference output signal $y_r(k)$ when $\theta = 0.1$, while from Figure 10.4, the insensitive controller is still quite insensitive to the controller coefficient variations, which shows the advantage of the proposed method.

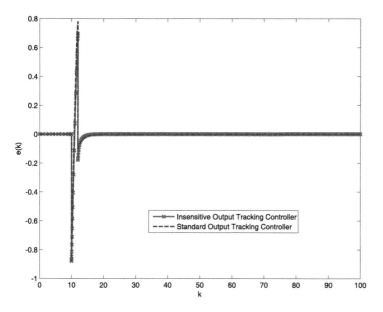

FIGURE 10.1
Comparison of the responses of the tracking error $e(k)$ via different controllers
for $\theta = 0$.

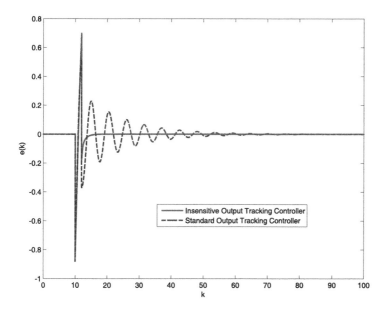

FIGURE 10.2
Comparison of the responses of the tracking error $e(k)$ via different controllers
for $\theta = 0.08$.

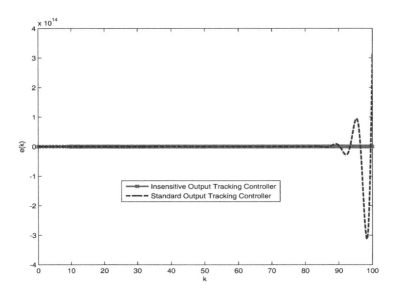

FIGURE 10.3
Comparison of the responses of the tracking error $e(k)$ via different controllers
for $\theta = 0.1$.

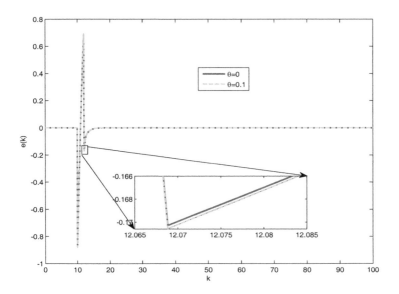

FIGURE 10.4
The responses of the tracking error $e(k)$ via insensitive output tracking controller when $\theta = 0$ and $\theta = 0.1$.

10.5 Conclusion

The coefficient sensitivity approach is employed to investigate the problem of designing insensitive H_∞ output tracking controllers for discrete-time systems with respect to controller coefficient variations based on delta operator models. The coefficient sensitivity approach discussed in the chapter suggests the possibility of an addition of a sensitivity measure to the original performance for the controlling problems to reduce the sensitivity of the methods to changes in controller coefficients. A novel real bounded lemma for delta operator systems, which has potential for less conservative systems with uncertainties, is presented. Furthermore, LMI-based conditions are obtained for the existence of admissible controllers with respect to controller coefficient variations. Finally, a numerical example based on a linearized model of F-18 aircraft is provided to demonstrate the effectiveness of the proposed methods.

11

Insensitive H_∞ Dynamic Output Feedback Control

11.1 Introduction

Coefficient sensitivity analysis is used widely to investigate the effects of coefficient change on the solution of mathematical models [124]. Therefore, *coefficient sensitivity analysis* is a powerful method for evaluating the effect of the controller/filter coefficient variations on the performance of the systems. In recent years, strictly convex conditions have been developed for the problem of insensitive H_∞ filter design by the authors [48, 146]. However, the problem of designing insensitive controllers with respect to *multiplicative/additive controller coefficient variations* is an non-deterministic polynomial (NP) optimization problem which is more difficult than the one considered in Chapters 8, 9, and 10. Therefore, it is necessary to develop new techniques and methods for solving this NP-hard problem.

Motivated by the above points, this chapter is concerned with the *multi-objective problem* of designing *insensitive H_∞ controllers* for *linear discrete-time systems*, which minimizes the controller coefficient sensitivity and meets the prescribed H_∞ norm constraint simultaneously. Two design methods are proposed to solve this *multi-objective control* problem. First, the H_∞ norms of the *coefficient sensitivity functions* are defined as the sensitivity of the transfer functions with respect to *multiplicative/additive controller coefficient variations*. Then, based on the sensitive measures, a sufficient and necessary condition for designing insensitive H_∞ dynamic output-feedback controllers with respect to multiplicative/additive coefficient variations is presented in terms of a set of *linear matrix inequalities* (LMIs) and additional coupling constraints which destroy the convexity of the overall problem. To solve such a problem, an LMI-based procedure which is a *sequential linear programming matrix method (SLPMM)* [85] is proposed. However, when the SLPMM algorithm acts on a module of very high dimension, the search for satisfactory solutions may be very difficulty. To overcome this difficulty, the *non-fragile controller design method* is adopted to obtain an initial solution for the SLPMM algorithm for the first time. Finally, the effectiveness and superiority of the proposed method is demonstrated by a numerical example.

11.2   Problem Statement

Consider a discrete-time system described as follows:

$$
\begin{aligned}
x(k+1) &= Ax(k) + B_1w(k) + B_2u(k) \\
z(k) &= C_1x(k) + D_{12}u(k) \\
y(k) &= C_2x(k) + D_{21}w(k)
\end{aligned}
\tag{11.1}
$$

where $x(k) \in R^n$ is the state, $w(k) \in R^r$ is an exogenous disturbance which belongs to gauss white noise signals, $z(k) \in R^m$ is the controlled output, $u(k) \in R^q$ is the control input, and $y(k) \in R^p$ is the measured output. A, B_1, B_2, C_1, C_2, D_{12}, and D_{21} are known constant matrices of appropriate dimensions.

Consider a linear time-invariant controller with a state-space representation

$$
\begin{aligned}
\xi(k+1) &= \tilde{A}\xi(k) + \tilde{B}y(k) \\
u(k) &= \tilde{C}\xi(k)
\end{aligned}
\tag{11.2}
$$

where $\xi(k) \in R^n$ is the controller's state, and $\tilde{A}$, $\tilde{B}$, and $\tilde{C}$ are the controller's matrices to be designed.

To facilitate the following sections, denote

$$
\tilde{A} = [\ \tilde{a}_{ij}\ ]_{n \times n}, \tilde{B} = [\ \tilde{b}_{ik}\ ]_{n \times p}, \tilde{C} = [\ \tilde{c}_{lj}\ ]_{q \times n},
$$
$$
\text{for } i = 1, \ldots, n; j = 1, \ldots, n; k = 1, \ldots, p; l = 1, \ldots, q.
$$

Applying the controller (11.2) to the system (11.1), we obtain the closed-loop system

$$
\begin{aligned}
x_{cl}(k+1) &= A_{cl}x_{cl}(k) + B_{cl}w(k) \\
z(k) &= C_{cl}x_{cl}(k)
\end{aligned}
\tag{11.3}
$$

where $x_{cl}(k) = \begin{bmatrix} x(k) \\ \xi(k) \end{bmatrix}$, $A_{cl} = \begin{bmatrix} A & B_2\tilde{C} \\ \tilde{B}C_2 & \tilde{A} \end{bmatrix}$, $B_{cl} = \begin{bmatrix} B_1 \\ \tilde{B}D_{21} \end{bmatrix}$, $C_{cl} = \begin{bmatrix} C_1 & D_{12}\tilde{C} \end{bmatrix}$.

The transfer matrix of the augmented system (11.3) from $w(k)$ to $z(k)$ is given by

$$
T(z, [\tilde{a}_{ij}]_{n \times n}, [\tilde{b}_{ik}]_{n \times p}, [\tilde{c}_{lj}]_{q \times n}) = C_{cl}(zI - A_{cl})^{-1}B_{cl}.
\tag{11.4}
$$

11.2.1   Sensitivity Function

Generally, *sensitivity theory* is always used to characterize the phenomenon of the trivial deviations, which motivates us to design insensitive controllers in the framework of coefficient sensitivity theory because the controller coefficient variations resulting from the limitation of the available computer memory are of trivial deviations.

In the following, $\theta^{\tilde{a}}_{ij}$, $\theta^{\tilde{b}}_{ik}$, and $\theta^{\tilde{c}}_{lj}$ (for $i, j = 1, \cdots, n; k = 1, \cdots, p; l = 1, \cdots, q$) are used to denote the magnitudes of the deviation of the controller's coefficients of $\tilde{A}$, $\tilde{B}$, and $\tilde{C}$, respectively. Then, based on Definition 8.1 and by means of the techniques developed in Gevers and Li [43] and Hilaire, Chevrel, and Trinquet [60], Lemma 11.1 is presented.

Lemma 11.1 *Let $T_e(z)$ be defined in (11.4), and $\tilde{a}_{ij}$, $\tilde{b}_{ik}$, and $\tilde{c}_{lj}$ are the elements of the controller's matrices $\tilde{A}$, $\tilde{B}$, and $\tilde{C}$, respectively. Then, the sensitivity functions of the transfer function with respect to the elements of the controller's matrices are given as follows:*

$$S(T(z, [\tilde{a}_{ij}]_{n \times n}, [\tilde{b}_{ik}]_{n \times p}, [\tilde{c}_{lj}]_{q \times n}), \tilde{a}_{ij})$$
$$= C_{cl}(zI - A_{cl})^{-1}N^{\tilde{a}}_{ij}(zI - A_{cl})^{-1}B_{cl}$$
$$= \begin{bmatrix} C_{cl} & 0 \end{bmatrix} \left(zI - \begin{bmatrix} A_{cl} & N^{\tilde{a}}_{ij} \\ 0 & A_{cl} \end{bmatrix} \right)^{-1} \begin{bmatrix} 0 \\ B_{cl} \end{bmatrix}$$
$$= C^{\tilde{a}}_{ij}(zI - A^{\tilde{a}}_{ij})^{-1}B^{\tilde{a}}_{ij}, i = 1, \ldots, n; j = 1, \cdots, n$$

$$S(T(z, [\tilde{a}_{ij}]_{n \times n}, [\tilde{b}_{ik}]_{n \times p}, [\tilde{c}_{lj}]_{q \times n}), \tilde{b}_{ik})$$
$$= C_{cl}(zI - A_{cl})^{-1}N^{\tilde{b}}_{ik}(zI - A_{cl})^{-1}B_{cl} + C_{cl}(zI - A_{cl})^{-1}\bar{N}^{\tilde{b}}_{ik}$$
$$= \begin{bmatrix} C_{cl} & 0 \end{bmatrix} \left(zI - \begin{bmatrix} A_{cl} & N^{\tilde{b}}_{ik} \\ 0 & A_{cl} \end{bmatrix} \right)^{-1} \begin{bmatrix} 0 \\ B_{cl} \end{bmatrix} + \tilde{C}(zI - A_{cl})^{-1}\bar{N}^{\tilde{b}}_{ik}$$
$$= \begin{bmatrix} C_{cl} & 0 & C_{cl} \end{bmatrix} \left(zI - \begin{bmatrix} A_{cl} & N^{\tilde{b}}_{ik} & 0 \\ 0 & A_{cl} & 0 \\ 0 & 0 & A_{cl} \end{bmatrix} \right)^{-1} \begin{bmatrix} 0 \\ B_{cl} \\ \bar{N}^{\tilde{b}}_{ik} \end{bmatrix}$$
$$= C^{\tilde{b}}_{ik}(zI - A^{\tilde{b}}_{ik})^{-1}B^{\tilde{b}}_{ik}, i = 1, \ldots, n; k = 1, \ldots, p$$

$$S(T(z, [\tilde{a}_{ij}]_{n \times n}, [\tilde{b}_{ik}]_{n \times p}, [\tilde{c}_{lj}]_{q \times n}), \tilde{c}_{lj})$$
$$= \bar{N}^{\tilde{c}}_{lj}(zI - A_{cl})^{-1}B_{cl} + C_{cl}(zI - A_{cl})^{-1}N^{\tilde{c}}_{lj}(zI - A_{cl})^{-1}B_{cl}$$
$$= \bar{N}^{\tilde{c}}_{lj}(zI - A_{cl})^{-1}B_{cl} + \begin{bmatrix} C_{cl} & 0 \end{bmatrix} \left(zI - \begin{bmatrix} A_{cl} & N^{\tilde{c}}_{lj} \\ 0 & A_{cl} \end{bmatrix} \right)^{-1} \begin{bmatrix} 0 \\ B_{cl} \end{bmatrix}$$
$$= \begin{bmatrix} \bar{N}^{\tilde{c}}_{lj} & C_{cl} & 0 \end{bmatrix} \left(zI - \begin{bmatrix} A_{cl} & N^{\tilde{c}}_{lj} & 0 \\ 0 & A_{cl} & 0 \\ 0 & 0 & A_{cl} \end{bmatrix} \right)^{-1} \begin{bmatrix} B_{cl} \\ 0 \\ B_{cl} \end{bmatrix}$$
$$= C^{\tilde{c}}_{lj}(zI - A^{\tilde{c}}_{lj})^{-1}B^{\tilde{c}}_{lj}, l = 1, \cdots, q; j = 1, \cdots, n$$

where

$$A_{ij}^{\tilde{a}} = \begin{bmatrix} A_{cl} & N_{ij}^{\tilde{a}} \\ 0 & A_{cl} \end{bmatrix}, A_{ik}^{\tilde{b}} = \begin{bmatrix} A_{cl} & N_{ik}^{\tilde{b}} & 0 \\ 0 & A_{cl} & 0 \\ 0 & 0 & A_{cl} \end{bmatrix}, A_{lj}^{\tilde{c}} = \begin{bmatrix} A_{cl} & N_{lj}^{\tilde{c}} & 0 \\ 0 & A_{cl} & 0 \\ 0 & 0 & A_{cl} \end{bmatrix}$$

$$B_{ij}^{\tilde{a}} = \begin{bmatrix} 0 & B_{cl}^T \end{bmatrix}^T, B_{ik}^{\tilde{b}} = \begin{bmatrix} 0 & B_{cl}^T(\bar{N}_{ik}^{\tilde{b}})^T \end{bmatrix}^T, B_{lj}^{\tilde{c}} = \begin{bmatrix} B_{cl}^T & 0 & B_{cl}^T \end{bmatrix}^T$$

$$C_{ij}^{\tilde{a}} = \begin{bmatrix} C_{cl} & 0 \end{bmatrix}, C_{ik}^{\tilde{b}} = \begin{bmatrix} C_{cl} & 0 & C_{cl} \end{bmatrix}, C_{lj}^{\tilde{c}} = \begin{bmatrix} \bar{N}_{lj}^{\tilde{c}} & C_{cl} & 0 \end{bmatrix} \quad (11.5)$$

$$N_{ij}^{\tilde{a}} = \begin{bmatrix} 0 & 0 \\ 0 & R_{ij}^{\tilde{a}} \end{bmatrix}, N_{ik}^{\tilde{b}} = \begin{bmatrix} 0 & 0 \\ R_{ik}^{\tilde{b}}C_2 & 0 \end{bmatrix}, \bar{N}_{ik}^{\tilde{b}} = \begin{bmatrix} 0 \\ R_{ik}^{\tilde{b}}D_{21} \end{bmatrix}$$

$$N_{lj}^{\tilde{c}} = \begin{bmatrix} 0 & B_2 R_{lj}^{\tilde{c}} \\ 0 & 0 \end{bmatrix}, \bar{N}_{lj}^{\tilde{c}} = \begin{bmatrix} 0 & D_{12}R_{lj}^{\tilde{c}} \end{bmatrix}$$

and

$$R_{ij}^{\tilde{a}} = e_i e_i^T \tilde{A} e_j e_j^T, \quad R_{ik}^{\tilde{b}} = e_i e_i^T \tilde{B} h_h h_h^T, \quad R_{lj}^{\tilde{c}} = g_l g_l^T \tilde{C} e_j e_j^T,$$

$$(11.6)$$

for the multiplicative form, and

$$R_{ij}^{\tilde{a}} = e_i e_j^T, \quad R_{ik}^{\tilde{b}} = e_i h_k^T, \quad R_{lj}^{\tilde{c}} = g_l e_j^T \quad (11.7)$$

for the additive form, respectively. In addition, $e_k \in R^n$, $h_k \in R^p$, and $g_k \in R^q$ denote the column vectors in which the kth element equals 1 and the others equal 0.

Remark 11.1 *In fact, in the course of the actualization of the controller, the coefficients of the controller may accrue some coefficient variations due to the existence of the parameter drift, accuracy problem, and other factors [77, 93]. Therefore, the designed controllers should be able to tolerate some level of controller coefficient variations $\Delta a_{f_{ij}}$, $\Delta b_{f_{ik}}$, and $\Delta c_{f_{lj}}$ which are described as follows.*

- *The multiplicative form [17, 47, 53, 148]:*

$$\Delta \tilde{a}_{ij} = \theta_{\tilde{a}_{ij}} \tilde{a}_{ij}, \Delta \tilde{b}_{ik} = \theta_{\tilde{b}_{ik}} \tilde{b}_{ik}, \Delta \tilde{c}_{lj} = \theta_{\tilde{c}_{lj}} \tilde{c}_{lj}, \quad (11.8)$$

$$\text{where } i, j = 1, \cdots, n; k = 1, \cdots, p; l = 1, \cdots, q.$$

- *The additive form [15, 16, 19, 26, 143, 147]:*

$$\Delta \tilde{a}_{ij} = \theta_{\tilde{a}_{ij}}, \Delta \tilde{b}_{ik} = \theta_{\tilde{b}_{ik}}, \Delta \tilde{c}_{lj} = \theta_{\tilde{c}_{lj}} \quad (11.9)$$

$$\text{where } i, j = 1, \cdots, n; k = 1, \cdots, p; l = 1, \cdots, q,$$

where $\theta_{ij}^{\tilde{a}}$, $\theta_{ik}^{\tilde{b}}$, and $\theta_{lj}^{\tilde{c}}$ (for $i, j = 1, \cdots, n; k = 1, \cdots, p; l = 1, \cdots, q$) are used to denote the magnitudes of the deviation of the controller's coefficients of $\tilde{A}$, $\tilde{B}$, and $\tilde{C}$, respectively, with the following form:

$$|\theta_{\tilde{a}_{ij}}| \leq \theta, i, j = 1, \cdots, n$$

$$|\theta_{\tilde{b}_{ik}}| \leq \theta, i = 1, \cdots, n, k = 1, \cdots, p \quad (11.10)$$

$$|\theta_{\tilde{c}_{lj}}| \leq \theta, l = 1, \cdots, q, j = 1, \cdots, n.$$

11.2.2 Sensitivity Measures

In this chapter, the coefficient sensitivity measures $M_{ij}^{\tilde{a}}$, $M_{ik}^{\tilde{b}}$, and $M_{lj}^{\tilde{c}}$ will be taken as

$$
\begin{aligned}
M_{ij}^{\tilde{a}} &= \|S(T(z, [\tilde{a}_{ij}]_{n \times n}, [\tilde{b}_{ik}]_{n \times p}, [\tilde{c}_{lj}]_{q \times n}), \tilde{a}_{ij})\|_\infty, i = 1, \cdots, n; j = 1, \cdots, n \\
M_{ik}^{\tilde{b}} &= \|S(T(z, [\tilde{a}_{ij}]_{n \times n}, [\tilde{b}_{ik}]_{n \times p}, [\tilde{c}_{lj}]_{q \times n}), \tilde{b}_{ik})\|_\infty, i = 1, \cdots, n; k = 1, \cdots, p \\
M_{lj}^{\tilde{c}} &= \|S(T(z, [\tilde{a}_{ij}]_{n \times n}, [\tilde{b}_{ik}]_{n \times p}, [\tilde{c}_{lj}]_{q \times n}), \tilde{c}_{lj})\|_\infty, l = 1, \cdots, q; j = 1, \cdots, n.
\end{aligned}
$$
(11.11)

The measures are the H_∞ norms of the sensitivity functions of the closed-loop transfer function with respect to the perturbations in controller parameters. The problem of designing *insensitive H_∞ controllers* now can be stated in the following section.

11.2.3 Insensitive H_∞ Control with Controller Coefficient Variations

Given positive scalars γ and β, design a controller described by (11.2) such that the closed-loop system (11.3) is *asymptotically stable* and keeps

$$
\|T(z, [\tilde{a}_{ij}]_{n \times n}, [\tilde{b}_{ik}]_{n \times p}, [\tilde{c}_{lj}]_{q \times n})\|_\infty < \gamma
$$

and in the meantime, satisfies $M_{ij}^{\tilde{a}} < \beta, M_{ik}^{\tilde{b}} < \beta$ and $M_{lj}^{\tilde{c}} < \beta$ for $i, j = 1, \cdots, n; k = 1, \cdots, p; l = 1, \cdots, q$.

11.3 Insensitive H_∞ Controller Design

In this section, we consider the problem of insensitive H_∞ output-feedback control for the system in (11.1). At first, a sufficient and necessary condition for the existence of insensitive controllers with respect to multiplicative/additive coefficient variations is presented. Then, the non-fragile controller design method is adopted to obtain an effective solution which is regarded as an effective initial solution for the SLPMM algorithm for the case where the search for satisfactory solutions may be very difficult when the SLPMM algorithm acts on a module of very high dimension.

11.3.1 Step 1: General Conditions for the Existence of Insensitive H_∞ Controllers

The following lemma can be obtained based on the Bounded Real Lemma and the definitions of the two types of sensitivity measures, respectively.

Lemma 11.2 *Consider the system of the form (11.1). For scalars $\gamma > 0$ and*

$\beta > 0$, the closed-loop system (11.3) is asymptotically stable and the conditions in (11.16) hold, if and only if there exist symmetric positive-definite matrices $P \in R^{2n \times 2n}$, $P_{ij}^{\tilde{a}} \in R^{4n \times 4n}$, $P_{ik}^{\tilde{b}} \in R^{6n \times 6n}$, and $P_{lj}^{\tilde{c}} \in R^{6n \times 6n}$ such that

$$
\begin{bmatrix}
-P & 0 & PA_{cl} & PB_{cl} \\
* & -I & C_{cl} & 0 \\
* & * & -P & 0 \\
* & * & * & -\gamma^2 I
\end{bmatrix} < 0, \tag{11.12}
$$

$$
\begin{bmatrix}
-P_{ij}^{\tilde{a}} & 0 & P_{ij}^{\tilde{a}} A_{ij}^{\tilde{a}} & P_{ij}^{\tilde{a}} B_{ij}^{\tilde{a}} \\
* & -I & C_{ij}^{\tilde{a}} & 0 \\
* & * & -P_{ij}^{\tilde{a}} & 0 \\
* & * & * & -\beta^2 I
\end{bmatrix} < 0, \tag{11.13}
$$

$$
\begin{bmatrix}
-P_{ik}^{\tilde{b}} & 0 & P_{ik}^{\tilde{b}} A_{ik}^{\tilde{b}} & P_{ik}^{\tilde{b}} B_{ik}^{\tilde{b}} \\
* & -I & C_{ik}^{\tilde{b}} & 0 \\
* & * & -P_{ik}^{\tilde{b}} & 0 \\
* & * & * & -\beta^2 I
\end{bmatrix} < 0, \tag{11.14}
$$

$$
\begin{bmatrix}
-P_{lj}^{\tilde{c}} & 0 & P_{lj}^{\tilde{c}} A_{lj}^{\tilde{c}} & P_{lj}^{\tilde{c}} B_{lj}^{\tilde{c}} \\
* & -I & C_{lj}^{\tilde{c}} & 0 \\
* & * & -P_{lj}^{\tilde{c}} & 0 \\
* & * & * & -\beta^2 I
\end{bmatrix} < 0, \tag{11.15}
$$

hold for $i, j = 1, \cdots, n; k = 1, \cdots, p; l = 1, \cdots, q$, where $A_{ij}^{\tilde{a}}, B_{ij}^{\tilde{a}}, C_{ij}^{\tilde{a}}$, $A_{ik}^{\tilde{b}}, B_{ik}^{\tilde{b}}, C_{ik}^{\tilde{b}}, A_{lj}^{\tilde{c}}, B_{lj}^{\tilde{c}}$, and $C_{lj}^{\tilde{c}}$ are defined in (11.5).

The following theorem based on Lemma 11.2 presents a sufficient and necessary condition under which the insensitive H_∞ control problem with respect to controller coefficient variations is solvable.

Theorem 11.1 *Consider the system of the form (11.1). For scalars $\gamma > 0$ and $\beta > 0$, the closed-loop system (11.3) is asymptotically stable and*

$$
\|T(z, [\tilde{a}_{ij}]_{n \times n}, [\tilde{b}_{ik}]_{n \times p}, [\tilde{c}_{lj}]_{q \times n})\|_\infty < \gamma, M_{ij}^{\tilde{a}} < \beta, M_{ik}^{\tilde{b}} < \beta, M_{lj}^{\tilde{c}} < \beta
$$
$$
where \; i, j = 1, \cdots, n; k = 1, \cdots, p; l = 1, \cdots, q
$$
$$\tag{11.16}$$

hold, if and only if there exist symmetric positive-definite matrices $P \in R^{2n \times 2n}$, $P_{ij}^{\tilde{a}} \in R^{4n \times 4n}$, $P_{ik}^{\tilde{b}} \in R^{6n \times 6n}$, $P_{lj}^{\tilde{c}} \in R^{6n \times 6n}$, $\hat{P} \in R^{2n \times 2n}$, $\hat{P}_{ij}^{\tilde{a}} \in R^{4n \times 4n}$, $\hat{P}_{ik}^{\tilde{b}} \in R^{6n \times 6n}$, and $\hat{P}_{lj}^{\tilde{c}} \in R^{6n \times 6n}$, and the matrices $\tilde{A} \in R^{n \times n}$,

$\tilde{B} \in R^{n \times p}$, and $\tilde{C} \in R^{q \times n}$ such that

$$\begin{bmatrix} -\hat{P} & 0 & A_{cl} & B_{cl} \\ * & -I & C_{cl} & 0 \\ * & * & -P & 0 \\ * & * & * & -\gamma^2 I \end{bmatrix} < 0 \tag{11.17}$$

$$\begin{bmatrix} -\hat{P}_{ij}^{\tilde{a}} & 0 & A_{ij}^{\tilde{a}} & B_{ij}^{\tilde{a}} \\ * & -I & C_{ij}^{\tilde{a}} & 0 \\ * & * & -P_{ij}^{\tilde{a}} & 0 \\ * & * & * & -\beta^2 I \end{bmatrix} < 0 \tag{11.18}$$

$$\begin{bmatrix} -\hat{P}_{ik}^{\tilde{b}} & 0 & A_{ik}^{\tilde{b}} & B_{ik}^{\tilde{b}} \\ * & -I & C_{ik}^{\tilde{b}} & 0 \\ * & * & -P_{ik}^{\tilde{b}} & 0 \\ * & * & * & -\beta^2 I \end{bmatrix} < 0 \tag{11.19}$$

$$\begin{bmatrix} -\hat{P}_{lj}^{\tilde{c}} & 0 & A_{lj}^{\tilde{c}} & B_{lj}^{\tilde{c}} \\ * & -I & C_{lj}^{\tilde{c}} & 0 \\ * & * & -P_{lj}^{\tilde{c}} & 0 \\ * & * & * & -\beta^2 I \end{bmatrix} < 0 \tag{11.20}$$

$$\hat{P}P = I, \hat{P}_{ij}^{\tilde{a}} P_{ij}^{\tilde{a}} = I, \hat{P}_{ik}^{\tilde{b}} P_{ik}^{\tilde{b}} = I, \hat{P}_{lj}^{\tilde{c}} P_{lj}^{\tilde{c}} = I \tag{11.21}$$

hold for $i, j = 1, \cdots, n; k = 1, \cdots, p; l = 1, \cdots, q$. In addition, A_{cl}, B_{cl}, and C_{cl} are defined in (11.3), and where $A_{ij}^{\tilde{a}}, B_{ij}^{\tilde{a}}, C_{ij}^{\tilde{a}}, A_{ik}^{\tilde{b}}, B_{ik}^{\tilde{b}}, C_{ik}^{\tilde{b}}, A_{lj}^{\tilde{c}}, B_{lj}^{\tilde{c}}$, and $C_{lj}^{\tilde{c}}$ are defined in (11.5) with

$$\begin{cases} R_{ij}^{\tilde{a}}, R_{ik}^{\tilde{b}}, \text{ and } R_{lj}^{\tilde{c}} \text{ being defined in (11.6) (multiplicative case)}, \\ R_{ij}^{\tilde{a}}, R_{ik}^{\tilde{b}}, \text{ and } R_{lj}^{\tilde{c}} \text{ being defined in (11.7) (additive case)}. \end{cases}$$

Then, if there exist solutions of these inequalities, the controller's matrices $\tilde{A}$, $\tilde{B}$, and $\tilde{C}$ can be obtained immediately.

Proof 11.1 *Sufficiency: Suppose that there exist symmetric positive-definite matrices $\hat{P}$, P, $\hat{P}_{ij}^{\tilde{a}}$, $P_{ij}^{\tilde{a}}$, $\hat{P}_{ik}^{\tilde{b}}$, $P_{ik}^{\tilde{b}}$, $\hat{P}_{lj}^{\tilde{c}}$, and $P_{lj}^{\tilde{c}}$ with constraints in (11.21) such that (11.17), (11.18), (11.19), and (11.20) hold.*
Substituting $\hat{P}P = I$ into (11.17), one has

$$\begin{bmatrix} -(P)^{-1} & 0 & A_{cl} & B_{cl} \\ * & -I & C_{cl} & 0 \\ * & * & -P & 0 \\ * & * & * & -\gamma^2 I \end{bmatrix} < 0. \tag{11.22}$$

Then, performing a congruence transformation to (11.22) by $diag\{P, I, I, I\}$ yields (11.41). At the same time, the equivalence between (11.18) and (11.13), (11.19) and (11.14), and (11.20) and (11.15) can be proved in the same way. Therefore, by Lemma 11.2, the closed-loop system (11.3) is asymptotically stable and the constraints in (11.16) are satisfied simultaneously.

Necessity: Suppose that the closed-loop system (11.3) is asymptotically stable and the constraints in (11.16) are satisfied simultaneously. Then, it follows from Lemma 11.2 that there exist symmetric positive-definite matrices P, $P_{ij}^{\tilde{a}}$, $P_{ik}^{\tilde{b}}$, and $P_{lj}^{\tilde{c}}$ such that (11.41), (11.13), (11.14), and (11.15) hold.

Performing a congruence transformation to (11.41) by $diag\{P^{-1}, I, I, I\}$, one obtains (11.22). Now, let us set $\hat{P} = P^{-1}$. Then, (11.17) can be immediately obtained from the inequality (11.22). Furthermore, we can also prove that the inequalities (11.13), (11.14), and (11.15) are equivalent to the conditions in Theorem 11.1. This ends the proof.

It is noted that the conditions in Theorem 11.1 are not all LMIs due to the additional matrix equation constraints in (11.21). Several approaches have been proposed to solve such non-convex feasibility problems, among which the cone complementarity linearization method (CCLM) [32] is the most commonly used one. Very recently, the sequential linear programming matrix method (SLPMM) was also proposed in Leibfritz [85]. In fact, the SLPMM algorithm may be interpreted as an improved version of the CCLM algorithm. As indicated in Leibfritz [85], the SLPMM algorithm is superior to the CCLM algorithm in that it always generates a sequence of iterates with strictly decreasing objective function values, and is globally convergent. In this chapter, we will adopt the SLPMM algorithm to solve the non-convex feasibility problem formulated above.

On the other hand, the coupling constraints in (11.21) can be weakened to the following well-known semidefinite programming (SDP) relaxation:

$$\begin{bmatrix} \hat{P} & I \\ I & P \end{bmatrix} \geq 0, \begin{bmatrix} \hat{P}_{ij}^{\tilde{a}} & I \\ I & P_{ij}^{\tilde{a}} \end{bmatrix} \geq 0, \begin{bmatrix} \hat{P}_{ik}^{\tilde{b}} & I \\ I & P_{ik}^{\tilde{b}} \end{bmatrix} \geq 0, \begin{bmatrix} \hat{P}_{lj}^{\tilde{c}} & I \\ I & P_{lj}^{\tilde{c}} \end{bmatrix} \geq 0,$$
$$(11.23)$$

and

$$trace(\hat{P}P) \geq 2n, trace(\hat{P}_{ij}^{\tilde{a}}P_{ij}^{\tilde{a}}) \geq 4n, trace(\hat{P}_{ik}^{\tilde{b}}P_{ik}^{\tilde{b}}) \geq 6n, trace(\hat{P}_{lj}^{\tilde{c}}P_{lj}^{\tilde{c}}) \geq 6n,$$

where the equations

$$trace(\hat{P}P) = 2n, trace(\hat{P}_{ij}^{\tilde{a}}P_{ij}^{\tilde{a}}) = 4n, trace(\hat{P}_{ik}^{\tilde{b}}P_{ik}^{\tilde{b}}) = 6n, trace(\hat{P}_{lj}^{\tilde{c}}P_{lj}^{\tilde{c}}) = 6n,$$

hold if and only if the equations in (11.21) hold. Thus, we further need to solve the following problem:

$$\min \quad trace(\hat{P}P) + \sum_{i=1}^{n}\sum_{j=1}^{n} trace(\hat{P}_{ij}^{\tilde{a}}P_{ij}^{\tilde{a}}) + \sum_{i=1}^{n}\sum_{k=1}^{p} trace(\hat{P}_{ik}^{\tilde{b}}P_{ik}^{\tilde{b}})$$
$$+ \sum_{l=1}^{q}\sum_{j=1}^{n} trace(\hat{P}_{lj}^{\tilde{c}}P_{lj}^{\tilde{c}})$$

subject to (11.23). Then, the constraints in (11.23) is a closed convex set and thus the SLPMM algorithm can be applied to find feasible solutions of Theorem 11.1, and the detailed SLPMM algorithm is presented as follows.

Algorithm 11.1

Step 1. *Fix the maximum number of iterations N, error bound ϵ, and the prescribed H_∞ performance γ beforehand.*

Step 2. *Obtain the initial solutions $(\hat{P}^{(\kappa)}, P^{(\kappa)}, \hat{P}_{ij}^{\tilde{a}(\kappa)}, P_{ij}^{\tilde{a}(\kappa)}, \hat{P}_{ik}^{\tilde{b}(\kappa)}, P_{ik}^{\tilde{b}(\kappa)}, \hat{P}_{lj}^{\tilde{c}(\kappa)}, P_{lj}^{\tilde{c}(\kappa)}, \beta^{(\kappa)}, \kappa = 0)$ satisfying the LMIs (11.17), (11.18), (11.19), (11.20), and (11.23). If there are none, exit.*

Step 3. *Solve the minimization problem*

$$\min_{s.t.(11.17),(11.18),(11.19),(11.20)\, and\,(11.23)} \varpi + \beta$$

for the matrix variables $\hat{P}$, P, $\hat{P}_{ij}^{\tilde{a}}$, $P_{ij}^{\tilde{a}}$, $\hat{P}_{ik}^{\tilde{b}}$, $P_{ik}^{\tilde{b}}$, $\hat{P}_{lj}^{\tilde{c}}$, $P_{lj}^{\tilde{c}}$, $\tilde{A}$, $\tilde{B}$, $\tilde{C}$, and the positive scalar β, where $\varpi = trace(\hat{P}P^{(\kappa)} + \hat{P}^{(\kappa)}P) + \sum_{i=1}^{n}\sum_{j=1}^{n} trace(\hat{P}_{ij}^{\tilde{a}}P_{ij}^{\tilde{a}(\kappa)} + \hat{P}_{ij}^{\tilde{a}(\kappa)}P_{ij}^{\tilde{a}}) + \sum_{i=1}^{n}\sum_{k=1}^{p} trace(\hat{P}_{ik}^{\tilde{b}}P_{ik}^{\tilde{b}(\kappa)} + \hat{P}_{ik}^{\tilde{b}(\kappa)}P_{ik}^{\tilde{b}}) + \sum_{l=1}^{q}\sum_{j=1}^{n} trace(\hat{P}_{lj}^{\tilde{c}}P_{lj}^{\tilde{c}(\kappa)} + \hat{P}_{lj}^{\tilde{c}(\kappa)}P_{lj}^{\tilde{c}})$ for $i,j = 1, \cdots, n; k = 1, \cdots, p; l = 1, \cdots, q$.

Step 4. *If*

$$|\varpi - 4n(2n^2 + 3np + 3qn + 1) + \beta^{(\kappa)} - \beta| \leq \epsilon$$

holds, then the controller gains of (11.2) can be obtained immediately. Exit.

Step 5. *If $\kappa > N$, exit.*

Step 6. *Calculate $\rho^* \in [0, 1]$ by solving*

$$\min_{s.t.\rho\in[0,1]} \varpi_1$$

where $\varpi_1 = trace((\hat{P}^{(\kappa)} + \rho(\hat{P} - \hat{P}^{(\kappa)}))(P^{(\kappa)} + \rho(P - P^{(\kappa)}))) + \sum_{i=1}^{n}\sum_{j=1}^{n} trace((\hat{P}_{ij}^{\tilde{a}(\kappa)} + \rho(\hat{P}_{ij}^{\tilde{a}} - \hat{P}_{ij}^{\tilde{a}(\kappa)}))(P_{ij}^{\tilde{a}(\kappa)} + \rho(P_{ij}^{\tilde{a}} - P_{ij}^{\tilde{a}(\kappa)})))$

$+ \sum_{i=1}^{n}\sum_{k=1}^{p} trace((\hat{P}_{ik}^{\tilde{b}(\kappa)} + \rho(\hat{P}_{ik}^{\tilde{b}} - \hat{P}_{ik}^{\tilde{b}(\kappa)}))(P_{ik}^{\tilde{b}(\kappa)} + \rho(P_{ik}^{\tilde{b}} - P_{ik}^{\tilde{b}(\kappa)})))$

$+ \sum_{l=1}^{q}\sum_{j=1}^{n} trace((\hat{P}_{lj}^{\tilde{c}(\kappa)} + \rho(\hat{P}_{lj}^{\tilde{c}} - \hat{P}_{lj}^{\tilde{c}(\kappa)}))(P_{lj}^{\tilde{c}(\kappa)} + \rho(P_{lj}^{\tilde{c}} - P_{lj}^{\tilde{c}(\kappa)})))$

$+ (\beta^{(\kappa)} + \rho(\beta - \beta^{(\kappa)})).$

Step 7. *Set*

$$
\begin{aligned}
\hat{P}^{(\kappa)} &= \hat{P}^{(\kappa)} + \rho^*(\hat{P} - \hat{P}^{(\kappa)}), & P^{(\kappa)} &= P^{(\kappa)} + \rho^*(P - P^{(\kappa)}), \\
\hat{P}_{ij}^{\tilde{a}(\kappa)} &= \hat{P}_{ij}^{\tilde{a}(\kappa)} + \rho^*(\hat{P}_{ij}^{\tilde{a}} - \hat{P}_{ij}^{\tilde{a}(\kappa)}), & P_{ij}^{\tilde{a}(\kappa)} &= P_{ij}^{\tilde{a}(\kappa)} + \rho^*(P_{ij}^{\tilde{a}} - P_{ij}^{\tilde{a}(\kappa)}), \\
\hat{P}_{ik}^{\tilde{b}(\kappa)} &= \hat{P}_{ik}^{\tilde{b}(\kappa)} + \rho^*(\hat{P}_{ik}^{\tilde{b}} - \hat{P}_{ik}^{\tilde{b}(\kappa)}), & P_{ik}^{\tilde{b}(\kappa)} &= P_{ik}^{\tilde{b}(\kappa)} + \rho^*(P_{ik}^{\tilde{b}} - P_{ik}^{\tilde{b}(\kappa)}), \\
\hat{P}_{lj}^{\tilde{c}(\kappa)} &= \hat{P}_{lj}^{\tilde{c}(\kappa)} + \rho^*(\hat{P}_{lj}^{\tilde{c}} - \hat{P}_{lj}^{\tilde{c}(\kappa)}), & P_{lj}^{\tilde{c}(\kappa)} &= P_{lj}^{\tilde{c}(\kappa)} + \rho^*(P_{lj}^{\tilde{c}} - P_{lj}^{\tilde{c}(\kappa)}) \\
\beta^{(\kappa)} &= \beta^{(\kappa)} + \rho^*(\beta - \beta^{(\kappa)}),
\end{aligned}
$$

and $\kappa = \kappa + 1$, *go to Step 3.*

Obviously, the second step of the algorithm and every Step 2 are simple LMI problems. Similar to El Ghaoui, Oustry, and AitRami [32] and Jiang et al. [72], the sequence ϖ is bounded by $4n(2n^2 + 3np + 3qn + 1)$ and decreasing. Thus, the sequence ϖ converges to some value $\varpi_{opt} \geq 4n(2n^2 + 3np + 3qn + 1)$. However, Algorithm 11.1 will act on a module of very high dimension, which may make the search for solutions very difficult depending on the numerical example such as the numerical example proposed in this chapter. To overcome this difficulty, we will propose a new initialization step for obtaining an initial solution because an iterative algorithm depends critically on the initialization step. Obviously, the initial solution should be insensitive to the variations of the controller coefficients. Then, instead of using initialization step as Step 2 of Algorithm 11.1, we propose a new initial solution of the SLPMM algorithm by using the non-fragile H_∞ controller design method.

11.3.2 Step 2: Non-Fragile H_∞ Controller Design with Interval-Bounded Controller Coefficient Variations

In the sequel, we present a design method of non-fragile H_∞ controllers based on matrix inequalities, which can be effectively solved by using the SLPMM algorithm involving convex optimization.

Consider a controller with controller coefficient variations described by

$$
\begin{aligned}
\xi(k+1) &= (\tilde{A} + \Delta\tilde{A})\xi(k) + (\tilde{B} + \Delta\tilde{B})y(k) \\
u(k) &= (\tilde{C} + \Delta\tilde{C})\xi(k)
\end{aligned}
\tag{11.24}
$$

where $\tilde{A}$, $\tilde{B}$, and $\tilde{C}$ are controller gain matrices to be designed. $\Delta\tilde{A}$, $\Delta\tilde{B}$, and $\Delta\tilde{C}$ represent the additive interval-bounded controller coefficient variations as in (11.9) and multiplicative interval-bounded controller coefficient variations as follows:

$$
\begin{aligned}
\Delta\tilde{A} &= \tilde{A}diag\{\theta_{11}^{\tilde{a}}, \theta_{22}^{\tilde{a}}, \cdots, \theta_{ii}^{\tilde{a}}, \cdots, \theta_{nn}^{\tilde{a}}\} \\
\Delta\tilde{B} &= \tilde{B}diag\{\theta_{11}^{\tilde{b}}, \theta_{22}^{\tilde{b}}, \cdots, \theta_{kk}^{\tilde{b}}, \cdots, \theta_{pp}^{\tilde{b}}\} \\
\Delta\tilde{C} &= \tilde{C}diag\{\theta_{11}^{\tilde{c}}, \theta_{22}^{\tilde{c}}, \cdots, \theta_{ii}^{\tilde{c}}, \cdots, \theta_{nn}^{\tilde{c}}\}.
\end{aligned}
\tag{11.25}
$$

Remark 11.2 *It should be noted that (11.25) can be viewed as a special form of (11.8), that is, (11.25) may bring some conservativeness.*

Obviously, $\Delta\tilde{A}$, $\Delta\tilde{B}$, and $\Delta\tilde{C}$ can further be described as:

$$\begin{cases} \Delta\tilde{A} = \sum_{i=1}^{n} \theta_{ii}^{\tilde{a}}\tilde{A}e_i e_i^T, \quad \Delta\tilde{B} = \sum_{k=1}^{p} \theta_{kk}^{\tilde{a}}\tilde{B}h_k h_k^T, \quad \Delta\tilde{C} = \sum_{i=1}^{n} \theta_{ii}^{\tilde{c}}\tilde{C}e_i e_i^T \\ \qquad\qquad\qquad\qquad\qquad\qquad\qquad\qquad\qquad\qquad \text{multiplicative case} \\ \Delta\tilde{A} = \sum_{i=1}^{n}\sum_{j=1}^{n} \theta_{ij}^{\tilde{a}}e_i e_j^T, \quad \Delta\tilde{B} = \sum_{i=1}^{n}\sum_{k=1}^{p} \theta_{ik}^{\tilde{b}}e_i h_k^T, \quad \Delta\tilde{C} = \sum_{i=1}^{n}\sum_{l=1}^{q} \theta_{lj}^{\tilde{c}}g_l e_j^T \\ \qquad\qquad\qquad\qquad\qquad\qquad\qquad\qquad\qquad\qquad \text{additive case.} \end{cases}$$

In order to use the techniques developed in Yang and Wang [147,148] for non-fragile filter design with norm-bounded uncertainty, the controller coefficient variations can further be rewritten as

• Multiplicative case

$$\Delta\tilde{A} = \tilde{A}H^{\tilde{a}}F_1(k)E^{\tilde{a}}, \Delta\tilde{B} = \tilde{B}H^{\tilde{b}}F_2(k)E^{\tilde{b}}, \Delta\tilde{C} = \tilde{C}H^{\tilde{c}}F_3(t)E^{\tilde{c}}$$

where

$$\begin{aligned} H^{\tilde{a}} &= \begin{bmatrix} H_1^{\tilde{a}} & H_2^{\tilde{a}} & \cdots & H_\mu^{\tilde{a}} & \cdots & H_n^{\tilde{a}} \end{bmatrix}, E^{\tilde{a}} = H^{\tilde{a}T} \\ H^{\tilde{b}} &= \begin{bmatrix} H_1^{\tilde{b}} & H_2^{\tilde{b}} & \cdots & H_\mu^{\tilde{b}} & \cdots & H_p^{\tilde{b}} \end{bmatrix}, E^{\tilde{b}} = H^{\tilde{b}T} \\ H^{\tilde{c}} &= \begin{bmatrix} H_1^{\tilde{c}} & H_2^{\tilde{c}} & \cdots & H_\mu^{\tilde{c}} & \cdots & H_n^{\tilde{c}} \end{bmatrix}, E^{\tilde{c}} = H^{\tilde{c}T} \end{aligned}$$

and

$$\begin{aligned} F_1(k) &= diag\{\theta_{11}^{\tilde{a}}, \theta_{22}^{\tilde{a}}, \cdots, \theta_{ii}^{\tilde{a}}, \cdots, \theta_{nn}^{\tilde{a}}\} \\ F_2(k) &= diag\{\theta_{11}^{\tilde{b}}, \theta_{22}^{\tilde{b}}, \cdots, \theta_{kk}^{\tilde{b}}, \cdots, \theta_{pp}^{\tilde{b}}\} \\ F_3(k) &= diag\{\theta_{11}^{\tilde{c}}, \theta_{22}^{\tilde{c}}, \cdots, \theta_{ii}^{\tilde{c}}, \cdots, \theta_{nn}^{\tilde{c}}\} \end{aligned}$$

with

$$\begin{aligned} H_\mu^{\tilde{a}} &= e_i, \text{ where } \mu = i, i = 1, \cdots, n \\ H_\mu^{\tilde{b}} &= h_k, \text{ where } \mu = k, k = 1, \cdots, p \\ H_\mu^{\tilde{c}} &= e_i, \text{ where } \mu = i, i = 1, \cdots, n. \end{aligned}$$

• Additive case

$$\Delta\tilde{A} = H^{\tilde{a}}F_1(k)E^{\tilde{a}}, \Delta\tilde{B} = H^{\tilde{b}}F_2(k)E^{\tilde{b}}, \Delta\tilde{C} = H^{\tilde{c}}F_3(t)E^{\tilde{c}}$$

where

$$\begin{aligned} H^{\tilde{a}} &= \begin{bmatrix} H_1^{\tilde{a}} & H_2^{\tilde{a}} & \cdots & H_\mu^{\tilde{a}} & \cdots & H_{n^2}^{\tilde{a}} \end{bmatrix} \\ E^{\tilde{a}T} &= \begin{bmatrix} E_1^{\tilde{a}T} & E_2^{\tilde{a}T} & \cdots & E_\mu^{\tilde{a}T} & \cdots & E_{n^2}^{\tilde{a}T} \end{bmatrix} \\ H^{\tilde{b}} &= \begin{bmatrix} H_1^{\tilde{b}} & H_2^{\tilde{b}} & \cdots & H_\mu^{\tilde{b}} & \cdots & H_{np}^{\tilde{b}} \end{bmatrix} \\ E^{\tilde{b}T} &= \begin{bmatrix} E_1^{\tilde{b}T} & E_2^{\tilde{b}T} & \cdots & E_\mu^{\tilde{b}T} & \cdots & E_{np}^{\tilde{b}T} \end{bmatrix} \\ H^{\tilde{c}} &= \begin{bmatrix} H_1^{\tilde{c}} & H_2^{\tilde{c}} & \cdots & H_\mu^{\tilde{c}} & \cdots & H_{qn}^{\tilde{c}} \end{bmatrix} \\ E^{\tilde{c}T} &= \begin{bmatrix} E_1^{\tilde{c}T} & E_2^{\tilde{c}T} & \cdots & E_\mu^{\tilde{c}T} & \cdots & E_{qn}^{\tilde{c}T} \end{bmatrix} \end{aligned}$$

and

$$
\begin{aligned}
F_1(k) &= diag\{\theta_{11}^{\tilde{a}}, \theta_{12}^{\tilde{a}}, \cdots, \theta_{ij}^{\tilde{a}}, \cdots, \theta_{nn}^{\tilde{a}}\} \\
F_2(k) &= diag\{\theta_{11}^{\tilde{b}}, \theta_{12}^{\tilde{b}}, \cdots, \theta_{ik}^{\tilde{b}}, \cdots, \theta_{np}^{\tilde{b}}\} \\
F_3(k) &= diag\{\theta_{11}^{\tilde{c}}, \theta_{12}^{\tilde{c}}, \cdots, \theta_{lj}^{\tilde{c}}, \cdots, \theta_{qn}^{\tilde{c}}\}
\end{aligned}
$$

with

$$
\begin{aligned}
H_\mu^{\tilde{a}} &= e_i, E_\mu^{\tilde{a}} = e_j^T, \text{ where } \mu = (i-1)n + j, i, j = 1, \cdots, n \\
H_\mu^{\tilde{b}} &= e_i, E_\mu^{\tilde{b}} = h_k^T, \text{ where } \mu = (i-1)p + k, i = 1, \cdots, n, k = 1, \cdots, p \\
H_\mu^{\tilde{c}} &= g_l, E_\mu^{\tilde{c}} = e_j^T, \text{ where } \mu = (l-1)n + j, l = 1, \cdots, q, j = 1, \cdots, n.
\end{aligned}
$$

From (11.10), we can obtain that $F_i(k)^T F_i(k) \leq \theta^2 I (i = 1, 2, 3)$. Then, Lemma 2.12 can be used to solve the controller coefficient variations. However, it should be mentioned that using Lemma 2.12 to solve the interval-bounded controller coefficient variations may bring some conservativeness.

Combining controller (11.24) with system (11.1), the closed-loop system can be obtained as follows:

$$
\left[\begin{array}{c|c} A_{cl} & B_{cl} \\ \hline C_{cl} & 0 \end{array} \right] = \left[\begin{array}{cc|c} A & B_2(\tilde{C} + \Delta\tilde{C}) & B_1 \\ (\tilde{B} + \Delta\tilde{B})C_2 & \tilde{A} + \Delta\tilde{A} & (\tilde{B} + \Delta\tilde{B})D_{21} \\ \hline C_1 & D_{12}(\tilde{C} + \Delta\tilde{C}) & 0 \end{array} \right].
$$

$$(11.26)$$

A sufficient condition for the existence of a feasible solution to the non-fragile H_∞ control problem is obtained in the following theorem.

Theorem 11.2 *Consider the system in (11.1). Let the scalar $\gamma > 0$ be given. Then, admissible non-fragile H_∞ controllers with respect to multiplicative controller coefficient variations and additive controller coefficient variations exist if the following conditions are satisfied, respectively.*
   • *Multiplicative case: For some positive scalars $\lambda_i(1 = 1, 2, 3)$, if there exist symmetric positive-definite matrices $X \in R^{n \times n}$, $Y \in R^{n \times n}$, $\tilde{H} \in R^{2n \times 2n}$, and $\bar{H} \in R^{2n \times 2n}$, and matrices $A_c \in R^{n \times n}$, $B_c \in R^{n \times p}$, and $C_c \in R^{q \times n}$, are such that the following equations hold:*

$$
\left[\begin{array}{ccc} \Sigma & \theta\Sigma_h & \Sigma_e^T \\ * & \Sigma_\lambda & 0 \\ * & * & \Sigma_\lambda \end{array} \right] < 0 \tag{11.27}
$$

$$
\left[\begin{array}{cc} \tilde{H} & diag\{Y, X\} \\ * & \Xi_4 \end{array} \right] \geq 0 \tag{11.28}
$$

$$
\tilde{H}\bar{H} = I \tag{11.29}
$$

where

$$\Sigma_e = \begin{bmatrix} 0 & 0 & 0 & \lambda_1 E^{\tilde{a}} & 0 & 0 \\ 0 & 0 & 0 & \lambda_2 E^{\tilde{b}} C_2 & \lambda_2 E^{\tilde{b}} C_2 & \lambda_2 E^{\tilde{b}} D_{21} \\ 0 & 0 & 0 & \lambda_3 E^{\tilde{c}} & 0 & 0 \end{bmatrix}$$

$$\Sigma = \begin{bmatrix} -\Xi_4 & 0 & \Xi_1 & \Xi_2 \\ * & -I & \Xi_3 & 0 \\ * & * & -\bar{H} & 0 \\ * & * & * & -\gamma^2 I \end{bmatrix}, \Sigma_h = \begin{bmatrix} 0 & 0 & B_2 C_c H^{\tilde{c}} \\ A_c H^{\tilde{a}} & B_c H^{\tilde{b}} & B_2 C_c H^{\tilde{c}} \\ 0 & 0 & D_{12} C_c H^{\tilde{c}} \\ 0 & 0 & 0 \\ 0 & 0 & 0 \\ 0 & 0 & 0 \end{bmatrix}$$

$$\Sigma_\lambda = diag\{-\lambda_1 I, -\lambda_2 I, -\lambda_3 I\}$$

with

$$\Xi_1 = \begin{bmatrix} A + B_2 C_c & A \\ A + B_c C_2 + B_2 C_c + A_c & A + B_c C_2 \end{bmatrix}, \Xi_2 = \begin{bmatrix} B_1 \\ B_1 + B_c D_{21} \end{bmatrix}$$

$$\Xi_3 = \begin{bmatrix} C_1 + D_{12} C_c & C_1 \end{bmatrix}, \Xi_4 = \begin{bmatrix} Y & X \\ X & X \end{bmatrix}.$$

Suppose that $(X, Y, \tilde{H}, \bar{H}, A_c, B_c, C_c)$ is a feasible solution, then the desired non-fragile controller gain matrices can be computed by

$$\tilde{A} = Y(X - Y)^{-1} A_c, \tilde{B} = Y(X - Y)^{-1} B_c, \tilde{C} = C_c. \tag{11.30}$$

• *Additive case: For some positive scalar λ, if there exist symmetric positive-definite matrices $X \in R^{n \times n} Y \in R^{n \times n}$ and S, and matrices $A_c \in R^{n \times n}$, $B_c \in R^{n \times p}$, and $C_c \in R^{q \times n}$ are such that the following equations hold:*

$$\begin{bmatrix} \Sigma & \theta \Sigma_h & \lambda \Sigma_e \\ * & -\lambda I & 0 \\ * & * & -\lambda I \end{bmatrix} < 0 \tag{11.31}$$

$$SX = I \tag{11.32}$$

where

$$\Sigma_e^T = \begin{bmatrix} 0 & 0 & 0 & E^{\tilde{a}} X & 0 & 0 \\ 0 & 0 & 0 & E^{\tilde{b}} C_2 X & E^{\tilde{b}} C_2 & E^{\tilde{b}} D_{21} \\ 0 & 0 & 0 & E^{\tilde{c}} X & 0 & 0 \end{bmatrix}$$

$$\Sigma = \begin{bmatrix} -W & 0 & \Xi_{\tilde{a}} & \Xi_{\tilde{b}} \\ * & -I & \Xi_{\tilde{c}} & 0 \\ * & * & -W & 0 \\ * & * & * & -\gamma^2 I \end{bmatrix}$$

$$\Sigma_h = \begin{bmatrix} 0 & 0 & B_2 H^{\tilde{c}} \\ (S-Y)H^{\tilde{a}} & (S-Y)H^{\tilde{b}} & YB_2 H^{\tilde{c}} \\ 0 & 0 & D_{12}H^{\tilde{c}} \\ 0 & 0 & 0 \\ 0 & 0 & 0 \\ 0 & 0 & 0 \end{bmatrix}$$

with

$$\Xi_{\tilde{a}} = \begin{bmatrix} AX+B_2 H & A \\ L & YA+GC_2 \end{bmatrix}, \Xi_{\tilde{b}} = \begin{bmatrix} B_1 \\ YB_1+GD_{21} \end{bmatrix}$$

$$W = W^T = \begin{bmatrix} X & I \\ I & Y \end{bmatrix} > 0, \Xi_{\tilde{c}} = \begin{bmatrix} C_1 X + D_{12}H & C_1 \end{bmatrix}.$$

Suppose that $(X, Y, S, A_{\tilde{c}}, B_c, C_c)$ is a feasible solution, then the desired non-fragile controller gain matrices can be computed by

$$\begin{bmatrix} \tilde{A} & \tilde{B} \\ \tilde{C} & 0 \end{bmatrix} = \begin{bmatrix} (S-X)^{-1} & -(S-X)^{-1}XB_2 \\ 0 & I \end{bmatrix} \begin{bmatrix} A_c - XAY & B_c \\ C_c & 0 \end{bmatrix}.$$

$$\begin{bmatrix} S & 0 \\ -C_2 & I \end{bmatrix}$$

$$\tag{11.33}$$

Proof 11.2 • *Multiplicative case: From the real bounded lemma, if (11.41) holds, then the closed-loop system (11.26) is asymptotically stable and satisfies the prescribed H_∞ performance.*
    Construct the following nonsingular matrices

$$J_1 = \begin{bmatrix} Y & I \\ V & 0 \end{bmatrix}, J_2 = \begin{bmatrix} I & X^{-1} \\ 0 & U \end{bmatrix}$$

and

$$P = \begin{bmatrix} X^{-1} & U^T \\ U & X_{22} \end{bmatrix}, P^{-1} = \begin{bmatrix} Y & V^T \\ V & Y_{22} \end{bmatrix} \tag{11.34}$$

where $X_{22} = -UYV^{-1}, Y_{22} = -U^{-T}X^{-1}V^T$.
    Because X, Y, U, V are invertible, the above definition about P is applicable. It is obvious that J_1 and J_2 are also invertible. From $PP^{-1} = I$, we infer

$$PJ_1 = J_2$$
$$J_1^T P J_1 = J_2^T J_1 = \begin{bmatrix} Y & I \\ I & X^{-1} \end{bmatrix} > 0.$$

Performing congruence transformations to (11.41) by $J_1 \oplus I \oplus J_1 \oplus I$ yields

$$\begin{bmatrix} -J_1^T P J_1 & 0 & J_1^T P A_{cl} J_1 & J_1^T P B_{cl} \\ * & -I & C_{cl} J_1 & 0 \\ * & * & -J_1^T P J_1 & 0 \\ * & * & * & -\gamma^2 I \end{bmatrix} < 0. \tag{11.35}$$

Denote matrices $T_1 = I \oplus X$ and $T_2 = Y^{-1} \oplus I$.

Then, performing congruence transformations to (11.35) by $T_1 \oplus I \oplus T_2 \oplus I$, we can obtain

$$\begin{bmatrix} -T_1^T J_1^T P J_1 T_1 & 0 & T_1^T J_1^T P A_{cl} J_1 T_2 & T_1^T J_1^T P B_{cl} \\ * & -I & C_{cl} J_1 T_2 & 0 \\ * & * & -T_2^T J_1^T P J_1 T_2 & 0 \\ * & * & * & -\gamma^2 I \end{bmatrix} < 0. \qquad (11.36)$$

From (11.30), we have $A_c = (X - Y)Y^{-1}\tilde{A}, B_c = (X - Y)Y^{-1}\tilde{B}, C_c = \tilde{C}$.

Let $\bar{H} = T_2^T J_1^T P J_1 T_2$ and $V = Y$, then, combining the closed-loop system (11.26), we obtain

$$T_1^T J_1^T P A_{cl} J_1 T_2 = \begin{bmatrix} \Psi_{m1} & A \\ \Psi_{m2} & A + B_c(I + H^{\tilde{b}} F_2(k)E^{\tilde{b}})C_2 \end{bmatrix}$$

$$T_1^T J_1^T P B_{cl} = \begin{bmatrix} B_1 \\ B_1 + B_c(I + H^{\tilde{b}} F_2(k)E^{\tilde{b}})D_{21} \end{bmatrix}$$

$$C_{cl} J_1 T_2 = \begin{bmatrix} C_1 + D_{12}C_c(I + H^{\tilde{c}} F_3(t)E^{\tilde{c}}) & C_1 \end{bmatrix}$$

where $\Psi_{m1} = A + B_2 C_c(I + H^{\tilde{c}} F_3(t)E^{\tilde{c}}), \Psi_{m2} = A + B_c(I + H^{\tilde{b}} F_2(k)E^{\tilde{b}})C_2 + B_2 C_c(I + H^{\tilde{c}} F_3(t)E^{\tilde{c}}) + A_c(I + H^{\tilde{a}} F_1(k)E^{\tilde{a}})$.

Recalling the definition of T_1 and J_1, we can infer that

$$\Xi_4 = \begin{bmatrix} Y & X \\ X & X \end{bmatrix} = \begin{bmatrix} I & 0 \\ 0 & X \end{bmatrix} \begin{bmatrix} Y & I \\ I & X^{-1} \end{bmatrix} \begin{bmatrix} I & 0 \\ 0 & X \end{bmatrix} = T_1^T J_1^T P J_1 T_1. \qquad (11.37)$$

From (11.29), (11.28) can be rewritten as

$$\begin{bmatrix} \bar{H}^{-1} & T_2^{-1} T_1 \\ T_1^T T_2^{-T} & \Xi_4 \end{bmatrix} \geq 0. \qquad (11.38)$$

Using the Schur complement lemma, (11.38) is equivalent to

$$\Xi_4 - T_1^T T_2^{-T} \bar{H} T_2^{-1} T_1 \geq 0$$

that is,

$$\bar{H} \leq T_2^T T_1^{-T} \Xi_4 T_1^{-1} T_2 = T_2^T J_1^T P J_1 T_2. \qquad (11.39)$$

Then, using the Schur complement lemma and Lemma 2.12, (11.36) is equivalent to (11.27). The first part of the proof is completed.

• *Additive case: Using the spirit of the works [24, 153], let the matrices P and P^{-1} be partitioned as*

$$P = \begin{bmatrix} Y & N \\ N^T & ? \end{bmatrix}, P^{-1} = \begin{bmatrix} X & M \\ M^T & ? \end{bmatrix}$$

where $X = X^T > 0$, $Y = Y^T > 0$, M, and N are invertible such that $MN^T + XY = I$ holds; the notations "?" denote blocks in these matrices with no importance for the derivations to be presented in the sequel.

Define the following invertible matrices

$$T_1 = \begin{bmatrix} X & I \\ M^T & 0 \end{bmatrix}, T_2 = \begin{bmatrix} I & Y \\ 0 & N^T \end{bmatrix}.$$

In order to solve the problem, let $M = X$ and $S = X^{-1}$, which infer that $N = X^{-1} - Y = S - Y$ where $SX = I$. Then, it is easy to obtain (11.31) by similar matrix transformation operations [24, 39]. For brevity, it is omitted here. The proof is completed.

Remark 11.3 *It should be noted that the conditions given in Theorem 11.2 with respect to multiplicative interval-bounded coefficient variations and additive interval-bounded coefficient variations are not all LMI conditions due to (11.29) and (11.32). However, by using the SLPMM algorithm, we can solve this non-convex feasibility problem by formulating it into a sequential optimization problem subject to LMI constraints.*

$$\begin{cases} \underset{s.t. \begin{bmatrix} \tilde{H} & I \\ I & \bar{H} \end{bmatrix} \geq 0, (11.27), (11.28)}{\min} & trace(\tilde{H}\bar{H}) + \gamma^2 \quad \text{Multiplicative Case} \\ \underset{s.t. \begin{bmatrix} S & I \\ I & X \end{bmatrix} \geq 0, (11.31)}{\min} & trace(SX) + \gamma^2 \quad \text{Additive Case.} \end{cases} \tag{11.40}$$

The steps are similar to Algorithm 11.1, although the above algorithm does not have the global convergence property. However, the proposed nonlinear minimization problem is easier to solve than the original non-convex feasibility problem.

Remark 11.4 *It should be mentioned that the constraints $M = X$ and $N = X^{-1} - Y$, which are different from those of De Oliveira, Geromel, and Bernussou [24] and Zhang and Yang [153], are introduced to solve the problem. However, the constraints will introduce conservativeness. In comparison with earlier works of Che and Yang [19] and Che and Wang [17], the number of LMI conditions involved in the design conditions is significantly decreased. At the same time, the proposed procedure is not only simple but also very easy to obtain the designed controller. Moreover, it is noted that the type of multiplicative coefficient variations in (11.25) is just a particular case of (11.8), that is, Theorem 11.2 and Che and Wang [17] do not really solve the problem of designing non-fragile H_∞ dynamic output feedback controllers with respect to multiplicative interval-bounded controller coefficient variations as in (11.8).*

At the same time, the following corollary presents a sufficient condition for the existence of the standard H_∞ dynamic output feedback controller.

Corollary 11.1 *Consider the linear discrete-time system (11.1). Given scalar $\gamma > 0$, the dynamic output feedback control problem is solvable if there exist matrices A_c, B_c, C_c, $X = X^T > 0$, and $Y = Y^T > 0$ such that the following inequality holds:*

$$
\begin{bmatrix}
-W & 0 & \Xi_{\tilde{a}} & \Xi_{\tilde{b}} \\
* & -I & \Xi_{\tilde{c}} & 0 \\
* & * & -W & 0 \\
* & * & * & -\gamma^2 I
\end{bmatrix} < 0
\tag{11.41}
$$

where

$$
\Xi_{\tilde{a}} = \begin{bmatrix} AX + B_2 C_c & A \\ A_c & YA + B_c C_2 \end{bmatrix}, \Xi_{\tilde{b}} = \begin{bmatrix} B_1 \\ YB_1 + B_c D_{21} \end{bmatrix}
$$

$$
W = W^T = \begin{bmatrix} X & I \\ I & Y \end{bmatrix} > 0, \Xi_{\tilde{c}} = \begin{bmatrix} C_1 X + D_{12} C_c & C_1 \end{bmatrix}.
$$

Then, if the above conditions is solvable, an appropriate dynamic output feedback controller can be obtained as follows:

$$
\begin{bmatrix} \tilde{A} & \tilde{B} \\ \tilde{C} & 0 \end{bmatrix} = \begin{bmatrix} N^{-1} & -N^{-1}YB_2 \\ 0 & I \end{bmatrix} \begin{bmatrix} A_c - YAX & B_c \\ C_c & 0 \end{bmatrix} \begin{bmatrix} M^{-T} & 0 \\ -C_2 X M^{-T} & I \end{bmatrix}
\tag{11.42}
$$

where $M \in R^{n \times n}$ and $N \in R^{n \times n}$ are any nonsingular matrices satisfying

$$
MN^T = I - XY.
$$

11.3.3   Summary of the Approach

The search for satisfactory solutions may be very difficult when the SLPMM algorithm acts on a module of very high dimension. Because the initialization step is important for an iterative algorithm, therefore, in order to overcome this difficulty, a new initialization step for obtaining an initial solution is proposed by combining it with the non-fragile controller design method.

Based on the above analysis, instead of using initialization step as Step 2 of Algorithm 11.1, we have the following algorithm to obtain the initial solution of the SLPMM algorithm.

Algorithm 11.2

Step 1. *By solving the optimization problem (11.40), the non-fragile controller gain matrices $\tilde{A}$, $\tilde{B}$, and $\tilde{C}$ can be obtained.*

Step 2. *Obviously, when $\tilde{A}$, $\tilde{B}$, and $\tilde{C}$ are given, the conditions in Lemma 11.2 become LMIs. Then, P, $P_{ij}^{\tilde{a}}$, $P_{ik}^{\tilde{b}}$, and $P_{lj}^{\tilde{c}}$ can be obtained by solving* *the LMI-based conditions. Let* $P^{(0)} = P$, $\hat{P}^{(0)} = P^{-1}$, $P_{ij}^{\tilde{a}(\kappa)} = P_{ij}^{\tilde{a}}$, $\hat{P}_{ij}^{\tilde{a}(\kappa)} =$ $(P_{ij}^{\tilde{a}})^{-1}$, $P_{ik}^{\tilde{b}(\kappa)} = P_{ik}^{\tilde{b}}$, $\hat{P}_{ik}^{\tilde{b}(\kappa)} = (P_{ik}^{\tilde{b}})^{-1}$, $P_{lj}^{\tilde{c}(\kappa)} = P_{lj}^{\tilde{c}}$, *and* $\hat{P}_{lj}^{\tilde{c}(\kappa)} = (P_{lj}^{\tilde{c}})^{-1}$ *as the initial value of Algorithm 11.1.*

Remark 11.5 *There have been a number of successful applications of the iterative algorithms to various controller design problems with the coupling conditions like (11.21) [see 32, 85, 114], and the references therein]. So it can be seen as a practical solution to the coupling conditions like (11.21), although a global optimum might be difficult to find in general due to the non-convexity of (11.21). Therefore, a good initial starting point for the iterative algorithm is very important. Particularly, when Algorithm 11.1 acts on a module of very high dimension, the search for satisfactory solutions may be very difficult. For the above reasons, Algorithm 11.2 is given to obtain the initial solutions, and the convergence speed of the proposed iterative algorithm may be accelerated. Then, the satisfactory solutions can be obtained by combining Algorithm 11.1 and Algorithm 11.2 and implementations can be done by using MATLAB YALMIP Toolbox [97]. Later in Section 4 we will illustrate, using a numerical example, that the above algorithm can provide satisfactory results.*

11.3.4 Insensitive H_∞ Control with Multiplicative Controller Coefficient Variations

First, the following *coefficient sensitivity measures* based on the average of the sensitivity functions are introduced. The coefficient sensitivity measure $M^{\tilde{a}}$ is taken as

$$
\begin{aligned}
M^{\tilde{A}} &= \|\frac{1}{n^2}\sum_{i=1}^{n}\sum_{j=1}^{n}S(T(z,[\tilde{a}_{ij}]_{n\times n},[\tilde{b}_{ik}]_{n\times p},[\tilde{c}_{lj}]_{q\times n}),\tilde{a}_{ij})\|_\infty \\
&= \|\frac{1}{n^2}\sum_{i=1}^{n}\sum_{j=1}^{n}C_{cl}(zI-A_{cl})^{-1}N_{ij}^{\tilde{a}}(zI-A_{cl})^{-1}B_{cl}\|_\infty \\
&= \|C_{cl}(zI-A_{cl})^{-1}N^{\tilde{a}}(zI-A_{cl})^{-1}B_{cl}\|_\infty \\
&= \|C^{\tilde{a}}(zI-A^{\tilde{a}})^{-1}B^{\tilde{a}}\|_\infty \\
M^{\tilde{B}} &= \|\frac{1}{np}\sum_{i=1}^{n}\sum_{k=1}^{p}S(T(z,[\tilde{a}_{ij}]_{n\times n},[\tilde{b}_{ik}]_{n\times p},[\tilde{c}_{lj}]_{q\times n}),\tilde{b}_{ik})\|_\infty \\
&= \|C^{\tilde{b}}(zI-A^{\tilde{b}})^{-1}B^{\tilde{b}}\|_\infty \\
M^{\tilde{C}} &= \|\frac{1}{qn}\sum_{l=1}^{q}\sum_{j=1}^{n}S(T(z,[\tilde{a}_{ij}]_{n\times n},[\tilde{b}_{ik}]_{n\times p},[\tilde{c}_{lj}]_{q\times n}),\tilde{c}_{lj})\|_\infty \\
&= \|C^{\tilde{c}}(zI-A^{\tilde{c}})^{-1}B^{\tilde{c}}\|_\infty
\end{aligned}
$$

where

$$A^{\tilde{a}} = \begin{bmatrix} A_{cl} & N^{\tilde{a}} \\ 0 & A_{cl} \end{bmatrix}, A^{\tilde{b}} = \begin{bmatrix} A_{cl} & N^{\tilde{b}} \\ 0 & A_{cl} \end{bmatrix} \oplus A_{cl}, A^{\tilde{c}} = \begin{bmatrix} A_{cl} & N^{\tilde{c}} \\ 0 & A_{cl} \end{bmatrix} \oplus A_{cl}$$

$$B^{\tilde{a}} = \begin{bmatrix} 0 & B_{cl}^T \end{bmatrix}^T, B^{\tilde{b}} = \begin{bmatrix} 0 & B_{cl}^T & \bar{N}_b^T \end{bmatrix}^T B^{\tilde{c}} = \begin{bmatrix} B_{cl}^T & 0 & B_{cl}^T \end{bmatrix}^T$$

$$C^{\tilde{a}} = \begin{bmatrix} C_{cl} & 0 \end{bmatrix}, C^{\tilde{b}} = \begin{bmatrix} C_{cl} & 0 & C_{cl} \end{bmatrix}, C^{\tilde{c}} = \begin{bmatrix} \bar{N}_c & C_{cl} & 0 \end{bmatrix} \quad (11.43)$$

$$N^{\tilde{a}} = \begin{bmatrix} 0 & 0 \\ 0 & \frac{1}{n^2}R^{\tilde{a}} \end{bmatrix}, N^{\tilde{b}} = \begin{bmatrix} 0 & 0 \\ \frac{1}{np}R^{\tilde{b}}C_2 & 0 \end{bmatrix}, \bar{N}_b = \begin{bmatrix} 0 \\ \frac{1}{np}R^{\tilde{b}}D_{21} \end{bmatrix}$$

$$N^{\tilde{c}} = \begin{bmatrix} 0 & \frac{1}{qn}B_2R^{\tilde{c}} \\ 0 & 0 \end{bmatrix}, \bar{N}_c = \begin{bmatrix} 0 & \frac{1}{qn}D_{12}R^{\tilde{c}} \end{bmatrix}$$

with

$$R^{\tilde{a}} = \tilde{A}, R^{\tilde{b}} = \tilde{B}, R^{\tilde{c}} = \tilde{C} \quad (11.44)$$

for the multiplicative form.

Compared with the coefficient sensitivity measures defined in (11.11), the advantage of the new measures is that the matrices $R^{\tilde{c}}, R^{\tilde{b}}$, and $R^{\tilde{c}}$ are compatible with the controller gain matrices. The feature is crucial in dealing with the control problem with respect to multiplicative coefficient variations.

In addition, Lemma 11.3 is introduced to reduce the conservativeness caused by the product terms between the Lyapunov matrix P and the closed-loop system matrices as in (11.41) by the introduction of an additional slack matrix variable.

Lemma 11.3 *[144] The system with its transfer function as $\|T(z, [\tilde{a}_{ij}]_{n \times n}, [\tilde{b}_{ik}]_{n \times p}, [\tilde{c}_{lj}]_{q \times n})\|_\infty < \gamma$ is said to be asymptotically stable with $\|T_e(z)\|_\infty < \gamma$ if there exist a symmetric matrix $P = P^T > 0$ and a matrix G such that*

$$\begin{bmatrix} P - G - G^T & 0 & G^T A_{cl} & G^T B_{cl} \\ * & -I & C_{cl} & 0 \\ * & * & -P & 0 \\ * & * & * & -\gamma^2 I \end{bmatrix} < 0. \quad (11.45)$$

Then, the following lemma can be obtained based on the above lemma and the definition of the new type of sensitivity measures.

Lemma 11.4 *Given positive scalars $\gamma > 0$ and $\beta > 0$, the insensitive dynamic output feedback H_∞ control problem for the system (11.1) with respect to multiplicative controller coefficient variations is solvable if there exist matrices $0 < P = P^T \in R^{2n \times 2n}$, $0 < P_{\tilde{a}} = P_{\tilde{a}}^T \in R^{4n \times 4n}$, $0 < P_{\tilde{b}} = P_{\tilde{b}}^T \in R^{6n \times 6n}$, $0 < P_{\tilde{c}} = P_{\tilde{c}}^T \in R^{6n \times 6n}$, $G_s \in R^{2n \times 2n}$, $G_{\tilde{a}} \in R^{4n \times 4n}$, $G_{\tilde{b}} \in R^{6n \times 6n}$, and*

$G_{\tilde{c}} \in R^{6n \times 6n}$ such that (11.45) and the following inequalities hold:

$$
M_{af} = \begin{bmatrix} P_{\tilde{a}} - He\{G_{\tilde{a}}\} & 0 & G_{\tilde{a}}^T A^{\tilde{a}} & G_{\tilde{a}}^T B^{\tilde{a}} \\ * & -I & C^{\tilde{a}} & 0 \\ * & * & -P_{\tilde{a}} & 0 \\ * & * & * & -\beta^2 I \end{bmatrix} < 0 \tag{11.46}
$$

$$
M_{bf} = \begin{bmatrix} P_{\tilde{b}} - He\{G_{\tilde{b}}\} & 0 & G_{\tilde{b}}^T A^{\tilde{b}} & G_{\tilde{b}}^T B^{\tilde{b}} \\ * & -I & C^{\tilde{b}} & 0 \\ * & * & -P_{\tilde{b}} & 0 \\ * & * & * & -\beta^2 I \end{bmatrix} < 0 \tag{11.47}
$$

$$
M_{cf} = \begin{bmatrix} P_{\tilde{c}} - He\{G_{\tilde{c}}\} & 0 & G_{\tilde{c}}^T A^{\tilde{c}} & G_{\tilde{c}}^T B^{\tilde{c}} \\ * & -I & C^{\tilde{c}} & 0 \\ * & * & -P_{\tilde{c}} & 0 \\ * & * & * & -\beta^2 I \end{bmatrix} < 0. \tag{11.48}
$$

Then, the closed-loop system is asymptotically stable and satisfies the performance constraints (11.49).

Then, based on the new sensitivity measures and Lemma 11.4, the following theorem is obtained.

Theorem 11.3 *Consider the system in (11.1) and let scalars $\gamma > 0$, $\beta > 0$ be given constants. Then, the closed-loop system (11.3) is asymptotically stable and satisfies the following constraints*

$$
\|T(z, [\tilde{a}_{ij}]_{n \times n}, [\tilde{b}_{ik}]_{n \times p}, [\tilde{c}_{lj}]_{q \times n})\|_\infty < \gamma, M_{\tilde{A}} < \beta, M_{\tilde{B}} < \beta, M_{\tilde{C}} < \beta \tag{11.49}
$$

if there exist matrices $A_c \in R^{n \times n}$, $B_c \in R^{n \times p}$, $C_c \in R^{q \times n}$, $S \in R^{n \times n}$, $0 < X \in R^{n \times n}$, $0 < Y \in R^{n \times n}$, and symmetric positive-definite matrices $0 < \hat{H}_s = \hat{H}_s^T \in R^{2n \times 2n}$, $0 < \hat{H}_{\tilde{a}} = \hat{H}_{\tilde{a}}^T \in R^{4n \times 4n}$, $0 < \hat{H}_{\tilde{b}} = \hat{H}_{\tilde{b}}^T \in R^{6n \times 6n}$, and $0 < \hat{H}_{\tilde{c}} = \hat{H}_{\tilde{c}}^T \in R^{6n \times 6n}$ such that

$$
\begin{bmatrix} \hat{H}_s - \Xi_4 & 0 & \Xi_1 & \Xi_2 \\ * & -I & \Xi_3 & 0 \\ * & * & -\bar{H}_s & 0 \\ * & * & * & -\gamma^2 I \end{bmatrix} < 0 \tag{11.50}
$$

$$
\begin{bmatrix} \hat{H}_{\tilde{a}} - I_2 \otimes \Xi_4 & 0 & \begin{bmatrix} \Xi_1 & \Xi_{af} \\ 0 & \Xi_1 \end{bmatrix} & \begin{bmatrix} 0 \\ \Xi_2 \end{bmatrix} \\ * & -I & \begin{bmatrix} \Xi_3 & 0 \end{bmatrix} & 0 \\ * & * & -\bar{H}_{\tilde{a}} & 0 \\ * & * & * & -\beta^2 I \end{bmatrix} < 0 \tag{11.51}
$$

$$\begin{bmatrix} \hat{H}_b - I_3 \otimes \Xi_4 & 0 & \begin{bmatrix} \Xi_1 & \Xi_{bf} & 0 \\ 0 & \Xi_1 & 0 \\ 0 & 0 & \Xi_1 \end{bmatrix} & \begin{bmatrix} 0 \\ \Xi_2 \\ \Xi_{b1f} \end{bmatrix} \\ * & -I & \begin{bmatrix} \Xi_3 & 0 & \Xi_3 \end{bmatrix} & 0 \\ * & * & -\bar{H}_b & 0 \\ * & * & * & -\beta^2 I \end{bmatrix} < 0 \qquad (11.52)$$

$$\begin{bmatrix} \hat{H}_c - I_3 \otimes \Xi_4 & 0 & \begin{bmatrix} \Xi_1 & \Xi_{cf} & 0 \\ 0 & \Xi_1 & 0 \\ 0 & 0 & \Xi_1 \end{bmatrix} & \begin{bmatrix} \Xi_2 \\ 0 \\ \Xi_2 \end{bmatrix} \\ * & -I & \begin{bmatrix} \Xi_{c1f} & \Xi_3 & 0 \end{bmatrix} & 0 \\ * & * & -\bar{H}_c & 0 \\ * & * & * & -\beta^2 I \end{bmatrix} < 0 \qquad (11.53)$$

$$\begin{bmatrix} \tilde{H}_s & Y \oplus X \\ * & \hat{H}_s \end{bmatrix} \geq 0 \qquad (11.54)$$

$$\begin{bmatrix} \tilde{H}_{\tilde{a}} & I_2 \otimes (Y \oplus X) \\ * & \hat{H}_{\tilde{a}} \end{bmatrix} \geq 0 \qquad (11.55)$$

$$\begin{bmatrix} \tilde{H}_{\tilde{b}} & I_3 \otimes (Y \oplus X) \\ * & \hat{H}_{\tilde{b}} \end{bmatrix} \geq 0 \qquad (11.56)$$

$$\begin{bmatrix} \tilde{H}_{\tilde{c}} & I_3 \otimes (Y \oplus X) \\ * & \hat{H}_{\tilde{c}} \end{bmatrix} \geq 0 \qquad (11.57)$$

$$\tilde{H}_s \bar{H}_s = I, \tilde{H}_{\tilde{a}} \bar{H}_{\tilde{a}} = I, \tilde{H}_{\tilde{b}} \bar{H}_{\tilde{b}} = I, \tilde{H}_{\tilde{c}} \bar{H}_{\tilde{c}} = I \qquad (11.58)$$

where

$$\Xi_1 = \begin{bmatrix} \Upsilon_1 & A \\ \Upsilon_2 & \Upsilon_3 \end{bmatrix}, \Xi_2 = \begin{bmatrix} B_1 \\ \Upsilon_4 \end{bmatrix}, \Xi_3 = \begin{bmatrix} \Upsilon_5 & C_1 \end{bmatrix}, \Xi_{af} = \begin{bmatrix} 0 & 0 \\ \frac{1}{n \times n} A_c & 0 \end{bmatrix}$$

$$\Xi_{bf} = \begin{bmatrix} 0 & 0 \\ \frac{1}{n \times p} B_c C_2 & \frac{1}{n \times p} B_c C_2 \end{bmatrix}, \Xi_{b1f} = \begin{bmatrix} 0 \\ \frac{1}{n \times p} B_c D_{21} \end{bmatrix}$$

$$\Xi_{cf} = \begin{bmatrix} 0 & \frac{1}{q \times n} B_2 C_c \\ 0 & \frac{1}{q \times n} B_2 C_c \end{bmatrix}, \Xi_{c1f} = \begin{bmatrix} \frac{1}{q \times n} D_{12} C_c & 0 \end{bmatrix}$$

$$\Xi_4 = \begin{bmatrix} He\{Y\} & \Upsilon_6 \\ * & He\{X\} \end{bmatrix}$$

with

$$\Upsilon_1 = A + B_2 C_c, \Upsilon_2 = A + B_c C_2 + B_2 C_c + A_c$$
$$\Upsilon_3 = A + B_c C_2, \Upsilon_4 = B_1 + B_c D_{21}$$
$$\Upsilon_5 = C_1 + D_{12} C_c, \Upsilon_6 = X + S^T + Y^T.$$

Moreover, if there exist solutions of these inequalities, the controller can be given by

$$\begin{bmatrix} \tilde{A} & \tilde{B} \\ \tilde{C} & 0 \end{bmatrix} = \begin{bmatrix} YS^{-1}A_c & YS^{-1}B_c \\ C_c & 0 \end{bmatrix}. \qquad (11.59)$$

Proof 11.3 *Lemma 11.4 shows that the closed-loop system (11.3) is asymptotically stable and satisfies the H_∞ norm performance requirement with the insensitive constraint simultaneously, if there exist positive matrices P_s, P_a, P_b, P_c, and matrices G_s, G_a, G_b, and G_c satisfying (11.41), (11.13), (11.14), and (11.15), respectively. In the following proof, we will show that Theorem 11.3 is sufficient to guarantee that (11.41), (11.13), (11.14), and (11.15) hold.*

Let the matrices G_s, G_a, G_b, and G_c satisfy the following constraint:

$$G_s = G, G_{\tilde{a}} = I_2 \otimes G, G_{\tilde{b}} = I_3 \otimes G, G_{\tilde{c}} = I_3 \otimes G. \tag{11.60}$$

Further, we partition G and G^{-1} as

$$G = \begin{bmatrix} X^{-1} & X_{12} \\ U & X_{22} \end{bmatrix}, G^{-1} = \begin{bmatrix} Y & Y_{12} \\ V & Y_{22} \end{bmatrix} \tag{11.61}$$

where G and G^{-1} have appropriate dimensions, and $X > 0$, $Y > 0$, $X = X^T \in R^{n \times n}$, $Y = Y^T \in R^{n \times n}$, $X_{12} \in R^{n \times n}$, $Y_{12} \in R^{n \times n}$, $Y_{22} \in R^{n \times n}$. We can assume without loss of generality that U and V have full row rank (see Gahinet [39] for details).

Construct the following matrices:

$$J_1 = \begin{bmatrix} Y & I \\ V & 0 \end{bmatrix}, J_2 = \begin{bmatrix} I & X^{-1} \\ 0 & U \end{bmatrix} \tag{11.62}$$

$$\begin{bmatrix} A_c & B_c \\ C_c & 0 \end{bmatrix} = \begin{bmatrix} X^T U^T & 0 \\ 0 & I \end{bmatrix} \begin{bmatrix} \tilde{A} & \tilde{B} \\ \tilde{C} & 0 \end{bmatrix} \begin{bmatrix} VY^{-1} & 0 \\ 0 & I \end{bmatrix}. \tag{11.63}$$

From $GG^{-1} = I$, we infer

$$GJ_1 = J_2 \tag{11.64}$$

$$J_1^T G^T J_1 = J_2^T J_1 = \begin{bmatrix} Y & I \\ U^T V + X^{-T} Y & X^{-T} \end{bmatrix} > 0. \tag{11.65}$$

For brevity, we only prove that the inequalities (11.51) and (11.55) with $\tilde{H}_a \bar{H}_a = I$ are equivalent to (11.13) with constraint $G_a = I_2 \otimes G$. Applying the congruence transformations to (11.13) by $(I_2 \otimes J_1) \oplus I \oplus (I_2 \otimes J_1) \oplus I$, we have

$$\begin{bmatrix} H_{\tilde{a}} - I_2 \otimes (J_1^T G J_1) & 0 & \Psi_A & \begin{bmatrix} 0 \\ J_1^T G^T B_{cl} \end{bmatrix} \\ * & -I & \begin{bmatrix} C_{cl} J_1 & 0 \end{bmatrix} & 0 \\ * & * & -H_{\tilde{a}} & 0 \\ * & * & * & -\beta^2 I \end{bmatrix} < 0 \tag{11.66}$$

where $\Psi_A = \begin{bmatrix} J_1^T G^T A_{cl} J_1 & J_1^T G^T N^{\tilde{a}} J_1 \\ 0 & J_1^T G^T A_{cl} J_1 \end{bmatrix}$, $H_{\tilde{a}} = (I_2 \otimes J_1^T) P_{\tilde{a}} (I_2 \otimes J_1)$.

Following from (11.3) and (11.5), we obtain

$$
G^T J_1^T A_{cl} J_1 = \begin{bmatrix} \Upsilon_{11} & A \\ \Upsilon_{12} & \Upsilon_{13} \end{bmatrix}, G^T J_1^T B_{cl} = \begin{bmatrix} B_1 \\ \Upsilon_{14} \end{bmatrix}, C_{cl} J_1 = \begin{bmatrix} \Upsilon_{15} & C_1 \end{bmatrix}
$$

$$
J_1^T G^T N^{\tilde{a}} J_1 = \begin{bmatrix} 0 & 0 \\ U^T R^{\tilde{a}} V & 0 \end{bmatrix}, He\{J_1^T G J_1\} = \begin{bmatrix} He\{Y\} & \Upsilon_{16} \\ * & He\{X^{-1}\} \end{bmatrix}
$$

where

$$
\Upsilon_{11} = AY + B_2 \tilde{C} V, \Upsilon_{12} = X^{-T} AY + U^T \tilde{B} C_2 Y + X^{-T} B_2 \tilde{C} V + U^T \tilde{A} V
$$
$$
\Upsilon_{13} = X^{-T} A + U^T \tilde{B} C_2, \Upsilon_{14} = X^{-T} B_1 + U^T \tilde{B} D_{21}
$$
$$
\Upsilon_{15} = C_1 Y + D_{12} \tilde{C} V, \Upsilon_{16} = I + V^T U + Y^T X^{-1}.
$$

Then, (11.66) can be rewritten as follows:

$$
\begin{bmatrix} H_{\tilde{a}} - diag\{\Xi_{14}, \Xi_{14}\} & 0 & \begin{bmatrix} \Xi_{11} & \Xi_{a1} \\ 0 & \Xi_{11} \end{bmatrix} & \begin{bmatrix} 0 \\ \Xi_{12} \end{bmatrix} \\ * & -I & \begin{bmatrix} \Xi_{13} & 0 \end{bmatrix} & 0 \\ * & * & -H_a & 0 \\ * & * & * & -\beta^2 I \end{bmatrix} < 0 \qquad (11.67)
$$

where

$$
\Xi_{11} = \begin{bmatrix} \Upsilon_{11} & A \\ \Upsilon_{12} & \Upsilon_{13} \end{bmatrix}, \Xi_{12} = \begin{bmatrix} B_1 \\ \Upsilon_{14} \end{bmatrix}, \Xi_{13} = \begin{bmatrix} \Upsilon_{15} & C_1 \end{bmatrix}
$$
$$
\Xi_{a1} = \begin{bmatrix} 0 & 0 \\ U^T R^{\tilde{a}} V & 0 \end{bmatrix}, \Xi_{14} = \begin{bmatrix} He\{Y\} & \Upsilon_{16} \\ * & He\{X^{-1}\} \end{bmatrix}.
$$

Denote

$$
T_{a1} = I \oplus X \oplus I \oplus X, T_{a2} = Y^{-1} \oplus I \oplus Y^{-1} \oplus I \qquad (11.68)
$$

Performing a congruence transformation to (11.67) by $T_{a1} \oplus I \oplus T_{a2} \oplus I$, we obtain

$$
\begin{bmatrix} T_{a1}^T H_{\tilde{a}} T_{a1} - diag\{\Xi_{24}, \Xi_{24}\} & 0 & \begin{bmatrix} \Xi_{21} & \Xi_{a2ij} \\ 0 & \Xi_{21} \end{bmatrix} & \begin{bmatrix} 0 \\ \Xi_{22} \end{bmatrix} \\ * & -I & \begin{bmatrix} \Xi_{23} & 0 \end{bmatrix} & 0 \\ * & * & -T_{a2}^T H_{\tilde{a}} T_{a2} & 0 \\ * & * & * & -\beta^2 I \end{bmatrix} < 0
$$
$$(11.69)$$

where

$$
\Xi_{21} = \begin{bmatrix} \Upsilon_{21} & A \\ \Upsilon_{22} & \Upsilon_{23} \end{bmatrix}, \Xi_{22} = \begin{bmatrix} B_1 \\ \Upsilon_{24} \end{bmatrix}, \Xi_{23} = \begin{bmatrix} \Upsilon_{25} & C_1 \end{bmatrix}
$$
$$
\Xi_{a2f} = \begin{bmatrix} 0 & 0 \\ X^T U^T \tilde{A} V Y^{-1} & 0 \end{bmatrix}, \Xi_{24} = \begin{bmatrix} He\{Y\} & \Upsilon_{26} \\ * & He\{X\} \end{bmatrix}
$$

with

$$\Upsilon_{21} = A + B_2\tilde{C}VY^{-1}, \Upsilon_{22} = A + X^TU^T\tilde{B}C_2 + B_2\tilde{C}VY^{-1} + X^TU^T\tilde{A}VY^{-1}$$
$$\Upsilon_{23} = A + X^TU^T\tilde{B}C_2, \Upsilon_{24} = B_1 + X^TU^T\tilde{B}D_{21}, \Upsilon_{25} = C_1 + D_{12}\tilde{C}VY^{-1}$$
$$\Upsilon_{26} = X + V^TUX + Y^T.$$

Let

$$\hat{H}_{\tilde{a}} = \hat{H}_{\tilde{a}}^T = T_{a1}^T H_{\tilde{a}} T_{a1} \tag{11.70}$$

$$\bar{H}_{\tilde{a}} \le T_{a2}^T H_{\tilde{a}} T_{a2} = T_{a2}^T T_{a1}^{-T} \hat{H}_{\tilde{a}} T_{a1}^{-1} T_{a2} \tag{11.71}$$

where (11.71) can be rewritten as

$$\hat{H}_{\tilde{a}} - T_{a1}^T T_{a2}^{-T} \bar{H}_{\tilde{a}} T_{a2}^{-1} T_{a1} \ge 0. \tag{11.72}$$

Using the Schur complement, (11.72) implies that

$$\begin{bmatrix} \bar{H}_{\tilde{a}}^{-1} & T_{a2}^{-1}T_{a1} \\ T_{a1}^T T_{a2}^{-T} & \hat{H}_{\tilde{a}} \end{bmatrix} \ge 0. \tag{11.73}$$

Substituting (11.68) to (11.73), we can obtain (11.55) with $\tilde{H}_a\bar{H}_a = I$.
In addition, (11.69) can be rewritten as

$$\begin{bmatrix} \hat{H}_{\tilde{a}} - diag\{\Xi_{24}, \Xi_{24}\} & 0 & \begin{bmatrix} \Xi_{21} & \Xi_{a2ij} \\ 0 & \Xi_{21} \end{bmatrix} & \begin{bmatrix} 0 \\ \Xi_{22} \end{bmatrix} \\ * & -I & \begin{bmatrix} \Xi_{23} & 0 \end{bmatrix} & 0 \\ * & * & -\bar{H}_{\tilde{a}} & 0 \\ * & * & * & -\beta^2 I \end{bmatrix} < 0. \tag{11.74}$$

In addition, it can be easily argued that there exists a linear transformation on the controller state such that

$$\hat{x}(k) = T\bar{x}(k) \tag{11.75}$$

where $T \in R^{n \times n}$ *is an invertible matrix. Then, a new representation form of the controller (11.2) can be obtained as follows:*

$$\dot{\hat{x}}(k) = T\tilde{A}T^{-1}\hat{x}(k) + T\tilde{B}y(k)$$
$$u(k) = \tilde{C}T^{-1}\hat{x}(k)$$

where $\hat{x}(k) \in R^n$ *is the controller state. Obviously, the linear transformation has no effect on the controller mapping from* $y(k)$ *to* $u(k)$. *Without loss of generality, we may set* $T = VY^{-1} = I$, *and* $S = X^TU^TV$. *Then, (11.63) is equivalent to*

$$\begin{bmatrix} A_c & B_c \\ C_c & 0 \end{bmatrix} = \begin{bmatrix} X^TU^T & 0 \\ 0 & I \end{bmatrix} \begin{bmatrix} T\tilde{A}T^{-1} & T\tilde{B} \\ \tilde{C}T^{-1} & 0 \end{bmatrix} \begin{bmatrix} VY^{-1} & 0 \\ 0 & I \end{bmatrix}$$
$$= \begin{bmatrix} X^TU^TT\tilde{A}T^{-1}VY^{-1} & X^TU^TTB_{Fe} \\ C_{Fe}T^{-1}VY^{-1} & 0 \end{bmatrix} \tag{11.76}$$
$$= \begin{bmatrix} SY^{-1}\tilde{A} & SY^{-1}\tilde{B} \\ \tilde{C} & 0 \end{bmatrix}$$

then, (11.59) can be easily obtained. Note that because of the invertibility of the matrices involved in the above proof process, the above proof process is a reversible process, therefore, (11.51) and (11.55) with $\tilde{H}_a \bar{H}_a = I$ are equivalent to (11.13) with constraint $G_a = I_2 \otimes G$.

Moreover, performing congruence transformations to (11.41) by $J_1 \oplus I \oplus J_1 \oplus I$ and to (11.14) and (11.15) by $(I_3 \otimes J_1) \oplus I \oplus (I_3 \otimes J_1) \oplus I$, respectively, and following similar steps as the proof process of (11.51) and (11.55), the inequalities (11.50), (11.52), (11.53), (11.56), (11.57), and (11.58) are equivalent to (11.41), (11.14), and (11.15) with constraint (11.60), respectively. This completes the proof.

Remark 11.6 *It is obvious that the conditions given in Lemma 11.4 are not LMIs due to the product of the variables G_s, G_a, G_b, and G_c with the controller matrices A_{Ce}, B_{Ce}, and C_{Ce}, respectively. On the other hand, the difficulty lies in that G_s, G_a, G_b, and G_c are interdependent with A_{Ce}, B_{Ce}, and C_{Ce}. Therefore, the constraints in (11.60) are needed.*

Remark 11.7 *Using the similar algorithm as in Algorithm 11.1, the original non-convex feasibility problem formulated in Theorem 11.3 can be converted to the following nonlinear minimization problem involving LMI conditions:*

$$
\begin{cases}
\min \ Trace(\tilde{H}_s \bar{H}_s + \tilde{H}_{\tilde{a}} \bar{H}_{\tilde{a}} + \tilde{H}_{\tilde{b}} \bar{H}_{\tilde{b}} + \tilde{H}_{\tilde{c}} \bar{H}_{\tilde{c}}) \\
s.t.(11.50) - (11.57), \begin{bmatrix} \tilde{H}_s & I \\ I & \bar{H}_s \end{bmatrix} \geq 0 \\
\begin{bmatrix} \tilde{H}_{\tilde{a}} & I \\ I & \bar{H}_{\tilde{a}} \end{bmatrix} \geq 0, \begin{bmatrix} \tilde{H}_{\tilde{b}} & I \\ I & \bar{H}_{\tilde{b}} \end{bmatrix} \geq 0, \begin{bmatrix} \tilde{H}_{\tilde{c}} & I \\ I & \bar{H}_{\tilde{c}} \end{bmatrix} \geq 0.
\end{cases}
\tag{11.77}
$$

If the solution of the above minimization problem is $18n$, that is, $Trace(\tilde{H}_s \bar{H}_s + \tilde{H}_{\tilde{a}} \bar{H}_{\tilde{a}} + \tilde{H}_{\tilde{b}} \bar{H}_{\tilde{b}} + \tilde{H}_{\tilde{c}} \bar{H}_{\tilde{c}}) = 18n$, then the conditions in Theorem 11.3 are solvable.

Remark 11.8 *It must be pointed out that, from the theory point of view, the new coefficient sensitivity measures $M^{\tilde{A}}$, $M^{\tilde{B}}$, and $M^{\tilde{C}}$ cannot guarantee the all concerning individual sensitivity. However, the concerning individual sensitivity can practically be reduced in many examples by using these new sensitivity measures, see the examples in Section 11.4.*

Remark 11.9 *It is worth mentioning that the techniques used in the derivation of Theorem 11.1 are quite different from those in Theorem 11.3, manifested by the following aspects:*

(a) *It is difficult to use the techniques used in the derivation of Theorem 11.3 to obtain a solution of the insensitive H_∞ control problem with respect to additive controller coefficient variations, while this problem can be resolved well by using Theorem 11.1.*

(b) Theorem 11.1 presents a sufficient and necessary condition for the insensitive H_∞ control problem, while Theorem 11.3 presents a sufficient condition for this issue. On the other hand, we have imposed certain structural restrictions on the slack matrices as (11.60) and the partitioned matrices G and G^{-1} as (11.61) in the proof of Theorem 11.3, while these restrictions are not needed in Theorem 11.1. Obviously, Theorem 11.1 is less conservative than Theorem 11.3.

(c) From the theory point of view, the conditions in Theorem 11.1 can guarantee all the concerning individual sensitivity. In contrast, the conditions in Theorem 11.3 cannot guarantee it, but may reduce the individual sensitivity for many systems practically.

(d) It is widely known that finding the feasible solution would be computationally prohibitive for an iterative algorithm due to the high dimension of the search space. Therefore, despite the above four advantages, Theorem 11.1 ($\varpi_{opt} \geq 4n(2n^2 + 3np + 3qn + 1)$, i.e., high dimension) may be more difficult to find a feasible controller than Theorem 11.3 ($\varpi_{opt} \geq 4n$). In addition, the number of LMI conditions involved in Theorem 11.1 is more numerous than those involved in Theorem 11.3, which means that Theorem 11.1 needs more CPU time.

Remark 11.10 *Obviously, the problem of designing insensitive controllers is equivalent to minimizing β by using Algorithm 11.1 (or Algorithm 11.2). However, the iterative algorithm may introduce some conservativeness. The actual measures in (11.11) can be computed directly when the controller's matrices $\tilde{A}$, $\tilde{B}$, and $\tilde{C}$ are given, where the largest one is taken as the actual sensitivity measure of the designed controller.*

11.4 Example

In this section, a numerical example is presented to illustrate the effectiveness of the proposed insensitive output-feedback H_∞ controller design methods. By the way, in this section, SC denotes the standard H_∞ controller, IC denotes the insensitive H_∞ controller, and NC denotes the non-fragile H_∞ controller for brevity.

Example 11.1 *Consider a linear discrete-time system of the form (11.1) with the following matrices:*

$$
A = \begin{bmatrix} 1 & -1 & 0.5 \\ 1 & -0.5 & 0.4 \\ 1 & -1 & 1 \end{bmatrix}, B_1 = \begin{bmatrix} -0.05 & 0 \\ 0 & -0.08 \\ 0.04 & -0.1 \end{bmatrix}, B_2 = \begin{bmatrix} 0.5 \\ 0.5 \\ 0 \end{bmatrix}
$$

$$C_1 = \begin{bmatrix} 1 & -1 & 0 \\ 0.9 & 0.4 & 0.2 \end{bmatrix}, D_{12} = \begin{bmatrix} -0.01 \\ 0.5 \end{bmatrix}$$

$$C_2 = \begin{bmatrix} -0.8 & 2 & -1 \end{bmatrix}, D_{21} = \begin{bmatrix} -0.5 & 0.8 \end{bmatrix}.$$

It should be noted that we always let $N = 100$, $\epsilon = 0.005$, and $\theta = 0.05$ in this section.

Solving the LMI given in Corollary 11.1, we obtain the standard controller with the optimal H_∞ performance index $\gamma = 0.1249$ as follows:

$$\tilde{A} = \begin{bmatrix} -0.1256 & 0.9249 & -0.2478 \\ 0.6367 & 0.4022 & 0.1523 \\ 0.9354 & 1.5002 & 0.0270 \end{bmatrix}, \tilde{B} = \begin{bmatrix} -0.0307 \\ 0.0339 \\ 0.2773 \end{bmatrix}$$

$$\tilde{C} = \begin{bmatrix} 2.5748 & -3.0726 & 0.8853 \end{bmatrix}.$$

Multiplicative Coefficient Variation Case

In the following, the case of multiplicative coefficient variations is considered. It deserves to be mentioned that Theorem 11.1 cannot be used to obtain the insensitive controller by using Algorithm 11.1 directly in this numerical example. Then, we combine Algorithm 11.1 and Algorithm 11.2 to obtain the insensitive H_∞ controller below.

First, by solving the optimization algorithm (11.40) utilizing the YALMIP toolbox with the SeDuMi solver, after $\kappa = 15$ times of iterations, we obtain the following gain matrices of the non-fragile controller:

$$\tilde{A} = \begin{bmatrix} 0.2633 & -0.8500 & 0.4472 \\ 0.4043 & -0.5436 & 0.3163 \\ 0.4003 & -0.4297 & 0.6615 \end{bmatrix}, \tilde{B} = \begin{bmatrix} 0.0180 \\ -0.1904 \\ -0.1999 \end{bmatrix}$$

$$\tilde{C} = \begin{bmatrix} -0.6183 & -0.3201 & 0.1226 \end{bmatrix}$$

and the obtained minimum guaranteed H_∞ performance is $\gamma = 0.2107$. Furthermore, by using the definition of sensitivity measure in (11.11), we can obtain the sensitivity performance index as $\beta = 0.1686$.

Then, combining Algorithm 11.1 and Algorithm 11.2 with $\gamma = 0.18$, after $\kappa = 9$ times of iterations, we can obtain the gain matrices and sensitivity performance index of the insensitive controller with respect to multiplicative controller coefficient variations as follows:

$$\tilde{A} = \begin{bmatrix} 0.5126 & -0.5889 & 0.3188 \\ 0.2885 & 0.0928 & 0.0524 \\ 0.0818 & -0.0606 & 0.3720 \end{bmatrix}, \tilde{B} = \begin{bmatrix} 0.0300 \\ -0.0168 \\ -0.0309 \end{bmatrix}.$$

$$\tilde{C} = \begin{bmatrix} -0.4384 & -0.1329 & 0.2295 \end{bmatrix}, \beta = 0.4277$$

Further, by solving the optimization problem (11.77), after $\kappa = 60$ times of iterations, the gain matrices and the sensitivity performance index of the

TABLE 11.1

Comparison of Performance Indexes for Various Controllers

Index	SC	NC	IC		
	Corollary 11.1	Theorem 11.2	Theorem 11.1		Theorem 11.3
γ	0.1278	0.2107	0.18	0.23	0.23
β	Infeasible	0.1686	0.0382	0.0158	0.0390

insensitive controller can be given as follows:

$$\tilde{A} = \begin{bmatrix} 0.4054 & -1.0229 & 0.4773 \\ 0.4035 & -0.7010 & 0.4460 \\ 0.2773 & -0.4089 & 0.9615 \end{bmatrix}, \tilde{B} = \begin{bmatrix} 0.0220 \\ -0.0543 \\ -0.0571 \end{bmatrix}.$$

$$\tilde{C} = \begin{bmatrix} -0.5976 & -0.8846 & 0.1414 \end{bmatrix}, \tilde{\beta} = 0.0503$$

In addition, it is necessary to point out that some conservativeness may be introduced during the design process. Fortunately, the actual sensitivity measures of the insensitive controllers, non-fragile controller, and standard controller, respectively, can be computed by using the method given in Remark 11.10. The results are listed in Table 11.1.

From Table 11.1, we can see that the sensitivity indexes of the insensitive controller and non-fragile controller are much lower than that of the standard controller, which shows the superiority of our proposed method. On the other hand, it is shown that different insensitive controllers can be obtained by the trade-off between the sensitivity index and standard H_∞ performance index. Furthermore, this table also shows that the proposed insensitive controller design method is less conservative than the non-fragile controller design method.

For simulation purposes, we take the disturbance $w^T(k) = \begin{bmatrix} w_1^T(k) & w_2^T(k) \end{bmatrix}$ as follows:

$$w_1(k) = w_2(k) = \begin{cases} -2\cos(k) & 10 \le k \le 20 \\ 0 & otherwise. \end{cases}$$

In addition, the initial conditions of the system and the controller are $x_0 = \begin{bmatrix} 0 & 0 & 0 \end{bmatrix}$ and $\xi_0 = \begin{bmatrix} 0 & 0 & 0 \end{bmatrix}$, respectively.

Figure 11.1 through Figure 11.4 show the regulated output responses of the closed-loop system for $\theta = 0$ and $\theta = 0.05$, respectively. From Figure 11.1 and Figure 11.3, one can see that the three controllers have a similar effect for $\theta = 0$, while when there exist controller coefficient variations, that is, $\theta = 0.05$ is considered, the standard controller is sensitive to the controller coefficient variations and the closed-loop system even becomes unstable. On the other hand, from Figure 11.2 and Figure 11.4, the insensitive controller and the non-fragile controller are superior in their ability to tolerate multiplicative coefficient variations as compared to the standard controller, which shows the superiority and effectiveness of the proposed design method.

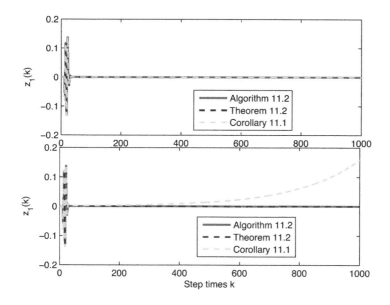

FIGURE 11.1
Comparison of the regulated output responses of $z_1(k)$ (the upper one for
$\theta = 0$ and the lower one for $\theta = 0.05$).

Additive Coefficient Variation Case

In this section, the case of additive coefficient variations is considered. Because Theorem 11.1 cannot be used to obtain the insensitive controller by use Algorithm 11.1 directly in this numerical example, we using Algorithm 11.2 to obtain the insensitive H_∞ controller with respect to additive controller coefficient variations below.

First, let $\lambda = 0.042$. After $\kappa = 5$ times of iterations, the H_∞ performance index and gain matrices of the non-fragile controller subject to additive controller coefficient variations are obtained by solving the optimization problem (11.40) as follows:

$$\tilde{A} = \begin{bmatrix} -0.1594 & -0.4267 & 0.2484 \\ -0.1855 & -0.3261 & 0.5132 \\ -0.1517 & -0.2218 & 0.5228 \end{bmatrix}, \tilde{B} = \begin{bmatrix} -0.0576 \\ -0.0694 \\ -0.1788 \end{bmatrix}$$
$$\tilde{C} = \begin{bmatrix} -0.2914 & -0.2427 & 0.2171 \end{bmatrix}, \gamma = 0.3851.$$

At the same time, the sensitivity performance index of this non-fragile controller is $\beta = 0.4509$ by using the definition of sensitivity measures in (11.11).

Likewise, by applying Algorithm 11.1 and Algorithm 11.2 to Theorem 11.1 with $\gamma = 0.2$, after $\kappa = 25$ times of iterations, the following results are

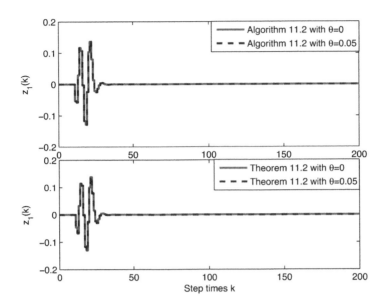

FIGURE 11.2
The regulated output responses of $z_1(k)$ by using insensitive controller and
non-fragile controller.

TABLE 11.2
Comparison of Performance Indexes for Various
Controllers

Index	SC	NC	IC
	Corollary 11.1	Theorem 11.2	Theorem 11.1
γ	0.1278	0.3851	0.2
β	6.7248	0.4509	0.2671

obtained:

$$\tilde{A} = \begin{bmatrix} 0.0389 & -0.0240 & 0.0044 \\ 0.0445 & 0.0474 & 0.1029 \\ 0.0936 & 0.0694 & 0.1298 \end{bmatrix}, \tilde{B} = \begin{bmatrix} -0.0211 \\ -0.0280 \\ -0.0713 \end{bmatrix}$$
$$\tilde{C} = \begin{bmatrix} 0.0653 & 0.1252 & 0.1232 \end{bmatrix}, \beta = 0.5447.$$

*Table 11.2 presents a further comparison of the sensitivity performance
indexes obtained by using the method proposed in Remark 11.10.*

This table shows similar results to those in Table 11.1.

*In order to show the effectiveness and superiority of the insensitive con-
troller design method and non-fragile controller design method more clearly,
maximum singular value curves of the closed-loop system's transfer functions
obtained via different controllers over the prespecified frequency interval are*

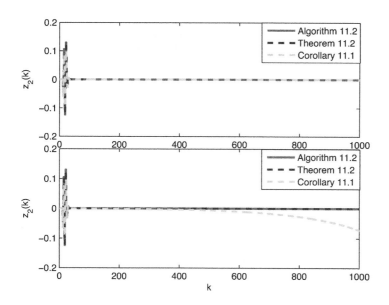

FIGURE 11.3
Comparison of the regulated output responses of $z_2(k)$ (the upper one for $\theta = 0$ and the lower one for $\theta = 0.05$).

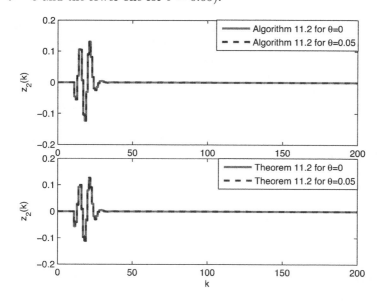

FIGURE 11.4
The regulated output responses of $z_2(k)$ by using an insensitive controller and non-fragile controller.

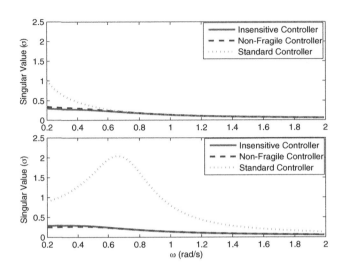

FIGURE 11.5
Maximum singular value curves of the closed-loop systems obtained via different controllers over the prespecified frequency interval (the upper one for $\theta = 0$ and the lower one for $\theta = 0.05$).

presented in Figure 11.5. From this figure, we can see that the additive controller coefficient variations can greatly influence the maximum singular value curve of the closed-loop systems obtained via the standard controller, while the curves obtained via the insensitive controller and standard controller are insensitive to the same variations.

Simulation also is carried out based on supposing that the disturbance inputs and the initial values are the same as in the last section. Figure 11.6 to Figure 11.9 show the regulated output responses of the closed-loop system, from which we can conclude that the insensitive and non-fragile controllers are insensitive to the additive controller coefficient variations, while the standard controller cannot guarantee the closed-loop stability when there exist the same controller coefficient variations, which demonstrate the superiority and effectiveness of the proposed method in this chapter.

11.5   Conclusion

This chapter studies the problem of designing multi-objective parameter insensitive H_∞ dynamic output feedback controllers for linear discrete-time systems, which is an NP-hard problem. Two different design methods with

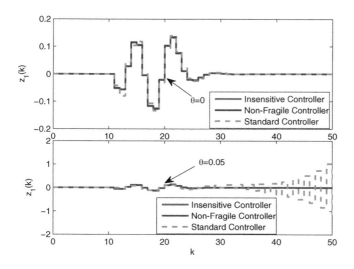

FIGURE 11.6
Comparison of the regulated output responses of $z_1(k)$ (the upper one for $\theta = 0$ and the lower one for $\theta = 0.05$).

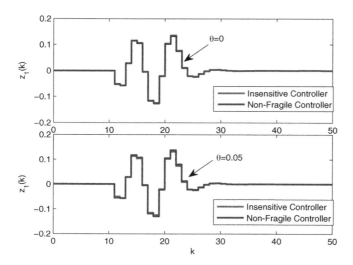

FIGURE 11.7
The regulated output responses of $z_1(k)$ by using an insensitive controller and non-fragile controller.

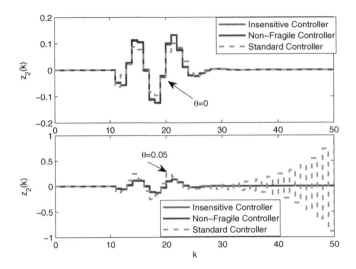

FIGURE 11.8
Comparison of the regulated output responses of $z_2(k)$ (the upper one for $\theta = 0$ and the lower one for $\theta = 0.05$).

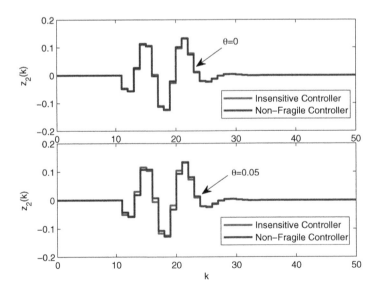

FIGURE 11.9
The regulated output responses of $z_2(k)$ by using an insensitive controller and non-fragile controller.

different degrees of conservativeness and computational complexity are proposed for this problem. The designed controllers are insensitive to the multiplicative/additive controller coefficient variations. The first method presents a necessary and sufficient condition for the existence of the insensitive controller. The problem of designing multi-objective dynamic output feedback controllers is a non-convex problem itself. An LMI-based procedure which is a sequential linear programming matrix method (SLPMM) is proposed to solve this non-convex problem. However, the search for satisfactory solutions may be very difficult when the SLPMM algorithm acts on a module of very high dimension. To overcome the above difficulty, the non-fragile controller design method is adopted to obtain an initial solution for the SLPMM algorithm for the first time. In the second method, a sufficient condition is provided for the multiplicative coefficient variation case based on a new type of sensitivity measures. Finally, the effectiveness of the proposed method is validated by numerical examples.

Bibliography

[1] J. Ackermann. *Sampled-data control systems: Analysis and synthesis, robust system design.* Springer-Verlag, New York, 1985.

[2] C. K. Ahn, S. Han, and W. H. Kwon. H_∞ FIR filters for linear continuous-time state-space systems. *IEEE Signal Process. Lett.*, 13(9):557–560, 2006.

[3] G. Amit and U. Shaked. Small roundoff noise realization of fixed-point digital filters and controllers. *IEEE Trans. Acoust. Speech Signal Process.*, 36(6):880–891, 1988.

[4] B. D. O. Anderson and J. B. Moore. *Optimal control: Linear quadratic methods*, volume 1. Prentice Hall, Englewood Cliffs, NJ, 1990.

[5] K. J. Åström and B. Wittenmark. *Computer-controlled systems: Theory and design.* Prentice Hall, Englewood Cliffs, NJ, 1997.

[6] B. R. Barmish. Necessary and sufficient conditions for quadratic stabilizability of an uncertain system. *J. Optimiz. Theory Appl.*, 46(4):399–408, 1985.

[7] P. Bernhard. H_∞*-optimal control and related minimax design problems: a dynamic game approach.* Birkhäuser Science, Germany, 1991.

[8] F. Blanchini, R. Lo Cigno, and R. Tempo. Control of atm networks: Fragility and robustness issues. In *Proc. American Control Conference*, volume 5, pp. 2847–2851. IEEE, 1998.

[9] F. Blanchini and R. Pesenti. Min-max control of uncertain multi-inventory systems with multiplicative uncertainties. *IEEE Trans. Autom. Control*, 46(6):955–960, 2001.

[10] P. Bolzern, P. Colaneri, and G. De Nicolao. Optimal robust filtering with time-varying parameter uncertainty. *Int. J. Control*, 63(3):557–576, 1996.

[11] S. Boyd, L. El Ghaoui, E. Feron, and V. Balakrishnan. Linear matrix inequalities in system and control theory. *SIAM Studies in Appl. Math.*, 1994.

[12] E. Castillo, J. M. Gutiérrez, and A. S. Hadi. Sensitivity analysis in discrete Bayesian networks. *IEEE Trans. Syst. Man Cybern. Part A Syst. Humans*, 27(4):412–423, 1997.

[13] E. Castillo, R. Minguez, and C. Castillo. Sensitivity analysis in optimization and reliability problems. *Reliab. Eng. Syst. Saf.*, 93(12):1788–1800, 2008.

[14] J. L. Chang. Dynamic output integral sliding-mode control with disturbance attenuation. *IEEE Trans. Autom. Control*, 54(11):2653–2658, 2009.

[15] X. H. Chang and G. H. Yang. Non-fragile H_∞ filtering of continuous-time Fuzzy systems. *IEEE Trans. Signal Process.*, 59(4):1528–1538, 2011.

[16] X. H. Chang and G. H. Yang. Non-fragile fuzzy H_∞ filter design for nonlinear continuous-time systems with D stability. *Signal Process.*, 92(2):575–586, 2012.

[17] W. W. Che and Y. L. Wang. Non-fragile dynamic output feedback h_∞ control for continuous-time systems with controller coefficient sensitivity consideration. In *2011 Chinese Control and Decision Conference*, pp. 2441–2446. IEEE, 2011.

[18] W. W. Che and G. H. Yang. Discrete-time quantized H_∞ filtering with quantizer ranges consideration. In *Proc. American Control Conference*, pp. 5659–5664. IEEE, 2009.

[19] W. W. Che and G. H. Yang. Non-fragile dynamic output feedback H_∞ control for discrete-time systems with FWL consideration. *Int. J. Control Autom. Syst.*, 9(5):993–997, 2011.

[20] S. Chen, J. Wu, R. H. Istepanian, J. Chu, and J. F. Whidborne. Optimising stability bounds of finite-precision controller structures for sampled-data systems in the δ-operator domain. *IEEE Proceedings: Control Theory and Applications*, 146(6):517–526, 1999.

[21] P. Chevrel and M. Yagoubi. A parametric insensitive H_2 control design approach. *Int. J. Robust Nonlinear Control*, 14(15):1283–1297, 2004.

[22] J. R. Corrado and W. M. Haddad. Static output feedback controllers for systems with parametric uncertainty and controller gain variation. In *Proc. American Control Conference*, volume 2, pp. 915–919. IEEE, 1999.

[23] J. B. Cruz Jr. *System sensitivity analysis*. Dowden, Hutchinson & Ross, Stroudsburg, PA, 1973.

[24] M. C. De Oliveira, J. C. Geromel, and J. Bernussou. Extended H_2 and H_∞ characterizations and controller parametrizations for discrete-time systems. *Int. J. Control*, 75(9):666–679, 2002.

[25] C. E. de Souza, A. Trofino, and K. A. Barbosa. Mode-independent H_∞ filters for Markovian jump linear systems. *IEEE Trans. Autom. Control*, 51(11):1837–1841, 2006.

[26] D. W. Ding, X. Li, Y. Yin, and C. Sun. Non-fragile H_∞ and H_2 filter designs for continuous-time linear systems based on randomized algorithms. *IEEE Trans. Ind. Electron.*, 59(11):4433–4442, 2012.

[27] P. Dorato. Non-fragile controller design: An overview. In *Proc. American Control Conference*, volume 5, pp. 2829–2831. IEEE, 1998.

[28] J. C. Doyle, K. Glover, P. P. Khargonekar, and B. A. Francis. State-space solutions to standard H_2 and H_∞ control problems. *IEEE Trans. Autom. Control*, 34(8):831–847, 1989.

[29] Z. Duan, J. Zhang, C. Zhang, and E. Mosca. Robust H_2 and H_∞ filtering for uncertain linear systems. *Automatica*, 42(11):1919–1926, 2006.

[30] Y. Ebihara and T. Hagiwara. New dilated LMI characterizations for continuous-time control design and robust multiobjective control. In *Proc. American Control Conference*, volume 1, pp. 47–52. IEEE, 2002.

[31] Y. Ebihara and T. Hagiwara. New dilated LMI characterizations for continuous-time multiobjective controller synthesis. *Automatica*, 40(11):2003–2009, 2004.

[32] L. El Ghaoui, F. Oustry, and M. AitRami. A cone complementarity linearization algorithm for static output-feedback and related problems. *IEEE Trans. Autom. Control*, 42(8):1171–1176, 1997.

[33] R. W. Eustace, B. A. Woodyatt, G. L. Merrington, and A. Runacres. Fault signatures obtained from fault implant tests on an F404 engine. *J. Eng. Gas Turbines Power*, 116:178, 1994.

[34] D. Famularo, P. Dorato, C. T. Abdallah, W. M. Haddad, and A. Jadbabaie. Robust non-fragile LQ controllers: The static state feedback case. *Int. J. Control*, 73(2):159–165, 2000.

[35] H. Fan, T. Soderstrom, M. Mossberg, B. Carlsson, and Y. Zou. Estimation of continuous-time AR process parameters from discrete-time data. *IEEE Trans. Signal Process.*, 47(5):1232–1244, 1999.

[36] H. H. Fan and P. De. High speed adaptive signal processing using the delta operator. *Digital Signal Process.*, 11(1):3–34, 2001.

[37] M. Fu, C. E. de Souza, and L. Xie. Quadratic stabilization and H_∞ control of discrete-time uncertain systems. In *Proc. Int. Symposium MTNS-91*, pp. 269–274, 1991.

[38] M. Fu, C. E. de Souza, and L. Xie. H_∞ estimation for uncertain systems. *Int. J. Robust Nonlinear Control*, 2(2):87–105, 1992.

[39] P. Gahinet. Explicit controller formulas for LMI-based H_∞ synthesis. *Automatica*, 32(7):1007–1014, 1996.

[40] P. Gahinet and P. Apkarian. A linear matrix inequality approach to H_∞ control. *Int. J. Robust Nonlinear Control*, 4(4):421–448, 1994.

[41] J. C. Geromel and M. C. De Oliveira. H_2 and H_∞ robust filtering for convex bounded uncertain systems. *IEEE Trans. Autom. Control*, 46(1):100–107, 2001.

[42] J. C. Geromel and G. Levin. Suboptimal reduced-order filtering through an LMI-based method. *IEEE Trans. Signal Process.*, 54(7):2588–2595, 2006.

[43] M. Gevers and G. Li. *Parameterizations in control, estimation and filtering problems: Accuracy aspects*. Springer Verlag, London, 1993.

[44] G. C. Goodwin, R. L. Leal, D. Q. Mayne, and R. H. Middleton. Rapprochement between continuous and discrete model reference adaptive control. *Automatica*, 22(2):199–207, 1986.

[45] G. C. Goodwin, R. H. Middleton, and H. V. Poor. High-speed digital signal processing and control. *Proc. of the IEEE*, 80(2):240–259, 1992.

[46] K. M. Grigoriadis and J. T. Watson Jr. Reduced-order H_∞ and $L_2 - L_\infty$ filtering via linear matrix inequalities. *IEEE Trans. Aerosp. Electron. Syst.*, 33(4):1326–1338, 1997.

[47] X. G. Guo and G. H. Yang. Non-fragile H_∞ filter design for delta operator formulated systems with circular region pole constraints: An LMI optimization approach. *Acta Autom. Sin.*, 35(9):1209–1215, 2009.

[48] X. G. Guo and G. H. Yang. H_∞ filter design for delta operator formulated systems with low sensitivity to filter coefficient variations. *IET Control Theory Appl.*, 5(15):1677–1688, 2011.

[49] X. G. Guo and G. H. Yang. H_∞ output tracking control for delta operator systems with insensitivity to controller coefficient variations. *Int. J. Syst. Sci.*, 2011 (DOI 10.108000207721.2011.617898).

[50] X.G. Guo and G.H. Yang. Low-sensitivity h_∞ filter design for linear delta operator systems with sampling time jitter. *Int. J. Control*, 85(4):397–408, 2012.

[51] W. M. Haddad and D. S. Bernstein. Robust, reduced-order, nonstrictly proper state estimation via the optimal projection equations with guaranteed cost bounds. *IEEE Trans. Autom. Control*, 33(6):591–595, 1988.

[52] W. M. Haddad and J. R. Corrado. Resilient dynamic controller design via quadratic lyapunov bounds. In *Proc. 36th IEEE Conf. Decision and Control*, volume 3, pp. 2678–2683. IEEE, 1997.

[53] W. M. Haddad and J. R. Corrado. Robust resilient dynamic controllers for systems with parametric uncertainty and controller gain variations. In *Proc. American Control Conference*, pp. 2837–2841. IEEE, 1998.

[54] W. M. Haddad and J. R. Corrado. Robust resilient dynamic controllers for systems with parametric uncertainty and controller gain variations. *Int. J. Control*, 73(15):1405–1423, 2000.

[55] S. Han and W. H. Kwon. Comparative study of finite wordlength effects in shift and delta operator parameterizations. *IEEE Trans. Autom. Control*, 38(5):803–807, 1993.

[56] S. Han and W. H. Kwon. L_2-EFIR smoothers for deterministic discrete-time state-space signal models. *IEEE Trans. Autom. Control*, 52(5):927–932, 2007.

[57] J. Hao and G. Li. Polynomial operator based sparse controller structures with stability consideration. *IET Proceedings Control Theory and Applications*, 152(5):521–530, 2005.

[58] J. X. Hao and G. Li. Polynomial operator-based digital controller structures of high stability performance and computation efficiency. In *Control, Automation, Robotics and Vision Conference, 2004. ICARCV 2004 8th*, volume 3, pp. 1604–1609. IEEE, 2004.

[59] T. Hilaire. Low-parametric-sensitivity realizations with relaxed l_2-dynamic-range-scaling constraints. *IEEE Trans. Circuits Syst.-II: Express Briefs*, 56(7):590–594, 2009.

[60] T. Hilaire, P. Chevrel, and Y. Trinquet. Designing low parametric sensitivity FWL realizations of LTI controllers/filters within the implicit state-space framework. In *Proc. 44th IEEE Conf. Decision & Control and the European Control Conference*, pp. 5192–5197. IEEE, 2005.

[61] T. Hilaire, P. Chevrel, and J. F. Whidborne. Finite wordlength controller realisations using the specialised implicit form. *Int. J. Control*, 83(2):330–346, 2010.

[62] T. Hinamoto and T. Kouno. Realization of state-estimate feedback controllers with minimum L 2-sensitivity. In *IEEE Int. Symposium on Circuits and Systems*, volume 2, pp. 837–840. IEEE, 2001.

[63] T. Hinamoto, T. Oumi, O. I. Omoifo, and W. S. Lu. Minimization of frequency-weighted l_2-sensitivity subject to l_2-scaling constraints for two-dimensional state-space digital filters. *IEEE Trans. Signal Process.*, 56(10):5157–5168, 2008.

[64] T. Hinamoto, Y. Zempo, Y. Nishino, and W. S. Lu. An analytical approach for the synthesis of two-dimensional state-space filter structures with minimum weighted sensitivity. *IEEE Trans. Circuits Syst.-I: Fundam. Theory Appl.*, 46(10):1172–1183, 1999.

[65] R. S. H. Istepanian, G. Li, J. Wu, and J. Chu. Analysis of sensitivity measures of finite-precision digital controller structures with closed-loop stability bounds. 145(5):472–478, 1998.

[66] R. S. H. Istepanian and J. F. Whidborne. *Digital controller implementation and fragility: A modern perspective.* Springer Verlag, London, 2001.

[67] T. Iwasaki and R. E. Skelton. All controllers for the general H_∞ control problem: LMI existence conditions and state space formulas. *Automatica*, 30(8):1307–1317, 1994.

[68] A. Jadbabaie, C. T. Abdallah, D. Famularo, and P. Dorato. Robust, non-fragile and optimal controller design via linear matrix inequalities. In *Proc. American Control Conference*, volume 5, pp. 2842–2846. IEEE, 1998.

[69] B. Jain. Guaranteed error estimation in uncertain systems. *IEEE Trans. Autom. Control*, 20(2):230–232, 1975.

[70] X. Jia, D. Zhang, X. Hao, and N. Zheng. Fuzzy H_∞ tracking control for nonlinear networked control systems in T-S fuzzy model. *IEEE Trans. Syst. Man Cybern. Part B Cybern.*, 39(4):1073–1079, 2009.

[71] Y. Jia. Alternative proofs for improved lmi representations for the analysis and the design of continuous-time systems with polytopic type uncertainty: A predictive approach. *IEEE Trans. Autom. Control*, 48(8):1413–1416, 2003.

[72] X. Jiang, Q. L. Han, S. Liu, and A. Xue. A New H_∞ stabilization criterion for networked control systems. *IEEE Trans. Autom. Control*, 53(4):1025–1032, 2008.

[73] S. J. Julier and J. K. Uhlmann. A new extension of the Kalman filter to nonlinear systems. In *Int. Symp. Aerospace/Defense Sensing, Simul. and Controls*, volume 3, p. 26. Spie, Bellingham, WA, 1997.

[74] D. Kaesbauer and J. Ackermann. How to escape from the fragility trap [control]. In *Proc. American Control Conference*, volume 5, pp. 2832–2836. IEEE, 1998.

[75] I. Kaminer, P. P. Khargonekar, and M. A. Rotea. Mixed control for discrete-time systems via convex optimization. *Automatica*, 29(1):57–70, 1993.

[76] S. Kanev, C. Scherer, M. Verhaegen, and B. De Schutter. Robust output-feedback controller design via local BMI optimization. *Automatica*, 40(7):1115–1127, 2004.

[77] L. H. Keel and S. P. Bhattacharyya. Robust, fragile, or optimal? *IEEE Trans. Autom. Control*, 42(8):1098–1105, 1997.

[78] L. H. Keel and S. P. Bhattacharyya. Stability margins and digital implementation of controllers. In *Proc. American Control Conference*, volume 5, pp. 2852–2856. IEEE, 1998.

[79] H. W. Knobloch, A. Isidori, and D. Flockerzi. *Topics in control theory.* Birkhäuser, Basel, Switzerland, 1993.

[80] P. V. Kokotovic. Method of sensitivity points in the investigation and optimization of linear control systems. *Autom. Remote Control*, 25(12):1670–1676, 1964.

[81] A. G. Kuznetsov, R. O. Bowyer, and D. W. Clark. Estimation of multiple order models in the delta-domain. *Int. J. Control*, 72(7):629–642, 1999.

[82] H. Kwakernaak and R. Sivan. *Linear optimal control systems*, volume 172. Wiley-Interscience, New York, 1972.

[83] P. Lancaster and L. Rodman. *Algebraic Riccati Equations*. Oxford University Press, USA, 1995.

[84] H. J. Lee, J. B. Park, and Y. H. Joo. Further refinement on LMI-based digital redesign: Delta-operator approach. *IEEE Trans. Circuits Syst.-II: Express Briefs*, 53(6):473–477, 2006.

[85] F. Leibfritz. An LMI-based algorithm for designing suboptimal static H_2/H_∞ output feedback controllers. *SIAM J. Control Optim.*, 39(6):1711–1735, 2000.

[86] B. Lennartson, R. Middleton, A. K. Christiansson, and T. McKelvey. Low order sampled data H_∞ control using the Delta operator and LMIs. In *Proc. 43rd IEEE Conf. Decis. Control*, volume 4, pp. 4479–4484, 2004.

[87] G. Li. Minimization of pole/zero sensitivity in digital filter design with sparse structure consideration. In *IEEE Int. Conf. Acoustics, Speech, and Signal Processing*, volume 3, pp. III/581–III/587. IEEE, 1994.

[88] G. Li. On pole and zero sensitivity of linear systems. *IEEE Trans. Circuits Syst.-I: Fundam. Theory Appl.*, 44(7):583–590, 1997.

[89] G. Li. On the structure of digital controllers with finite word length consideration. *IEEE Trans. Autom. Control*, 43(5):689–693, 1998.

[90] G. Li. A polynomial-operator-based DFIIt structure for IIR filters. *IEEE Trans. Circuits Syst.-II: Express Briefs*, 51(3):147–151, 2004.

[91] G. Li, B. D. O. Anderson, M. Gevers, and J. E. Perkins. Optimal FWL design of state-space digital systems with weighted sensitivity minimization and sparseness consideration. *IEEE Trans. Circuits Syst.-I: Fundam. Theory Appl.*, 39(5):365–377, 1992.

[92] G. Li, J. Wu, and S. Chen. Sparse controller realisation with small roundoff noise. 151(2):246–251, 2004.

[93] L. Li and Y. Jia. Non-fragile dynamic output feedback control for linear systems with time-varying delay. *IET Control Theory Appl.*, 3(8):995–1005, 2009.

[94] C. Lin, Q. G. Wang, and T. H. Lee. H_∞ output tracking control for nonlinear systems via T-S fuzzy model approach. *IEEE Trans. Syst. Man Cybern. Part B Cybern.*, 36(2):450–457, 2006.

[95] C. Lin, Q. G. Wang, and T. H. Lee. A less conservative robust stability test for linear uncertain time-delay systems. *IEEE Trans. Autom. Control*, 51(1):87–91, 2006.

[96] R. Lin, F. Yang, and Q. Chen. Design of robust non-fragile H_∞ controller based on Delta operator theory. *J. Control Theory Appl.*, 5(4):404–408, 2007.

[97] J. Lofberg. YALMIP: A toolbox for modeling and optimization in MATLAB. In *2004 IEEE Int. Symposium on Computer Aided Control Systems Design*, pp. 284–289. IEEE, 2004.

[98] J. G. Lu and D. J. Hill. Impulsive synchronization of chaotic lur'e systems by linear static measurement feedback: An lmi approach. *IEEE Trans. Circuits Syst.-II: Express Briefs*, 54(8):710–714, 2007.

[99] W. J. Lutz and S. L. Hakimi. Design of multi-input multi-output systems with minimum sensitivity. *IEEE Trans. Circuits Syst.*, 35(9):1114–1122, 1988.

[100] A. G. Madievski, B. Anderson, and M. Gevers. Optimum realizations of sampled-data controllers for FWL sensitivity minimization. *Automatica*, 31(3):367–379, 1995.

[101] M. S. Mahmoud. *Resilient control of uncertain dynamical systems*, volume 303. Springer Verlag, London, 2004.

[102] M. S. Mahmoud. Resilient linear filtering of uncertain systems. *Automatica*, 40(10):1797–1802, 2004.

[103] M. S. Mahmoud. Resilient $L_2 - L_\infty$ filtering of polytopic systems with state delays. *IET Control Theory Appl.*, 1(1):141–154, 2007.

[104] P. Martin, S. Devasia, and B. Paden. A different look at output tracking: Control of a VTOL aircraft. *Automatica*, 32(1):101–107, 1996.

[105] I. Masubuchi, A. Ohara, and N. Suda. LMI-based controller synthesis: A unified formulation and solution. *Int. J. Robust Nonlinear Control*, 8(8):669–686, 1998.

[106] R. M. Palhares and P. L. D. Peres. LMI approach to the mixed H_2/H_∞ filtering design for discrete-time uncertain systems. *IEEE Trans. Aerosp. Electron. Syst.*, 37(1):292–296, 2001.

[107] I. R. Petersen. A stabilization algorithm for a class of uncertain linear systems. *Syst. Control Lett.*, 8(4):351–357, 1987.

[108] I. R. Petersen, B. Anderson, and E. A. Jonckheere. A first principles solution to the non-singular H_∞ control problem. *Int. J. Robust and Nonlinear Control*, 1(3):171–185, 1991.

[109] I. R. Petersen and C. V. Hollot. A riccati equation approach to the stabilization of uncertain linear systems. *Automatica*, 22(4):397–411, 1986.

[110] I. R. Petersen and D. C. McFarlane. Optimal guaranteed cost control and filtering for uncertain linear systems. *IEEE Trans. Autom. Control*, 39(9):1971–1977, 1994.

[111] J. Qiu, Y. Xia, H. Yang, and J. Zhang. Robust stabilisation for a class of discrete-time systems with time-varying delays via delta operators. *IET Control Theory Appl.*, 2(1):87–93, 2008.

[112] J. R. Ragazzini and G. F. Franklin. *Sampled-data control systems*. McGraw-Hill, New York, 1958.

[113] C. Scherer, P. Gahinet, and M. Chilali. Multiobjective output-feedback control via LMI optimization. *IEEE Trans. Autom. Control*, 42(7):896–911, 1997.

[114] C. W. Scherer. A full block S-procedure with applications. In *Proc. 36th IEEE Conf. Decis. Control*, volume 3, pp. 2602–2607. IEEE, 1997.

[115] U. Shaked and C. E. de Souza. Robust minimum variance filtering. *IEEE Trans. Signal Process.*, 43(11):2474–2483, 1995.

[116] M. H. Shor and W. R. Perkins. Reliable control in the presence of sensor/actuator failures: A unified discrete/continuous approach. In *Proc. 30th IEEE Conf. Decis. Control*, pp. 1601–1606, 1991.

[117] J. M. A. D. Silva, C. Edwards, and S. K. Spurgeon. Linear matrix inequality based dynamic output feedback sliding mode control for uncertain plants. In *Proc. American Control Conference*, pp. 763–768. IEEE, 2009.

[118] W. Slob. Uncertainty analysis in multiplicative models. *Risk Anal.*, 14(4):571–576, 1994.

[119] T. Song, E. G. Collins Jr, and R. H. Istepanian. Improved closed-loop stability for fixed-point controller implementation using the delta operator. In *Proc. American Control Conference*, volume 6, pp. 4328–4332. IEEE, 1999.

[120] G. Tadmor. Worst-case design in the time domain: The maximum principle and the standard H_∞ problem. *Math. Control, Signals Syst.*, 3(4):301–324, 1990.

[121] V. Tavsanoglu and L. Thiele. Optimal design of state-space digital filters by simultaneous minimization of sensitivity and roundoff noise. *IEEE Trans. Circuits Syst.*, 31(10):884–888, 1984.

[122] D. E. Torfs, R. Vuerinckx, J. Swevers, and J. Schoukens. Comparison of two feedforward design methods aiming at accurate trajectory tracking of the end point of a flexible robot arm. *IEEE Trans. Control Syst. Technol.*, 6(1):2–14, 1998.

[123] H. D. Tuan, P. Apkarian, and T. Q. Nguyen. Robust and reduced-order filtering: New LMI-based characterizations and methods. *IEEE Trans. Signal Process.*, 49(12):2975–2984, 2001.

[124] T. Turányi. Sensitivity analysis of complex kinetic systems. Tools and applications. *J. Math. Chem.*, 5(3):203–248, 1990.

[125] K. Uesaka and M. Kawamata. Synthesis of low coefficient sensitivity digital filters using genetic programming. In *IEEE Int. Symposium on Circuits and Systems*, volume 3, pp. 307–310. IEEE, 1999.

[126] R. J. Veillette. Reliable linear-quadratic state-feedback control. *Automatica*, 31(1):137–143, 1995.

[127] Y. L. Wang and G. H. Yang. Output tracking control for networked control systems with time delay and packet dropout. *Int. J. Control*, 81(11):1709–1719, 2008.

[128] Y. L. Wang and G. H. Yang. Output tracking control for discrete-time networked control systems. In *Proc. American Control Conference*, pp. 5109–5114. IEEE, 2009.

[129] J. F. Whidborne, R. S. H. Istepanian, and J. Wu. Reduction of controller fragility by pole sensitivity minimization. *IEEE Trans. Autom. Control*, 46(2):320–325, 2001.

[130] N. Wong and T. S. Ng. A generalized direct-form delta operator-based IIR filter with minimum noise gain and sensitivity. *IEEE Trans. Circuits Syst.-II: Analog Digital Signal Process.*, 48(4):425–431, 2001.

[131] J. Wu, S. Chen, G. Li, and J. Chu. Global optimal realizations of finite precision digital controllers. In *Proc. 41th IEEE Conf. Decision and Control*, volume 3, pp. 2941–2946. IEEE, 2002.

[132] J. Wu, S. Chen, G. Li, R. H. Istepanian, and J. Chu. An improved closed-loop stability related measure for finite-precision digital controller realizations. *IEEE Trans. Autom. Control*, 46(7):1162–1166, 2001.

[133] J. Wu, G. Li, R. H. Istepanian, and J. Chu. Shift and delta operator realisations for digital controllers with finite word length considerations. *IEE Proceedings Control Theory and Applications*, 147(6):664–672, 2000.

[134] L. Wu and D. W. C. Ho. Reduced-order $L_2 - L_\infty$ filtering for a class of nonlinear switched stochastic systems. *IET Control Theory Appl.*, 3(5):493–508, 2009.

[135] Y. Xia, M. Fu, H. Yang, and G. P. Liu. Robust sliding-mode control for uncertain time-delay systems based on delta operator. *IEEE Trans. Ind. Electron.*, 56(9):3646–3655, 2009.

[136] C. S. Xiao, P. Agathoklis, and D. J. Hill. Coefficient sensitivity and structure optimization of multidimensional state-space digital filters. *IEEE Trans. Circuits Syst.-I: Fundam. Theory Appl.*, 45(9):993–998, 1998.

[137] L. Xie. Guaranteed cost control of uncertain discrete-time systems. *Control Theory Adv. Technol.*, 10(4):1235–1251, 1995.

[138] L. Xie, L. Lu, D. Zhang, and H. Zhang. Improved robust H_2 and H_∞ filtering for uncertain discrete-time systems. *Automatica*, 40(5):873–880, 2004.

[139] L. Xie and Y. C. Soh. Robust Kalman filtering for uncertain systems. *Syst. Control Lett.*, 22(2):123–129, 1994.

[140] S. Yamaki, M. Abe, and M. Kawamata. On the absence of limit cycles in state-space digital filters with minimum l2-sensitivity. *IEEE Trans. Circuits Syst.-II: Express Briefs*, 55(1):46–50, 2008.

[141] W. Y. Yan and J. B. Moore. On L_2-sensitivity minimization of linear state-space systems. *IEEE Trans. Circuits Syst.-I: Fundam. Theory Appl.*, 39(8):641–648, 1992.

[142] G. H. Yang and W. W. Che. Non-fragile H_∞ controller design with sparse structure. In *Proc. 2007 IEEE Int. Conf. Control Automation*, pp. 57–62. IEEE, 2007.

[143] G. H. Yang and W. W. Che. Non-fragile H_∞ filter design for linear continuous-time systems. *Automatica*, 44(11):2849–2856, 2008.

[144] G. H. Yang and W. W. Che. Discrete-time non-fragile dynamic output feedback H_∞ controller design. *Proc. American Control Conference*, 32(7):5126–5131, 2009.

[145] G. H. Yang and X. G. Guo. Insensitive H_∞ filter design for discrete-time systems: An LMI optimization approach. In *Proc. 48th IEEE Conf. Decision and Control and 28th Chinese Control Conference*, pp. 7157–7162. IEEE, 2009.

[146] G. H. Yang and H. Wang. Fault detection for a class of uncertain state-feedback control systems. *IEEE Trans. Control Syst. Technol.*, 18(1):201–212, 2010.

[147] G. H. Yang and J. L. Wang. Non-fragile H_∞ control for linear systems with multiplicative controller gain variations. *Automatica*, 37(5):727–737, 2001.

[148] G. H. Yang and J. L. Wang. Robust nonfragile Kalman filtering for uncertain linear systems with estimator gain uncertainty. *IEEE Trans. Automat. Control*, 46(2):343–348, 2001.

[149] G. H. Yang, J. L. Wang, and C. Lin. H_∞ control for linear systems with additive controller gain variations. *Int. J. Control*, 73(16):1500–1506, 2000.

[150] G. H. Yang, J. L. Wang, and Y. C. Soh. Guaranteed cost control for discrete-time linear systems under controller gain perturbations. *Linear Algebra and its Applications*, 312(1-3):161–180, 2000.

[151] H. Yang, Y. Xia, and P. Shi. Observer-based sliding mode control for a class of discrete systems via delta operator approach. *J. Franklin Inst.*, 347(7):1199–1213, 2010.

[152] X. Yang, M. Kawamata, and T. Higuchi. Coefficient sensitivity analysis of periodically time-varying state-space digital filters. *IEEE Trans. Circuits Syst.-II: Analog Digital Signal Process.*, 41(8):526–536, 1994.

[153] X. N. Zhang and G. H. Yang. Dynamic output feedback control synthesis with mixed frequency small gain specifications. *Acta Autom. Sin.*, 34(5):551–557, 2008.

[154] K. Zhou, J. C. Doyle, and K. Glover. *Robust and optimal control*, Prentice Hall, Englewood Cliffs, NJ, 1996.

Index